SOCIAL SCIENCE FOR WHAT?

SOCIAL SCIENCE FOR WHAT?

BATTLES OVER PUBLIC FUNDING FOR THE "OTHER SCIENCES" AT THE NATIONAL SCIENCE FOUNDATION

MARK SOLOVEY

The MIT Press
Cambridge, Massachusetts
London, England

This book was set in Bembo Book MT Pro by Westchester Publishing Services. Printed and bound in the United States of America.

Library of Congress Cataloging-in-Publication Data

Names: Solovey, Mark, 1964- author.
Title: Social science for what? : Battles over public funding for the "other sciences" at the National Science Foundation / Mark Solovey.
Description: Cambridge, Massachusetts : MIT Press, [2020] | Includes bibliographical references and index.
Identifiers: LCCN 2019040890 | ISBN 9780262539050 (paperback)
Subjects: LCSH: National Science Foundation (U.S.) | Social sciences--Research--United States--History. | Research--United States--Finance--History. | Endowment of research--United States--History.
Classification: LCC H62.5.U5 S653 2020 | DDC 300.72/073--dc23
LC record available at https://lccn.loc.gov/2019040890

10 9 8 7 6 5 4 3 2 1

For the individuals and institutions who support historical inquiry

CONTENTS

ACKNOWLEDGMENTS

While working on this project I have been fortunate to receive encouragement and support from many people. This group includes Roger Backhouse, Alex Csiszar, Christian Daye, David Engerman, Peter Galison, Christopher Green, Pieter Huistra, Joel Isaac, Noortje Jacobs, David Kaiser, Alice Kehoe, Donald McGraw, Neil McLaughin, Wade Pickren, Alexandra Rutherford, Thomas Teo, Krist Vaesen, Jessica Wang, and Nadine Weidman. I am grateful to those who provided me with valuable feedback on earlier versions of one or more chapters: Janet Abbate, Robert Adcock, Michael Bernstein, Howard Brick, Jamie Cohen-Cole, Matthew Farish, Philippe Fontaine, Daniel Geary, Emily Gibson, Thomas Gieryn, Hunter Heyck, Sarah Igo, Andrew Jewett, Michael Pettit, Jeff Pooley, Ted Porter, and Joy Rohde. My friend Juan Ilerbaig and my colleague Marga Vicedo deserve special thanks for reading, commenting on, and discussing the entire manuscript with me—and more than once. Juan assisted me by putting together the funding and organizational charts as well. Marga, who is also my dear wife, has helped me to become a better scholar in many ways, which has also helped to make this a much better study.

It is a pleasure to acknowledge my debt to those who facilitated access to certain published and unpublished materials: Richard Atkinson, Ann Bushmiller, Sonja Gardner-Clarke, Onaona Guay, Elise Lipkowitz, Marc Rothenberg, Leo Slater, and Brock Temanson. Thanks as well to those who helped with permissions to use archival collections and documents: Janice Goldblum (National Academy of Sciences), Daniel Meyer (University of Chicago Library Special Collections Research Center), Mary M. Clark (Columbia Center for Oral History Research), Mary Ann Quinn (Rockefeller Archive Center). Others deserve my gratitude for helping me to find and secure permission to use images of individuals who I thought should be seen, not just discussed: Emma Alpert, Geoffrey Alpert, Spencer Alpert, Roberta Balstad, Christina Bartlett-Whitcom, Steve Branch, Janice Davis, Angela Fritz, Michael

Gaetani, Jennifer Hadley, Randall Hagadorn, Fred Harris, Elizabeth Hogan, Lisa Marine, Eric Marshall, Julianne Mattera, Erica Mosner, Wendy Naus, Sherry Ortner, Erica Raskin, Jamie Raskin, Annie Riecken, Gilson Riecken, Susan Riecken, and Erica Weingartner.

My editor at MIT Press, Katie Helke, has been wonderful. Ever since I first told her about this project, back at the 2017 annual meeting of the History of Science Society, she has given me all of the encouragement, guidance, and reassurance that I could have reasonably hoped for. Her assistant, Laura Keeler, has provided superb support as well. At Westchester Publishing Services, Helen Wheeler made sure that the production process went smoothly, which included responding in a timely manner to my questions and concerns.

I am happy to recognize the institutions that have provided me with gainful employment and research funding. Without their support, this study would have been much harder to carry out properly: the Canadian Social Sciences and Humanities Research Council; and the University of Toronto, especially the Institute for the History and Philosophy of Science, Victoria College, Victoria University, and the College of Arts and Sciences. And a few of my Toronto colleagues have supplied much appreciated inspiration and fellowship: Hakob Barseghyan, Anne-Emanuelle Birn, David Cook, Angela Estherhammer, Craig Fraser, Paul Gooch, Eric Jennings, Paul Kingston, Nikolai Krementsov, Vanina Leschziner, Kanta Murali, Valentina Napolitano, Will Robins, and Denis Walsh. I also had the good fortune of spending a year in the congenial environment of the Institute for Advanced Study in Princeton, New Jersey, which was great for getting a good chunk of writing done on the first full draft.

Finally, for adding richness, fun, movies, and occasionally a bit of drama to my life, a big hug to my two families in the U.S. and Spain, to all of my good friends near and far, and to the amazing HAT.

INTRODUCTION

> [NSF social science funding] should not come at the expense of areas of science—math, engineering, computer science, physics, chemistry and biology—that are most likely to produce breakthroughs that will save lives, create jobs, and promote economic growth.
> —U.S. Republican Representative Lamar Smith, 2014[1]

> [The NSF is the] flagship of the [U.S.] social science enterprise.
> —Roberta B. Miller, leader of the NSF's Social and Economic Sciences Division, 1988[2]

> The social sciences have prospered best in the federal government where they have been included under broad umbrella classifications of the scientific disciplines. ... In close company with scientific areas which enjoy the prestige and status of biological or physical sciences, the social sciences have enjoyed a protection and nourishment which they normally do not have when they are identified as such and stand exposed, "naked and alone."
> —Harry Alpert, sociologist and the NSF's first social science policy architect, 1960[3]

In the contexts of U.S. federal science policy and American political culture, the social sciences have often found themselves under attack. Recently, the most persistent and strongest critics have come from conservative quarters and the Republican Party. Their criticisms have informed repeated efforts to slash public funding for the social sciences at the U.S. National Science Foundation (NSF), one of the country's premier federal science agencies.

In 2009, two rising stars in the Republican Party, John Boehner from Ohio and Eric Cantor from Virginia, sent President Obama a proposal to decrease NSF funding for social and behavioral science (SBS) research awards by 50 percent.[4] In 2011, Tom Coburn, a Republican senator from Oklahoma, suggested eliminating the NSF unit that funded SBS altogether.[5] The following year, in 2012, Republican Representative Jeff Flake from Arizona argued that NSF funding for political science "might satisfy the curiosities of a few academics," including one research grant for a study about "why political candidates make vague statements," but this was an inappropriate use of taxpayer money. Curiously, Flake himself had a master's degree in political science from Brigham Young University.[6] And in 2012, though Coburn's more ambitious proposal to eliminate SBS funding entirely had not been implemented, he did succeed in getting a restriction passed that limited NSF political science funding to studies that promised to enhance national security or economic growth, thus rendering a wide range of other research pursued by political scientists ineligible for support.[7]

Although that restriction did not remain in force for long, challenges from the right kept coming. In 2014, Texas Republican Lamar Smith said he supported NSF funding for "worthy" social science studies. However, far more important was funding for "math, engineering, computer science, physics, chemistry and biology" because, in his view, these fields—not the social sciences—were "most likely to produce breakthroughs that will save lives, create jobs, and promote economic growth."[8] Smith chaired the House science committee with jurisdiction over the NSF budget. Thus, his skepticism had considerable influence on congressional deliberations over social science funding at the agency. More recently, during Donald Trump's first year in the White House, the president's budget proposal to Congress included a 10.4 percent decrease in funding for the NSF Directorate for Social, Behavioral and Economic Sciences (SBE Directorate).[9]

This series of attacks merits serious attention because the NSF has had a central responsibility for maintaining the health of American social science. In 2016, the SBE Directorate received less than 5 percent of the agency's entire research budget. Yet, this same directorate provided two-thirds of all federal funding for basic research in the social sciences—not including basic research in psychology, which received extensive funding from other agencies, especially the National Institutes of Health and the Department of Defense. NSF support for basic research has aimed, first and foremost, to strengthen the academic foundations, methodological resources, and knowledge contributions of the social sciences. During the past few decades, the practical benefits of social research have also been of considerable interest to the agency. But the goals of strengthening the knowledge contributions and the knowledge-producing capacities of the social sciences have remained central.[10]

In addition, the NSF has an impressive track record of funding social scientists who receive high honors in their respective fields. Take the case of economics, where fifty-four scholars who received NSF funding at some point during their careers also won the Nobel Prize in Economics. And every single Economic Nobel Laureate from the U.S. between 1998 and 2016 received NSF funding.[11]

Moreover, social scientists and their advocates argue that by funding basic research and by strengthening social science at other levels (i.e., through the support of advanced training fellowships, workshops, conferences, and scientific resources such as computers), the NSF fortifies its ability to promote human welfare. According to a 2016 report from the Consortium of Social Science Associations (COSSA), despite persistent questions about the benefits of federally funded social science research, such research "makes meaningful contributions to nearly every aspect of American life." NSF-funded research has led, among other things, to many "discoveries" that have

> helped to improve public health, enhance the safety of troops in combat zones, understand how to prepare for and respond to natural and human-made disasters, reduce violence among our youth, improve the effectiveness of the criminal justice system, and generate billions of dollars for the U.S. Treasury with the creation of the telecommunications spectrum auctions.[12]

It must be emphasized as well that the scholarly oriented basic social research supported by the NSF has had little chance of attracting strong, broad-based, and sustained support from other funding sources in the philanthropic, commercial, and public sectors, which, by and large, have placed much greater emphasis on achieving practical goals. So, not only has NSF support been of considerable importance, but it is also not easily replaceable.

In light of these observations, why has the agency's work in this area been such a regular focal point of controversy? Note that social science funding is a thin slice of the total NSF research budget—less than 5 percent. So, why have Republican critics singled out this rather minor component of the agency's overall efforts for intense scrutiny and budgetary cuts? What is all the fuss about? Is this simply a matter of political grandstanding, where politicians eager to demonstrate their strident commitment to "responsible" budget making seek out potentially vulnerable items that they can pinpoint in the immense federal budget?

Surely, this political dynamic is part of the answer. However, there is also much more to the story that deserves our attention. To understand the recent events sketched above and to appreciate their broader significance for the social sciences, federal science policy, American political culture, and for the NSF itself, we require a deeper historical perspective.

Indeed, the NSF's engagements with the social sciences were problematic right from the beginning. In his famous national science policy report, *Science—The Endless Frontier* (1945), Vannevar Bush proposed a new and comprehensive science agency as the centerpiece of a greatly expanded postwar national science system. His report suggested that the task of creating such an agency should concentrate on supporting the natural sciences, while the social sciences could be set aside for later consideration. Shortly after Bush transmitted his report to President Truman in July 1945, a landmark national science debate over competing legislative proposals for a new science agency ensued. Finally, in 1950, the federal government passed enabling legislation for an independent and thus officially nonpartisan National Science Foundation, which would be located in the Executive Branch and dependent on Congress for annual budgetary appropriations.[13]

Led by a full-time director and a twenty-four-member governing board (called the National Science Board), the NSF's primary mandate was "to promote the progress of science" and, in doing so, to advance the "health, prosperity, and welfare" of the nation and "secure the national defense."[14] The NSF also had a specific directive to fund basic science, which, throughout the agency's history, has played a crucial role in shaping its mission and activities. The idea was that support for basic scientific investigation, rather than research with more practical aims in view, would advance the frontiers of scientific inquiry. Scientific advances, in turn, would lead, somewhere down the road, to powerful applications and practical benefits. Thus, in the long term, federal support for basic science would result in major contributions to national well-being. Along with this basic science focus, a strong commitment to funding the "best possible science" defined the agency's character from early on.[15] In pursuit of these objectives, it awarded research grants to scientific project proposals approved through a competitive, multilayered evaluation process that placed great weight on the judgment of scientific peer reviewers, who were instructed to focus principally on a proposal's scientific merit.[16]

Efforts to place the social sciences in the NSF faced an uphill battle from the beginning. As noted earlier, Vannevar Bush suggested they were not as important as the natural sciences. In addition, many passages in the 1950 NSF charter specifically mentioned the physical and biological sciences, thereby guaranteeing that the agency would have a strong natural science orientation. Yet the charter did not mention the social sciences—not even once. Yet the charter did include passages stating that the agency could support "other sciences" beyond those specifically mentioned, thereby leaving open the possibility that the social sciences would eventually be included and supported.

Not surprisingly, social scientists and their advocates in the nation's political and scholarly communities recognized a major opportunity here. Being included under the NSF umbrella would give them a place alongside the more firmly established, more highly valued, and more generously funded natural sciences. Inclusion would also mean at least some funding for scholars with basic social science research projects judged, by the agency's evaluation process, to be of high scientific caliber.[17]

As it turned out, the young NSF decided to include the social sciences and eventually became a patron of singular importance for them. In the decades that followed, the NSF provided extensive funding to advance scientific inquiry carried out by the nation's sociologists, political scientists, anthropologists, economists, and scholars in related fields of inquiry at a wide range of universities, colleges, and research institutes. In 1988, Roberta B. Miller, the head of the NSF Social and Economic Sciences Division at the time, recognized the agency as the flagship of American social science, attesting to the far-reaching significance it had acquired since its founding. However, as Miller herself noted, that positive assessment regarding the NSF's special role also came during a decade filled with serious difficulties, beginning with a 1981 proposal to drastically cut NSF social science funding put forth by the newly established Republican administration led by President Reagan.[18] The story of the social sciences at this agency was never an easy one.

***

In the chapters that follow, I examine the contentious story of the NSF's social science efforts, starting from the beginning. The principal time period of analysis runs from the mid-to-late 1940s, marked by the end of World War II and the onset of the Cold War, up through the end of the Reagan presidency in the late 1980s.[19] In the final piece of this study, we will see how insights gleaned from that time span provide an essential foundation for understanding more recent developments and the continuing controversy over the NSF's engagements with the social sciences leading up to the present day.

Telling this story involves four levels of analysis. The first one focuses on the NSF's social science efforts in relationship to the agency's broader history. Here, we will consider many basic issues of interest, including the evolution of the agency's social science policies and programs within the context of the agency's mission, leadership, structure, and budget; changes in the organizational standing, representation, and status of the social sciences inside the agency; and the trajectory of NSF funding for these sciences. The second level of analysis concerns the agency's position within the federal science establishment and the national science funding system more generally. This is necessary to understand how this one agency, which at the outset was rather small and rather cautious in approaching the social sciences, nevertheless became a leading federal patron of academically oriented social science.

The third level investigates the importance of NSF social sciences as a regular focal point for discussions about their nature and value within American political culture, especially within Congress. As hinted at already, those discussions have often had a strong partisan underpinning. The fourth and final level of analysis centers on the agency's relevance to the social sciences themselves. Topics of particular concern here include the agency's participation in longstanding debates regarding the scientific identity and practical value of the social sciences; its role in providing valuable support for social science research, theory, and methodology; and efforts by social scientists and social science organizations to support, reform, and sometimes critique NSF social science.

THE CENTRAL ARGUMENT

The primary goal is to establish the NSF's considerable yet underappreciated importance in the recent history of American social science. With this in mind, my central argument consists of two distinct but closely related claims.

The first is that the particular conditions and developments in American society, politics, and science that shaped the NSF's legislative origins and formative years had a powerful influence on the establishment of what I call a *scientistic* framework for understanding, evaluating, and supporting the social sciences. The conditions and developments of greatest relevance from the mid-1940s through the late-1950s include the following: the enormous presence of the natural sciences, especially the physical sciences, in the national science policy arena and federal science establishment; the corresponding second-class position of the social sciences in those spheres; intense partisan conflict over the social sciences, including conservative criticisms that raised sharp doubts about their scientific status, practical value, and political involvement; widespread (although by no means universal) agreement among social science leaders that gaining inclusion in the NSF had paramount importance in the nuclear age; and salient features of the new agency itself, especially its basic science mandate and natural science orientation.

At the NSF, the scientistic strategy rested on a number of principles or assumptions. Above all, this strategy supposed that the social sciences were part of a putatively unified scientific enterprise. In addition, it assumed (at least most of the time) that these sciences were relatively immature, junior partners compared to the allegedly more rigorous and more advanced natural sciences. Furthermore, this strategy postulated that, just as in the natural sciences, the key to progress in the social sciences depended on advances in basic research, at the levels of data, methodology, and theory. Such work would, presumably, yield reliable, value-neutral, and nonpartisan knowledge about fundamental features of society and basic social processes. In language commonly found in NSF documents, the agency committed itself to supporting work at the "hard-core" end of the social research continuum, defined as social inquiry that satisfied (allegedly) universal and rigorous scientific criteria, including objectivity, verifiability, and generalizability.

Furthermore, the agency's scientistic framework assumed and promoted a sharp distinction between social science and a variety of other things that the agency sought to distance its work from. These included ideology, philosophy, social criticism, and humanistic inquiry. In general, the agency took a strong stance against any viewpoint that posited fundamental differences between the social sciences and the natural sciences that seemed to threaten the presumed underlying unity of the sciences. Such a view, which would have implied the need for an alternative to the hard-core emphasis and scientistic strategy, never gained much support inside the agency.

In making these distinctions between legitimate social science and other areas of human activity that don't merit recognition as such, the agency became deeply engaged in what sociologist of science Thomas Gieryn and other scholars have referred to as scientific boundary work.[20] The agency thus became an important participant in the debate over scientific identity, a debate that has played a crucial role in the development of American social science ever since the initial period of professionalization and disciplinary formation in the late nineteenth and early twentieth centuries.[21]

My second claim is that the NSF's role in scientific boundary work and its scientistic framework had deep implications for developments in other areas. These areas correspond to the four lines of analysis identified above: concerning the NSF, the federal science establishment, American political culture and partisan politics, and the social sciences.

To begin with, at the NSF, the scientistic strategy was a pivotal rhetorical move as well as the basis for the agency's efforts to promote the social sciences in specific ways. This took place through the elaboration and implementation of agency policies, programs, priorities, and funding criteria for these sciences. In addition, the scientistic strategy both reflected and reinforced their rather weak position inside the agency, as will become evident by considering the presence of the social sciences in the agency's top leadership positions, the organizational standing of NSF social science programs, and the share of the NSF budget allocated to the social sciences. Over the decades, the agency grew in size considerably, as did the extent and scope of its programs aimed at strengthening the nation's social science enterprise. In addition, in 1968, the passage of landmark legislation known as the Daddario amendment moved the agency beyond its original basic science focus by giving it a new responsibility for supporting research relevant to national problems, including problem-oriented and applied research in both the natural and social sciences. Nevertheless, throughout this period and continuing into later decades, the agency maintained an unwavering commitment to a scientistic strategy for advancing the social sciences, reinforced by doubts about their scientific credentials raised by top natural science administrators, as well as an abiding quest for scientific legitimacy and funding within the social sciences themselves.

That strategy, with its assumptions about scientific unity, scientific hierarchy, and natural science superiority, helped to carve out a space for the social sciences at the agency. Equally important, it did so in a way that ensured their marginality.

The scientistic strategy also played a crucial role in establishing the significance of the NSF within the federal science establishment. No doubt, the dominant presence of the natural sciences, especially the physical sciences at leading government science agencies and in top science policy circles, influenced the ways in which the NSF engaged with the social sciences. However, as the agency itself grew in size and importance, it became known as the premier federal patron for academically oriented science, including academically oriented social science research. As a result, the agency's particular approach to the social sciences became a central point of reference in national science policy discussions. Moreover, the agency's scientistic strategy was considered the only appropriate one by other influential nodes within the federal science establishment, especially at the National Academy of Sciences. This situation made it extremely difficult to mount a case for federal funding based on an alternative and broader vision of the social sciences. These circumstances, in turn, raised the question of whether the nation needed a separate social science agency, an option that did in fact receive extensive attention in the late 1960s.

The NSF's engagements with the social sciences became a major focal point in discussions and debates about their intellectual character, practical value, and social relevance within American political culture and partisan politics as well. Liberal politicians, especially liberal Democrats, provided the strongest and most consistent support. For them, the agency contributed to the overall health and vitality of American social science through its commitment to first-class scholarship, its insistence on high scientific standards, and its support for objective, nonpartisan research whose results deserve respect regardless of one's politics or ideology. At the same time, liberals often anticipated that social science research and expertise would be useful in advancing causes that they supported, such as tackling the problems of racism and poverty. On the other side, the strongest and most frequent criticisms came from conservative politicians and the Republican Party. The reasons here surely include the common conservative worry that a great deal of social science has been aligned with liberal and more decidedly leftist causes, as seen, for example, in the extensive involvement of social scientists in the New Deal of the 1930s and, three decades later, in the Great Society of the 1960s. Another major line of conservative criticism focused on curbing NSF social science funding on the grounds that research in this area simply isn't very important to the national welfare and is much less valuable than research in the hard sciences. This partisan undercurrent is striking throughout the agency's entire history.

Last but not least, the evolution of the NSF and its commitment to a unified scientific enterprise, wherein the natural sciences were often taken as the gold standard, gave the agency far-reaching importance for the social sciences themselves. For social scientists, for professional social science associations, for social science research organizations, and for the nation's research universities, the NSF became a major source of public funding. On top of this, the agency's scientistic outlook made it directly relevant to the long-standing quest for scientific legitimacy. Sometimes, the agency's importance in this area concerned the social sciences broadly, as will be seen in strong NSF support for quantitative methods of study, in its insistence on particular epistemological ideals such as value neutrality, objectivity, and generalizability, and in its promotion, starting in the 1970s, of large-scale social science databases and other projects akin to "big science" in the natural sciences. At other times, the agency took up questions about scientific methodology and epistemology with respect to a particular social science discipline or interdisciplinary field of study, as will be seen in NSF backing of behavioralism in political science starting in the 1960s or in the agency's recognition of economics as the most scientifically rigorous of all the major social science disciplines and hence especially deserving of support. These are a few examples that shed light on the sort of social science projects that the agency funded and, in the process, bestowed a measure of financial support, scholarly prestige, and scientific legitimacy to developments in the social sciences that meshed well with the agency's scientistic outlook.

THE POLITICS–PATRONAGE–SOCIAL SCIENCE NEXUS IN COLD WAR AMERICA AND BEYOND

By examining the development and impact of science patronage, historians have learned a lot about the nature of the sciences and their place in the broader society. This point provides a basic motivation for the present study of NSF social science. In addition, focusing on the NSF addresses a significant gap in the scholarly literature. To clarify the nature of this gap, a few observations about the state of scholarly inquiry will be helpful.

In the past three decades or so, historians have produced a rich body of literature in this area, including a small mountain of studies on American science, patrons, and society during World War II and the Cold War era. Numerous books and articles have explored the development of private and public funding bodies, funding priorities, and relationships between the sciences and their patrons across a wide arrange of disciplines and interdisciplinary fields of study, including the physical, biological, medical, engineering, earth, psychological, and social sciences.[22] This literature attests to a widespread interest in making sense of the tremendous growth and heightened presence of the sciences in national and global affairs during the nuclear age. It also reveals considerable concern about the growing dependence of scientists, scientific projects, scientific infrastructure, research universities, and other scientific institutions on extra-university sources of funding, especially the federal government, including military and intelligence agencies.

In the case of the social sciences, historians have concentrated on work deemed relevant to the great social, economic, and political problems of the times, and the considerable importance of private and public patronage in supporting and shaping such work. Focusing on the complex and often problematic relationships between the social sciences and the national security state, scholars have charted the evolution of funding from military and intelligence agencies and their support for major research centers, such as the Special Operations Research Office and RAND. They have also analyzed the mid-to-late 1960s' debate over military patronage set off by the furor over Project Camelot, a U.S. Army–sponsored study of the revolutionary process that aimed to produce knowledge for counterinsurgency purposes. Other historical works have examined the enormous impact of the large private foundations (i.e., the Ford Foundation, Rockefeller Foundation, and Carnegie Corporation) on the development of area studies programs (Russian studies, Latin American studies, Southeast Asian studies, etc.) and on the evolution of support for social science research on related topics such as modernization, development, and population control.[23]

But historians have devoted much less attention to social science funding channeled through civilian agencies, including science agencies. Furthermore, work in this area has focused mainly on funding for research closely tied to specific policy issues regarding pressing social problems such as poverty.[24] Consequently, the question of how civilian agencies helped to shape and how they fit into the social science patronage system in the U.S. during the Cold War era has suffered from relative neglect. This lack of analysis is especially problematic in the case of the NSF.

I began to establish the NSF's importance as a central patron for American social science in my previous book, *Shaky Foundations: The Politics–Patronage–Social Science Nexus in Cold War America* (2013). There, I examined the development, from the mid-1940s to the early 1960s, of a largely new extra-university funding system for the social sciences, which included the NSF, the U.S. military, and the Ford Foundation. In certain respects, these funding bodies were very different from one another, for example, in the critical distinctions between private and public patrons, as well as between civilian and military patrons. Nevertheless, I showed that all three of these funding bodies supported two central commitments, to scientism and to social engineering, which makes it reasonable to consider them as complementary funding sources within a single, albeit loosely integrated, funding system. As suggested by the book's title—*Shaky Foundations*—my analysis also identified vulnerabilities in this funding system, including pointed challenges to scientism and social engineering, which rendered the system unstable by the mid-1960s. In the book's conclusion, I suggested that those criticisms are crucial for understanding the subsequent emergence, during the mid-to-late 1960s and continuing into the following decades, of impassioned debates about how powerful patrons (i.e., the military, intelligence agencies, and private foundations) threatened to compromise the intellectual, professional, and ethical integrity of the social sciences.[25]

The present study singles out the NSF for more extensive attention and covers a much longer period of time. Doing so deepens our appreciation of this agency's broad and enduring importance in the history of American social science during the Cold War era and beyond. This analysis also shows that to understand the evolution of the politics–patronage–social science nexus more fully, we need to recognize the special role of this civilian science agency and the broad significance of its scientistic strategy.

NOTE ON SCIENTISM

Throughout this study, I will use the term *scientism* in a specific sense. Readers may be familiar with a broader meaning of the word than I intend: the notion that scientific inquiry provides the only source of trustworthy knowledge in any realm.[26] Here, I use the term, instead, to refer to the notion that the social sciences are part of a unified scientific enterprise, wherein the natural sciences are often considered more rigorous, more objective, and more advanced, and hence following their lead seems to be a valuable—and perhaps even an essential—strategy for making progress in the social sciences.

This notion still leaves considerable room for interpretation, however. It has never been obvious what exactly following the natural sciences should entail at various levels of the social science enterprise, including the subject matters, methods, and aims of social science research; the investigative stance of social researchers; arrangements concerning the social, institutional, and financial supports for social research; the social practices and organization of the social science community; and relationships between the social sciences and other spheres of society, including the political arena. In the course of this study, we will see how scientism gained traction as a general strategy at the NSF, how it was interpreted with regard to those more specific issues noted above, how it was implemented at various levels, and how the agency's social science policies, priorities, practices, and programs developed, encountered various challenges, and were modified in response to changing circumstances.

Note as well that the individuals in this study whose views I characterize as scientistic, including many natural scientists and social scientists, did not use this term to define their position. Typically, they claimed that modern science, especially recent science, had made tremendous progress; that such progress depended on developments in scientific epistemology and scientific practice as well as in the growth of particular types of social, political, and financial supports that facilitated and protected scientific research from corrupting factors; that the greatest progress had occurred in the natural sciences; and that to the extent that the social sciences failed to pursue a similar path, they remained insufficiently rigorous and perhaps did not even qualify as legitimate sciences. But proponents of this viewpoint simply said that they supported real scientific progress in the social sciences, rather than saying that they favored a particular viewpoint that I am referring to as scientism.

In addition, authors who have been critical of a natural science model for the social sciences sometimes deployed the term *scientism* with negative connotations. To take just one example, the famous libertarian political philosopher and economist Friedrich Hayek did so, first in a series of essays from the early 1940s and subsequently in his classic 1952 book, *The Counter-Revolution of Science: Studies in the Abuse of Reason*.[27] Here, then, is a good reason why advocates of the unity of sciences, who in many cases also urged the social sciences to follow in the footsteps of the allegedly more advanced natural sciences, would not have been inclined to use the term *scientism* to describe their own position.

A ROAD MAP

This book has ten chapters. The first one examines the problematic status of the social sciences in the landmark postwar national science policy debate that led to the NSF's founding. In light of the enormous expansion in federal science funding and the dramatic power of scientific weaponry during World War II, the initial legislative proposals for a new science agency focused, as Vannevar Bush did in *Science—The Endless Frontier*, on establishing adequate support for the natural sciences. Soon, however, a movement supported by leaders at the U.S. Social Science Research Council sought to include the social sciences based on the claim that they were part of a unified scientific enterprise. The case for inclusion also acquired significant support from liberal Democratic senator Harley Kilgore. In this context, controversy over the social sciences' scientific identity, political meaning, and practical relevance erupted, bolstered by considerable and mainly conservative opposition from influential figures in the nation's political and scientific communities.

Chapter 2 examines the development of a small but viable social science program at the young NSF during the early to mid-1950s. At this time, anti-communist, McCarthyite attacks created major headaches for social science scholars, organizations, and patrons. At the natural science–oriented NSF, responsibility for crafting a viable policy framework for social science funding rested largely with sociologist Harry Alpert. At the heart of Alpert's efforts lay a cautious, scientistic strategy, as he emphasized support for studies at the hard-core end of the social research continuum, in a manner that positioned the social sciences as junior partners to the natural sciences within a unified scientific enterprise and promised to keep the agency's social science efforts free from unwanted political criticism.

From the late 1950s to late 1960s, soaring federal science budgets following the national panic over Sputnik, together with the resurgence of reform liberalism during the Kennedy and Johnson administrations, provided a much more favorable environment for NSF social science activities. Chapter 3 explores a deepened commitment to the social sciences during these years, marked by growth in NSF funding and the landmark establishment of a social science division, which had the same organizational status (although not the same level of funding) as the divisions for the physical and biological sciences. At the same time, persistent worries about the scientific status of the social sciences and political disapproval kept expansion of the agency's efforts within carefully restricted bounds, as seen in the development of NSF funding for political science and a bold new program for social science curriculum building called MACOS—Man, A Course of Study.

The agency's growing importance as a major social science patron on the national stage is the subject of chapter 4, told through a tale of two contrasting legislative initiatives. In the House of Representatives, liberal Democrat Emilio Daddario put forth a successful bill—the 1968 Daddario amendment—that gave the agency responsibility for supporting relevant science and applied science, including such work in the social sciences. That amendment also made the agency's authority for supporting the social sciences more explicit than it had been. Meanwhile, over in the Senate, Fred Harris, another liberal Democrat, put forth a more ambitious but ultimately unsuccessful legislative proposal that would have given the social sciences an agency of their own—a National Social Science Foundation. Harris, along with many supporters of his initiative, argued that these sciences needed a separate agency within the federal science establishment, where they would not be watched over by the natural sciences and where a wider range of social research, including humanistic inquiry and social criticism, would be encouraged and supported, something that the established NSF was unwilling to do.

This brings us to the period from the late 1960s to late 1970s, which saw changes in American politics and national science policy that curtailed the expansionary era, ushered in sharp criticisms of the social sciences from various quarters, and had far-reaching implications for the NSF in particular. Chapter 5 analyzes three controversies that posed major challenges for the social sciences. One controversy concerned a new NSF program known as RANN—Research Applied to National Needs—and centered on skepticism about the agency's ability to support applied social research and, equally important, skepticism about the practical value of such research more generally. Another controversy arose from the charge, advanced most famously by the maverick Democratic senator William Proxmire, that various federal agencies were wasting hard-earned taxpayer money on esoteric research projects, including some NSF-funded studies in the social and psychological sciences. The third controversy involved sharp attacks from conservative quarters directed against MACOS, which by the mid-1970s had emerged, by a wide margin, as the NSF's most important social science curriculum project.

Chapter 6 shows how a series of changes inside the agency brought about a major repositioning and retreat of NSF social science activities by the late 1970s. Even so, those changes did not halt mounting displeasure by both critics and supporters of those efforts inside and outside the agency. Also of special interest, these troubles unfolded while the NSF was led by Richard C. Atkinson, an accomplished mathematical psychologist. Atkinson was the first social or behavioral scientist to hold the influential position of director—the previous ones had all been physicists or biologists.

The last period covered in detail features Ronald Reagan's years in the White House and a newly energized conservative movement, which had a profound impact on American political culture, national science policy, and the social sciences. Chapter 7 begins in the early 1980s, when the first Reagan administration sought to decimate NSF funding for these sciences. Although the administration claimed this was simply a prudent budget-tightening measure, informed commentators pointed out that conservative journalists, scholars, think tanks, and politicians regularly portrayed a great deal of social science as a politicized, ideological, and leftist enterprise that should be defunded. This situation provoked a widespread sense of crisis in American social science and galvanized social science leaders and organizations into action. It was in this context that the Consortium of Social Science Associations first emerged as a strong advocate for the social sciences, for federal social science funding, and for NSF social science in particular.

Chapter 8 returns to developments inside the NSF, exploring how and why the social sciences fared rather poorly here throughout the decade. Led by physicists and engineers during these years (the psychologist Atkinson had left in 1980), the agency provided strong support for the Reagan administration's position that federal science policy should focus on strengthening the economy and national defense but not on solving social problems, especially not through social engineering grounded in social science expertise. Under these conditions, the social sciences suffered dwindling support, although economics fared better than the other major disciplines. Equally important, my analysis shows that the troubles faced by the NSF and the social sciences did not entail an ideological reorientation of the sort that occurred in the burgeoning world of conservative think tanks and policy institutes. Thus, the longstanding NSF commitment to funding what Harry Alpert had long ago called hard-core social research, which presumably lay beyond partisan and ideological influences, remained strong.

The ninth chapter shifts gears to consider salient challenges to the notion that the social sciences should take the natural sciences as the gold standard and hence as an investigative model. Whereas Senator Harris's National Social Science Foundation proposal in the late 1960s had challenged the scientistic strategy within the NSF and the federal science establishment more generally, developments in American society and thought nourished a number of anti-scientistic currents. Although a wide-ranging examination of these developments lies beyond the scope of this study, consideration of a few currents with special relevance will further our understanding of the agency's importance as well as its limitations as a major social science patron.

The final chapter begins with a summary of the central argument before proposing that the main insights established in the preceding chapters can be extended in two directions. First, we will see how those insights are crucial for understanding the evolution and continuing importance of NSF social science in the post-Reagan years, all the way to the present day. Second, I offer an assessment of the agency's engagements with the social sciences that has special pertinence in our current "posttruth" age. I end with a call for fresh thinking about the need for a new social science agency—a National Social Science Foundation. *Author's note:* For a timeline of events occurring in the book, visit https://mitpress.mit.edu/books/social-science-what.

***

But we should not get ahead of ourselves. Any lessons for the present that might be gleaned from this study need to come later, as they rest on the material presented in the following chapters.

So, let's start at the beginning. In 1945, Vannevar Bush suggested that questions about social science funding could be left aside for the time being. Within the scientific community, more serious criticisms concerning the scientific status and social implications of the social sciences surfaced during the postwar NSF debate. The first order of business, then, is to examine how at the outset the question of whether or not the social sciences should be included in the proposed agency became so problematic.

1

TO BE OR NOT TO BE INCLUDED: UNCOVERING THE ROOTS OF THE NSF'S SCIENTISTIC APPROACH

> Today we are faced with the preeminent fact that, if civilization is to survive, we must cultivate the science of human relationships—the ability of all peoples, of all kinds, to live together and work together, in the same world, at peace.
> —President Franklin D. Roosevelt, 1944[1]

> I asked an able scientist yesterday if he would define social science. I had been worrying about that. He said in his definition, "In the first place I would not call it science. What is commonly called social science is one individual or group of individuals telling another group how they should live."
> —U.S. Democratic Senator J. William Fulbright, 1946[2]

From World War II through the early Cold War years, American science underwent a dizzying expansion, accompanied by a much greater federal presence in scientific affairs than ever before. As countless movies, documentaries, and scholarly studies remind us, the natural sciences and especially the physical sciences had special prominence during those years. The development and wartime uses of radar, computers, and, of course, atomic weapons acquired legendary status. According to one common viewpoint, those events even mark the emergence of a new era in human history: the nuclear age. The ensuing battle for global supremacy between the Soviet Union and the United States ensured that the physical sciences remained in the national spotlight, as their work was inextricably linked to the nuclear arms race, intercontinental ballistic missiles, Sputnik, and atomic spying. Under these conditions, fears about the destructive power of modern science-based weapons and associated worries about the character of modern science itself became pervasive. Still, nobody doubted that the physical sciences had made spectacular progress at the levels of research and theory. Nor did anybody doubt that breathtaking technologies associated with those scientific advances had revolutionary implications for human affairs.[3]

As for the social sciences, they surely participated in the general expansion of scientific activity. World War II gave them opportunities to pursue a variety of investigations of an urgent nature, from studies of German and Japanese national characters to research on psychological warfare, communications, propaganda, mental health, and the logistical management of wartime operations. In April 1945, President Roosevelt prepared a speech that included the remarkable passage quoted above, about the dire need to "cultivate the science of human relationships—the ability of all peoples, of all kinds, to live together and work together, in the same world, at peace."

In addition, "the Second World War was much more than a symbolic dividing line, for it generally strengthened the social sciences, laying the foundations for their postwar expansion," as Roger Backhouse and Philippe Fontaine, historians of the social sciences, have explained.[4] With the onset and intensification of the Cold War, psychologists, sociologists, political scientists, and scholars from related fields acquired an expansive array of new opportunities, starting with the task of constructing an American-friendly alternative to the Marxist model of development for so-called underdeveloped, traditional, or Third World societies. In the meantime, unprecedented expansion of the American university system supported rapid growth in the social science professions and their disciplinary associations, just as it did in the physical and life sciences.[5]

Yet in the nuclear age, social scientists often found their work under attack. Conservative opposition to the New Deal and to the Roosevelt administration's wartime propaganda efforts inspired disparagement of the social sciences, whose extensive involvement in those developments was well known.[6] Heightened political and cultural tensions associated with the postwar Red Scare and McCarthyism gave additional impetus to critics who claimed that leftist scholars and intellectuals, including many social scientists, were subverting the American way of life.[7] No doubt atomic scientists faced McCarthyite pressures as well. The notorious case in which the Manhattan Project's scientific leader, J. Robert Oppenheimer, lost his security clearance is just the tip of that iceberg. However, whereas during World War II and the Cold War physical scientists enjoyed generous levels of funding, high-level institutional representation, and substantial policy influence within a burgeoning federal science establishment, the position of the social sciences was considerably weaker. Furthermore, leading physical scientists were often skeptical and sometimes downright dismissive of the social sciences. On one occasion, Edward Teller, the "father" of the H-bomb, declared that they were no more scientific than Christian science.[8]

These challenges—concerning the social sciences' scientific identity, their practical value, their involvement in partisan efforts, and their standing in federal science affairs—all came into clear view during a postwar national science policy debate over proposals for a major new science agency—the National Science Foundation (NSF). Among the central questions in this debate, which took place from the mid-to-late 1940s, was whether the social sciences even deserved to be included in the proposed agency. Historical studies of American science and science policy since have had surprisingly little to say about the contested position of the social sciences in the NSF debate, although the episode itself has long been recognized as a landmark in twentieth-century American science policy.[9] Accounts of the social and psychological sciences themselves more regularly include some discussion. But most authors have concentrated on developments and debates regarding particular disciplines, thus leaving aside broader analysis of the meaning of the NSF debate for the social science enterprise.[10]

The first section of this chapter focuses on early developments in the debate that stimulated partisan conflict over competing legislative proposals for a major new science agency and implicated the social sciences in that conflict. Worried that the social sciences might be left out of the proposed agency altogether, the Social Science Research Council (SSRC) and its scholars developed a case for inclusion based on a scientistic outlook, as discussed in the second section. The third section explores how mounting pressure from a coalition of conservative politicians and scientists combined with lackluster support from the liberal camp to undermine the case for inclusion. In addition, the SSRC's position on the fundamental unity of the social and natural sciences faced sharp criticism from a left-liberal perspective, as discussed in the fourth and final section. These lines of analysis will reveal how and why the NSF debate became pivotal in shaping the problematic status of the social sciences in the federal science establishment and in American political culture as the nuclear age got under way.[11]

DEFINING THE STAKES

Beginnings are important. This tidbit of common wisdom goes a long way in illuminating the contested position of the social sciences, first in the postwar NSF debate and later during the agency's early years. The initial piece of this story involves the emergence of competing legislative proposals for a major new science agency in a context where partisan struggle was rife, where the lion's share of attention focused on the natural sciences, and where Vannevar Bush, one of the nation's leading science-policy figures and a crucial figure in this particular episode, sought to place the social sciences on the sidelines.

On July 19, 1945, Bush presented President Truman with a landmark science policy document, *Science—The Endless Frontier* (*SEF*). Bush's report had been commissioned the previous fall (in November 1944) by Truman's predecessor in the White House, Franklin Roosevelt. But Bush had more conservative views on political and economic matters than either of these liberal Democratic presidents. Bush was especially wary of the expansion of federal power and control over realms of the economy and society that he believed were best left to the private sector and individual initiative. His views became embedded in the postwar science policy agenda advanced in *SEF*. Also, on July 19, Washington Democratic Senator Warren Magnuson, who had been consulting closely with Bush, introduced a bill (S. 1285) to create an agency based on *SEF*'s outlook. Although on balance Magnuson was more of a moderate liberal, his bill adhered to the conservatively oriented agenda presented by Bush.[12]

A few days later, West Virginia Democratic Senator Harley Kilgore introduced a competing bill (S. 1297) with a decidedly liberal, even left-liberal, bent. A vigorous defender of the New Deal and a good friend of President Truman, Kilgore's initiative conflicted with Magnuson's *SEF*-inspired legislation on a handful of questions regarding the leadership, mandate, and scope of the proposed agency. Who would choose the agency's leader, and how responsive would that leader be to political demands and especially to the president's will? Would the agency support basic research or also scientific studies with more practical aims? Would the agency seek to distribute its funds across the nation's regions and states, or would it look to support the best science regardless of geographical considerations? Would the public have free access to the results of agency-sponsored studies, and would the government control patents based on those results? And, most important here, would the social sciences be included and, if so, how? Bills similar to those from Magnuson and Kilgore were soon introduced into the House of Representatives, thus establishing the basis for an extensive national science policy debate.[13]

Crucial for understanding the historical importance of the emerging NSF debate as well as its special relevance to the social sciences is the fact that the liberal and conservative proposals both called for the creation not just of a new science agency but of what Vannevar Bush in *SEF* referred to as an "over-all agency."[14] Early debate over these proposals assumed that such an agency would be comprehensive and would serve as the centerpiece of a greatly expanded postwar federal science establishment.

But neither Magnuson nor Kilgore had included the social sciences. Why were they left out at this crucial early stage?

One of the principal reasons concerns the overwhelming emphasis placed on the natural sciences during the wartime years. In *SEF*, Bush stated that at this critical juncture in world history, it would be "folly" to develop a new program for expanded support in the natural and medical sciences "at the cost of the social sciences, humanities, and other studies so essential to national well-being." Yet he also suggested leaving the social sciences (and humanities) aside for the time being, so that the nation could focus on the supposedly more important and urgent task of establishing an agency to deal with the natural sciences.[15] Consistent with this position, Magnuson's bill did not discuss the social sciences at all. On the other side, Kilgore's bill called for the provision of science funding as well as support for "related economic and industrial studies," but it did not mention the social sciences more broadly.[16]

On closer inspection, it is clear that the position suggested by Bush and then followed by Magnuson was intentionally unfavorable to these sciences. This is a tricky issue to parse because Bush sometimes discussed them in a favorable light. For example, in a 1946 letter about the state of the social sciences, he noted that while researchers often had to depend on observational methods rather than experimental techniques, so did astronomers. He also stated that there were "of course many branches of the social sciences where experimentation under controlled conditions is just as fully present as it is in many of the natural sciences."[17] Thus, it is not fair to say, as one important account does, that Bush was in general "an opponent of the social sciences."[18] Yet Bush certainly had substantial worries about them. At another point in that 1946 letter, he declared that "a tremendous number of charlatans work under the social science tent, which is of course something sound social scientists regret and combat, but which still persists."[19]

Furthermore, whatever he said about the social sciences at certain moments, Bush had purposely excluded them from *SEF*—which provided the basis for their subsequent omission in Magnuson's bill. In a letter to President Truman and in various public statements, Bush did not make his role in excluding them evident. This left the impression that he had carried out the task of developing a national science policy report simply by following the guidelines given to him by President Roosevelt. But the truth, evident in unpublished correspondence, is that Bush had helped shape Roosevelt's request in the first place, such that the resulting report would confine "its attention to the natural sciences." In a letter to D. C. Josephs from the Carnegie Corporation, Bush also suggested that if "a similar inquiry

Figure 1.1
President Truman, Vannevar Bush, and National Defense Research Committee members, January 20, 1947. Seated left to right: James Conant, Harvard University president; President Harry S. Truman; and Alfred N. Richards, National Academy of Sciences president. Standing left to right: Karl T. Compton, MIT president; Lewis H. Weed, National Research Council Division of Medical Sciences chairman; Vannevar Bush, Office of Science and Development director; Frank B. Jewett, National Academy of Sciences former president; J. C. Hunsaker, National Advisory Committee for Aeronautics chairman; Roger Adams, University of Illinois chemist; A. Baird Hastings, Harvard Medical School; and A. R. Dochez, Columbia University. Copyright Unknown, Courtesy of Harry S. Truman Library and Museum.

into the social sciences" were desired, it "should be done through other channels."[20]

More important in the bigger picture, however, is the fact that Bush had reasons for excluding the social sciences that reflected prominent concerns in conservative quarters. As the historian of American science Nathan Reingold has pointed out, Bush firmly opposed the more ambitious planning components of the New Deal along with the expanding "welfare state and its centralizing tendencies."[21] He thus found the extensive involvement of social scientists in a wide array of New Deal agencies and programs troubling. One of the most important of those agencies was the National Resources Planning Board (NRPB), which included among its leaders the Columbia economist Wesley Clair Mitchell and the Chicago political scientist Charles Merriam. To the present day, the NRPB stands out as the only comprehensive planning agency in the nation's history.[22] Although attacks by conservatives in the political and business communities led to the NRPB's demise in 1943, many liberal economists, political scientists, and other scholars who had participated in the board's activities continued to advocate national planning initiatives for the post–World War II era.[23] Not only did Bush disagree with their politics at a fundamental level, but he also tried to curb their influence in postwar federal science policy. His effort to place consideration of the social sciences on the backburner in *SEF* and then in the NSF debate has to be understood in this broader context of partisan struggle over the future of American politics and society.

★★★

Throughout the summer of 1945, it seemed that social scientists themselves might not have much of a voice in this episode. Nevertheless, leading scholars at the SSRC were already considering the relevant policy issues among themselves. Founded in 1923 and with its headquarters in New York since 1927, the council was a private, nonprofit scholarly society whose members included representatives from the major professional associations for anthropology, economics, history, political science, psychology, sociology, and statistics. Council activities were carried out through its many committees, while the Problems and Policy Committee, board of directors, and president took care of general policy making and program oversight. The SSRC's influence extended in many directions. Heavily dependent on private patrons for support, especially the enormous Rockefeller and Carnegie philanthropies, the council served as a key intermediary organization that linked big philanthropy to the universities, extra-university research centers, scholars, and projects. The Council worked hard to promote scientific knowledge that promised practical payoffs. Its efforts to strengthen social science methodology acquired special renown, as did its support for interdisciplinary work, for example, in the study of personality and culture pursued by psychologists, psychiatrists, sociologists, and anthropologists.[24]

By 1945, the SSRC had established itself as the nation's leading organization responsible for the entire field of social science activity, in contrast to other scholarly associations that focused more narrowly on a particular discipline such as sociology or economics. In addition, the council helped bridge the worlds of academia and government. During the 1930s, council committees on social security and public administration worked closely with New Deal agencies. And in the early 1940s, the council established an office in the nation's capital to promote social science contributions to the war effort. Thus, well before the unveiling of competing legislative proposals for a new postwar science agency, council leaders were intimately familiar with the changing relationship between the federal government and the social sciences.

The council was especially concerned about the impact of wartime developments on the support, conduct, and uses of social research. To address these matters, in 1944, the council created a Committee on Federal Government and Research (CFGR), chaired by the economist Wesley Mitchell. Mitchell was also a former NRPB member, a long-time director of the National Bureau of Economic Research (1920–1945), and one of the original founders of the SSRC. Other CFGR members included the psychologist Robert Yerkes, famous for promoting intelligence testing during World War I and for his investigations with great apes; the sociologist William Ogburn, well known for his pioneering studies about the social implications of technological change and the concept of cultural lag; and the economist Edwin Nourse, who in 1946 would become the first chairman of the nation's new Council of Economic Advisors.[25]

SSRC discussions about the federal government's massive wartime role in supporting scientific research revealed deep worries about the potential for political control and corruption. In April 1945, Mitchell noted, at a Problems and Policy Committee meeting, that the members of his committee opposed the "general subsidization of research institutions" by the federal government. A dense, single-spaced, six-page draft report called "The

Figure 1.2
Edwin Nourse Reports to President Truman's Cabinet, October 3, 1947.
Front row, left to right: Acting Secretary of State Robert Lovett, Secretary of Defense James Forrestal, President Harry S. Truman, Secretary of the Treasury John Snyder, and Attorney General Tom C. Clark. Standing, left to right: Vice Chairman of the Economic Advisory Council Leon Keyserling, Postmaster General Robert Hannegan, Assistant to the President John Steelman, Secretary of Commerce Averell Harriman, Secretary of Agriculture Clinton Anderson, Chairman of the Economic Advisory Council Edwin Nourse, Secretary of the Interior Julius Krug, Secretary of Labor Lewis Schwellenbach, and Federal Works Administrator Philip Fleming.
Copyright Unknown. Courtesy of Harry S. Truman Library and Museum.

Federal Government and Research" identified many concerns that informed his committee's consensus on this point, among them doubts about "the feasibility of guarding subsidies against the intrusion of restraints upon freedom of inquiry" and "the great importance of maintaining the social scientist's complete independence with respect to many delicate problems of government policy and action." In addition, the dangers associated with "any tendency toward centralized control" over the work in any particular field led Mitchell's committee to emphasize the dire importance of maintaining and fostering a "diversity of sources of support for research."[26]

Yet a few months later, following the publication of *SEF* and the introduction of competing legislative bills by Magnuson and Kilgore, worries about government control of research were overshadowed by mounting concerns within the SSRC about the possible exclusion of the social sciences from the proposed science agency. A different interpretation appears in a major history of the council by the sociologist Donald Fisher, who claims that persistent fears about the impact of massive governmental support became the "major cause" for the council accepting the social sciences' exclusion from the legislative proposals. He also says that council leaders were not convinced that healthy scientific growth required substantial federal support in the first place.[27] However, my analysis of archival documents reveals that ongoing discussions within the council led to a collapse of the previous consensus. It's true that some scholars continued to express strong concerns about the potential dangers associated with large-scale federal support. But other scholars had concluded, by late July, that the earlier oppositional stance had been "too hesitant and fearful," that governmental support in the postwar era would continue to increase, and that the social sciences had a "greater need for federal support" than the physical sciences. These scholars also suggested that mechanisms for regulating such support could be established without "any disadvantages from centralized control."[28]

This more optimistic stance regarding federal support acquired greater appeal when, in an unexpected twist, the nation's president brought the social sciences into the picture. In early September, in an event widely covered in the press, Truman presented his reconversion address to Congress—about moving the nation from a wartime to a peacetime footing. As for postwar science planning, Truman, who in general favored Kilgore's approach, called for the establishment of a "single Federal research agency" that, among other things, would "promote and support research in the basic sciences and in the social sciences."[29]

Determining why Truman decided to include these sciences is difficult, probably because there was no single reason but a few that all pointed in the same direction. He may well have wanted to extend the close relationship between the federal government and the social sciences established during Roosevelt's presidency. Truman was also surely familiar with Roosevelt's undelivered Jefferson Day address from April 1945 that noted the need to "cultivate the science of human relationships." Furthermore, shortly before the president's September reconversion address, the Bureau of the Budget director Harold Smith told Truman that an "unbalanced research program" marked by much greater support for the physical sciences would leave the social sciences "overshadowed by the impact of the atomic bomb."[30]

In any case, besides inserting the social sciences into the national discussion in a highly visible way, Truman's address also linked them to the liberal science policy agenda presented by Kilgore. That, in turn, placed them in the midst of simmering partisan conflict.

Back at the SSRC, a board of directors meeting held shortly after Truman's address revealed a new consensus on the point that only the federal government could provide the "augmented funds" needed to allow social science research to "proceed at the maximum feasible rate." As the sociologist Samuel Klausner pointed out a few decades later, "any bashfulness on the part of the SSRC about plunging into the federal arena receded." To be sure, an array of difficult issues remained—regarding protections against political control of federally funded research to ensure the freedom of investigators and the integrity of research. Still, views at the council had undergone a decisive shift away from the earlier oppositional stance. A member of Mitchell's committee summed up the dominant new viewpoint by noting that the SSRC now seemed "forced to approve of social science participation or else lose any influence it might have" in the emerging science policy debate.[31]

★★★

If the developments noted above suggested that the bid to include the social sciences would be controversial, the nature of the controversy came more fully into view that fall during the first round of legislative hearings, held by the Senate, on the competing proposals. Altogether, Congress heard from nearly 100 witnesses from various sectors of American society, including government, higher education, business, and labor. The distribution of opinion revealed a reasonable level of support for the social sciences. Thirty-seven of the forty-five witnesses who commented on this issue supported their inclusion, while the others either favored inclusion though with some reservations or supported the creation of a separate social science agency.[32]

Yet Vannevar Bush presented a cautious position in his testimony. As in *SEF*, he recommended that consideration of the social sciences be put aside for the time being. Later, after the agency was established, another science policy study could determine the basis for "effective integration and partnership between the natural and social sciences."[33] Here, Bush did not directly challenge the status of the social sciences. Yet he implicitly suggested that they might be quite different from the natural sciences and thus required separate consideration.

In addition, some of Bush's close allies in the scientific community explained why the social sciences seemed so different. This distinguished group included the physicists Karl Compton and Isadore Rabi, the chemists Bradley Dewey and Roger Adams, the engineer Boris Bakhmeteff, the geographer Isaiah Bowman, and the medical scientists Henry Simms and Morris Fishbein. All of them supported Bush's science policy views. All of them favored the science agency proposal in Magnuson's legislation. And all of them conveyed distinct reservations about the social sciences.[34]

One set of concerns focused on the social sciences' shaky scientific status. According to the chemist Adams, social science inquiry differed so much from natural science inquiry that the former did not really qualify as scientific. Furthermore, this harsh judgment was not merely his personal opinion. The main professional society in his discipline, the American Chemical Society, opposed the inclusion of social science on the grounds that "the methods of approach to the study of its problems, the complete lack of any fundamental laws, [and] the necessity of analyzing vast bodies of facts, often unrelated, place this subject in the field of the humanities."[35] Others put forth similar concerns, with the Nobel Laureate physicist Isadore Rabi claiming that whereas natural scientists arrived at "quite objective" results through controlled experimentation, the work of social scientists was often "controversial" because they had great trouble proving that their findings were sound.[36]

A second set of concerns, which were often linked to the problem of scientific identity, involved the social sciences' engagement with politics and other matters involving value judgments and perhaps ideology, which the natural sciences were supposedly free of. As historian J. Merton England's account of the agency's legislative origins points out, scientists who favored Bush's views and opposed the effort to include social science typically found the latter's contributions to "governmental planning" troubling.[37] The medical scientist Henry Simms conveyed this concern with flare, as he warned that including social science might lead to the "the promotion of crackpot schemes for altering the form of government."[38] A statement from the Union of American Biological Societies declared that the social sciences "should not be included" because they "imply activities which are not purely objective and, therefore, may introduce disharmony through divergence of political viewpoints."[39]

These two related types of reservations carried additional bite because they dovetailed with the basic science ideology that Vannevar Bush presented in *SEF* and that many other scientists subscribed to. Basic science, as Bush explained it, is research "performed without thought of practical end. It results in general knowledge and an understanding of nature and its laws." These results then become the basis for "practical applications," thus serving as the "pacemaker of technological progress." In this way, basic science is crucial to the nation's "health, prosperity, and security."[40] In the NSF debate, and in postwar science policy discussions more broadly, these ideas provided a common framework for thinking about the nature of science, the relationship between science and technology, and the link between scientific advance and national well-being. An influential group of Bush's allies in the scientific, engineering, and medical communities suggested that the social sciences did not fit within this framework, however. Including these sciences would thus be problematic. Consider Bradley Dewey's position. Although he was the nephew of John Dewey and recognized his uncle as "the greatest living philosopher," they did not see eye to eye on this matter. Whereas Dewey the pragmatist philosopher often objected to strong distinctions between the social and natural sciences, Dewey the chemist saw them as worlds apart. In the latter's memorable comment, the natural sciences and the social sciences needed to be kept apart just like "hair and butter."[41]

In sum, events during the summer and fall of 1945 revealed the high stakes for the social sciences. SSRC discussions had led to the position that inclusion would be crucial for their future health. However, the early unfolding of the NSF debate highlighted a number of obstacles that lay in the way: the overwhelming emphasis at the time on the natural sciences; Vannevar Bush's maneuvering to place the social sciences on the sidelines; the call by President Truman to include them as part of Kilgore's proposal, which further stoked conservative fears; and the critique of the social sciences' scientific status and political valence presented to Congress by a small but powerful group of conservatively minded scientists who supported Bush's policy recommendations.

That fall, a group of leading social scientists from the SSRC were called on to present their case to Congress—giving them a chance to respond to the critics.

PITCHING SCIENTISM

Shortly after Truman's reconversion address, the SSRC received an invitation to testify at the fall hearings. The invitation came from Senator Magnuson, who noted that he and Senator Kilgore were seeking advice about the social sciences and special challenges, such as the greater difficulty in making "the distinction between fact and opinion" than in the physical sciences, and the problem of establishing sufficient measures to ensure the "objectivity" of federally supported social research. Recognizing the council's vital national role in representing "all of the diverse social science groups," Magnuson asked it to study and report on the social sciences' position regarding the proposed bills.[42] The council thus became responsible for constructing a coherent position on behalf of the American social science enterprise.

In November, a group of SSRC scholars presented their views during the first round of NSF hearings. By this point, Vannevar Bush and his allies who were opposed to including social science had testified. So, making a strong case for inclusion required challenging the critical position already established by others.

The scholars selected by the council included the economists Wesley Mitchell and Edwin Nourse, the psychologist Robert Yerkes, the sociologist William Ogburn, the political scientist John Gauss, and the anthropologist John Cooper. Many of them were well-known advocates of the view that the social sciences needed to establish more rigorous scientific foundations, marked by the search for objective truth, and that to do so, they should strive to emulate the more mature natural sciences. For instance, in his 1929 presidential address to the American Sociological Society (later renamed the American Sociological Association), Ogburn famously declared that "scientific sociology" was not interested in improving the world. Instead, as with the rest of the scientific enterprise, scientific sociology was "interested directly in one thing only, to wit, discovering new knowledge."[43] Another good example comes from the psychologist Yerkes, who, in a 1947 article on the "scope of science," remarked that "if the natural phenomena designated by such terms [behavior, experience, social relations, and institutions] are not amenable to systematic study by procedures which are in principle the same as those employed in the physical sciences, hosts of us who now are classified as scientists actually are self-deceived workers."[44]

At the November legislative hearings, Mitchell informed Congress that the SSRC agreed with President Truman's science policy recommendations and presented a council memorandum that argued for including social science. By this point, council discussions, as well as communications received from the broader social science community, underscored a conviction that extensive federal support would be essential for realizing the great potential value of the social sciences to society and government. To be sure, the dangers of government control and corruption still deserved careful consideration, as duly noted in the council memorandum:

> Restraints on objectivity of inquiry by any agency, public or private, tend to debase research into the preparation of apologia and rationalizations for the courses of those under whose dominance research is undertaken. Social science, under these circumstances, ceases to be scientific in spirit.[45]

Nevertheless, SSRC discussions had also concluded that "sufficient safeguards" could be established, by keeping in mind the need to preserve "freedom of research" and by maintaining a "diversity" of public and private sources of support, thus enabling researchers and institutions to "resist domination by any one interest."[46] The economist Nourse reinforced this point in his testimony, as he rejected the notion that large-scale federal support would necessarily "result in an impairment of the intellectual freedom of the social science worker."[47] Furthermore, the fact that the social sciences were immature compared to the natural sciences was, according to the council memorandum, "the strongest possible reason for advancing their development by every effective means."[48]

More fundamentally, Mitchell and his peers asserted that the social sciences were basically no different from the natural sciences. The alleged unity among the sciences thus became another key point in the case for inclusion.[49] Whereas critics from the natural sciences, medicine, and engineering claimed they saw crucial differences between the social and natural sciences, Nourse spoke of the "inherent unity of science."[50] "Traditional lines of demarcation between the natural and the social sciences have little meaning when confronted with the research problems involved in the safeguarding of the human aspects of every major problem of national interest," added the council memorandum.[51] In a similar vein, Yerkes remarked that "existing demarcations or barriers among the several sciences or principal areas of science are entirely artificial."[52]

The relationship between knowledge production and practical application was remarkably similar across the sciences as well, asserted these scholars. Whereas detractors had charged that social science knowledge claims were often biased and thus not trustworthy, the SSRC memorandum offered a more optimistic view: when "properly trained research workers imbued with scientific detachment and integrity" had "adequate resources" to carry out "earnest and objective investigations," they could produce "results of inestimable practical value."[53] Nourse added that all sciences aimed to "get fuller and more accurate knowledge" about the "materials" and "forces" in our world. When done properly, such work, whether in the social or natural sciences, produced unbiased, reliable knowledge that could be used, in instrumental fashion, to control materials and forces such that "mankind may have a safer and more satisfying existence."[54]

As such remarks about scientific detachment, objective investigation, and practical application suggest, the council's social scientists were also eager to address the worry that social research could be infected with personal opinions, social values, and ideological agendas. The political scientist Gauss, for one, acknowledged that the social sciences were sometimes suspect because their investigations seemed to be "too greatly influenced by the prejudices of their participants to constitute true sciences, presumably characterized by dispassionate inquiry." Yet he believed that when done correctly, social science research did not traffic in value judgments.[55]

The idea here—that scientific results were value free and thus had no particular political or ideological valence—appeared over and over again, as in this succinct formulation by Mitchell: "science properly considered does not undertake to say what ought to be done."[56] Ogburn agreed: although not long ago, the social sciences were "greatly confused by the mixing in of values with the consideration of knowledge," the distinction between scientific knowledge and value-laden viewpoints could and should be "kept clear." A disturbing example from the natural sciences reinforced this point that all of the sciences needed to respect the difference between the sphere of scientific inquiry and objective knowledge, on one hand, and the sphere of value judgments and practical action on the other: although it might be the scientist's task to produce a poison gas, it was not his job as a scientist "to say whether that gas shall be used for spraying fruit trees or for killing human beings."[57] The same point could be made for individual disciplines, with the anthropologist Cooper noting that his was "an empirical science, not a normative one." Thus, anthropologists did not "sponsor any specific philosophy of life."[58]

Among this group of scholars, there were some significant political differences, which would have made the points about value neutrality and nonpartisanship useful in maintaining a united front during the NSF hearings. The left-leaning Mitchell had voted for the socialist candidate Norman Thomas in the 1930 presidential election, was involved in various New Deal planning efforts, and favored the extension of New Deal initiatives during the postwar era.[59] Ogburn, who was more conservative, voted Republican, although he also believed social science knowledge could provide the basis for rational, technocratic planning.[60] And for many years, Nourse was a central figure at the Brookings Institution at a time when this nonprofit, think tank opposed New Deal planning efforts. Consistent with that stance, Nourse advocated solutions to economic problems mainly through reforms in the private sector and marketplace, not through national economic planning.[61]

During the NSF hearings, these scholars advanced a scientistic position for strategic reasons as well, as part of a group effort orchestrated by the SSRC. Not only was the unity of the sciences central to the council memorandum to Congress, but the council did not include any contrary voices. Of particular interest—and as will be discussed further in a later section—the council ignored the views of well-known scholars on the liberal left, such as the aforementioned pragmatist philosopher John Dewey, as well as three prominent social scientists: the political scientist and historian Charles Beard and the sociologists Robert Lynd and Louis Wirth. All of them believed value-laden inquiry oriented toward the realization of democratic aims and social betterment should be welcomed in the social sciences. As Heather Douglas has shown, well into the post–World War II years, the value-free ideal was also a contested issue among American philosophers of science.[62]

By failing to include any contrary voices, the SSRC was able to advance a consistent view that placed the social sciences under a wider scientific umbrella and promised to protect them against doubts regarding their scientific status and political character. Nevertheless, conservative critics in the scientific community who advanced those charges had significant weight. Moreover, as the NSF debate continued into 1946, these critics would gain valuable political allies eager to keep the social sciences at a safe distance.

INFLUENTIAL CRITICS ON THE RIGHT

After the 1945 fall hearings, the social sciences continued to receive significant attention. Initially, the bid to include them proceeded with continued support from the liberal camp. But conservative opposition soon hardened.

In the late fall, a group of scientists established the Committee Supporting the Bush Report to the President, also known as the Bowman Committee, named so for its leader Isaiah Bowman. One of the nation's leading geographers and a member of the prestigious National Academy of Sciences, Bowman had been a key adviser on foreign policy issues for presidents Wilson and Roosevelt. Yet Bowman disliked the New Deal because he, similar to his close associate Bush, believed this involved a dangerous expansion of the federal government and its powers. Regarding the social sciences, Bowman's own scholarship, somewhat curiously, included work on the social and political dimensions of geography. In the 1930s, he had also been a SSRC board member. Yet Bowman found this part of his discipline to be on uncertain ground compared to what he took to be physical geography's much sturdier foundations. In his 1934 book *Geography in Relation to the Social Sciences*, Bowman wrote that not all areas of geographical inquiry could be studied "scientifically in a strict sense," especially those involving "unpredictable variables—human beings organized in societies."[63]

Including the "social sciences (sociology, political science, economics, law & c.)" in the proposed agency "would be a serious mistake," declared the Bowman Committee in a public letter to President Truman. This letter, published in *Science* and the *New York Times*, recognized that the social sciences deserved federal funding. But it said this should be channeled through "a separate body."[64] The committee did not elaborate much on the nature of the mistake to be avoided or the reason for its seriousness, although familiarity with the wider context would have suggested the relevant concerns—the questionable scientific status and political character of the social sciences.

The following spring, during another round of legislative hearings and this time held by the House of Representatives, Bowman underscored the gravity of those issues. He recognized that certain fields within the social sciences already had a "sound methodology" and said that it would be appropriate to provide funding in the proposed agency for "factual and definite studies." But he opposed the idea of making social science an "integral part" of the agency (e.g., through a social science division) because doing so would open the door to political pressures "for the support of all sorts of hare-brained studies about things not capable of objective study, things that in the end have to be determined by individual opinion."[65]

Meanwhile, Kilgore's initiative had acquired backing from a different group of natural and social scientists led by the astronomer Harlow Shapley and the chemist Harold Urey. Other members from the physical sciences included three luminaries, Albert Einstein, Enrico Fermi, and J. Robert Oppenheimer. The group's social science members included Gaus, Mitchell, Nourse, Ogburn, and Yerkes, as well as the famous cultural anthropologist Ruth Benedict and two accomplished sociologists, Louis Wirth and Talcott Parsons.[66] The Urey-Shapley group, whose leadership and membership leaned to the left, aroused conservative suspicion. For example, Shapley, who in 1945 helped found the Federation of American Scientists, an organization that advocated for the international control of atomic energy and its peaceful uses, later came under investigation by the anti-communist House Un-American Activities Committee.[67]

While the Shapley-Urey group joined Truman, Kilgore, and the SSRC in supporting the social sciences, conservatives remained obstinate. In January 1946, the opposing sides came together to create a compromise science bill (S. 1850) that initially called for a social science division. But including such a division worried conservative figures. In late February, a report issued by the Senate subcommittee considering S. 1850 stated that, while the social sciences should not be excluded, the proposed agency should not include them because their "specific research needs" had not yet been examined carefully. Until adequate planning studies had been carried out, the agency should only support social science "studies of the impact of scientific discovery on the general welfare and studies required in connection with other projects supported by the foundation."[68] And Kilgore soon backed down on this issue.[69]

Conservative political opposition mounted as well that spring and summer. Consider the perspective offered by Ohio Republican Representative Clarence Brown. A leader in Warren G. Harding's successful 1920 presidential campaign and a persistent opponent of big government and communism, Brown provided one of the most colorful and damning passages from this entire episode. In a warning directed mainly at natural scientists, he declared that if congressmen believed the agency under consideration would include "a lot of short-haired women and long-haired men messing into everybody's personal affairs and lives ... you are not going to get your legislation."[70]

Over in the Senate, the Republican Thomas Hart from Connecticut joined the attack. A retired U.S. Navy admiral whose military service extended from the Spanish-American War in 1898 through World War II, Hart claimed that

> no agreement has been reached with reference to what social science really means. It may include philosophy, anthropology, all the racial questions, all kinds of economics, including political economics, literature, perhaps religion, and various kinds of ideology.

Such work therefore had "no place" in an agency focused on fundamental research in the natural sciences. Hart saw "no connection between the social sciences, a very abstract field, and the concrete field which constitutes the other subjects to be dealt with by the proposed science foundation."[71]

Piling on, the New Jersey Senator H. Alexander Smith, a deeply religious man and yet another conservative Republican, asserted that the science legislation under consideration aimed to support "research in pure science," with the goal of fostering "the discovery of truth," which had "nothing to do with the theory of life ... with history ... with law ... with sociology."[72] Smith spoke with some authority about these fields as well, for he had studied law and political science under Woodrow Wilson at Princeton University, obtained a law degree from Columbia University, and then taught international relations and foreign policy in Princeton's Department of Politics from 1927 to 1930.

In July, conservative legislators moved to seal the fate of the social sciences. Senator Hart took the lead by proposing an amendment that would have effectively eliminated them in the compromise legislation (S. 1850) still under consideration. After a brief discussion, the Hart amendment passed, with forty-six in favor and twenty-six opposed—the other twenty-four senators did not vote, most simply because they were absent.[73] The new version of the compromise bill then gained approval from Congress.[74] As it turned out, that wasn't the final word because President Truman refused to sign the bill into law. Nevertheless, the barrage of attacks in the House and Senate together with the Hart amendment's approval demonstrated how powerful the opposition to social science had become.

Furthermore, such criticism was not confined to the NSF debate but reflected and reinforced wider partisan battles. Conservative figures in the private and public sectors had real doubts about the scientific standing and practical implications of the social sciences with respect to a broad array of matters. Consider, for example, the raging controversy over national economic policy during the same years—1945 and 1946—when the position of these sciences in the NSF debate was still receiving serious attention.

During the New Deal and continuing during World War II and the postwar years, Keynesian-style macromanagement of the national economy acquired a solid base of liberal adherents, but it was anathema to conservatives who disliked anything that smacked of central planning. Partisan wrangling flared up during the legislative debate that culminated in the 1946 Employment Act, which, among other things, created the nation's Council of Economic Advisors. Social scientists and especially economists associated with the by-then-defunct National Resources Planning Board helped draft early versions of this legislation, which would have committed the federal government to the goal of achieving national full employment. When liberal senators Robert Wagner from New York and James Murray from Montana introduced this legislation, conservatives including Robert Taft, the most powerful Republican legislator of the day, protested fiercely. Besides warning of a socialist takeover, they successfully attacked passages that would have committed the government to maintaining full employment. Their influence was reflected in a key modification to the legislation's title: originally called the Full Employment Bill of 1945, the legislation that passed into law was called the Employment Act of 1946.[75] Other topics of partisan dispute involving the social sciences concerned their relevance to race relations, sex roles, labor laws, family life, religion, education, cultural norms, social security and other social welfare programs, the exercise and distribution of political power, and morality.

Equally important for understanding the course and significance of the NSF debate is the fact that support for social science from liberal and more progressive quarters turned out to be modest and receding. Senator Kilgore, who at first appeared to be a strong ally, opted to leave the social sciences aside when, as noted above, he backed off following negotiations over the 1946 compromise bill. The efforts of Democratic Arkansas Senator J. William Fulbright, although a more steadfast ally than Kilgore, are also revealing because of their ineffectiveness. Best remembered for his liberal internationalist philosophy and for creating the Fulbright Exchange Program (to promote international student education and scholarly exchanges), Fulbright acknowledged that the effort to include the social sciences faced daunting obstacles in the scientific and political communities. Shortly before the vote on the Hart amendment in July 1946, Fulbright reported—as seen in the opening quote by him—that one able scientist had bluntly told him that he "would not call it [social science] science."

Two years later, after an amendment Fulbright put forth to insert a social science division into one of the NSF bills then under consideration was voted down, he commented that some of his congressional colleagues still wondered if the proposed division meant he supported socialism. Unsuccessful in bringing them over to his side, a dismayed Fulbright explained that, no, what he actually had in mind was the study of such matters as "economics and politics," two vital fields that he believed the country was more "deficient" in than "mathematics, physics, or engineering science."[76]

With the controversy showing no signs of waning, the Truman White House backed down from its early pro–social science stance as well, as revealed by the Steelman Report, a major three-volume science policy study conceived of as a liberal alternative to Bush's *SEF* and published in 1947. The report's main architect, John R. Steelman, had been a labor relations mediator during the New Deal and then director of the Office of War Mobilization and Reconversion in the Truman administration. In 1946, Truman appointed him chief of staff (at the time, this position was officially called Assistant to the President), giving Steelman important responsibilities for the coordination of federal science policies and programs. Steelman was a social scientist by training as well, with graduate studies in sociology, first at Harvard, which he left because at the time Harvard did not offer a PhD in sociology, and then at the University of North Carolina–Chapel Hill, where he completed a PhD with a dissertation on mob violence including lynchings in the South. That, together with his work as a professor of sociology and economics for six years at Alabama College in Montevalla (1928–1934), meant that Steelman was well qualified to examine questions concerning federal policies regarding the social sciences. One might have reasonably expected the Steelman Report to make a strong case on their behalf. But, instead, it merely suggested that federal support for them should be considered elsewhere at a later time—exactly the position Vannevar Bush advocated.[77]

The tide having turned against them, the social sciences were never again invited by Congress to present their case. Their marginalization reflected many factors, including the overwhelming emphasis, present from the beginning, on creating a new agency dedicated to advancing the natural sciences as well as swelling conservative opposition. For a short while, the case for inclusion had backing from the liberal camp, starting with Truman's reconversion address and Kilgore's interest in them, followed by support from the Shapley-Urey group. But as conservative attacks from an alliance of scientists and politicians grew, liberal support receded. After the July 1946 vote on the Hart amendment, and despite sporadic efforts to bring the social sciences back into the picture (i.e., Fulbright's unsuccessful initiative in 1948), they no longer received much attention.

LEFT-LIBERAL CRITICS SIDELINED

We have seen that a conservative coalition of scientists and politicians opposed the social sciences based on intertwined worries about their scientific and political characters. This combination of charges suggested that they were not part of a unified scientific enterprise led by the natural sciences and thus should not be included in the proposed agency, or at least not in an integral way and not until a study could be undertaken after the agency itself was established. The case presented by leading social scientists from the SSRC, however ineffectual, countered those charges with the claim that the social sciences actually had a great deal in common with the natural sciences, while granting that the former were comparatively immature. Remarkably, no major proponent of inclusion discussed so far questioned this scientistic strategy. However, behind the scenes, and in wider discussions during the early postwar years, this strategy received trenchant criticism from a prominent group of left-liberal scholars.

Back in the 1930s—and thus well within recent memory at the time—a left-leaning critique of scientism had a significant presence in American intellectual circles, supported by, among others, John Dewey, Robert Lynd, Charles Beard, and Louis Wirth. Some European scholars supporting this position, including the sociologist Karl Mannheim and the economist Gunnar Myrdal, were also well known in the U.S. Although this critique lost substantial steam during the 1940s and 1950s, these figures continued to promote it. Consider the view presented by Dewey in his 1947 article "Liberating the Social Scientist." Eighty-eight years old at the time and still the country's leading pragmatist thinker, Dewey charged that "a narrowness, a restraint, [and] a constriction imposed upon the social sciences by their present 'frame of reference,' i.e., the axioms, terms, and boundaries under which they function today," had become worrisome. Scholarship in this vein, "instead of resulting in liberation from conditions previously fixed," tended "to give scientific warrant, barring minor changes, to the status quo—or the established order, a matter especially injurious in the case of economic inquiry." To strengthen social science contributions to progressive social change, Dewey proposed that they needed to embrace what was so often excluded, namely, "critical examination" of "the very order that is the nominal subject of investigation."[78]

Regarding the NSF debate, it was the left-liberal sociologist Louis Wirth who became the main proponent of this position. Although not nearly as well remembered today as Dewey, Beard, or Lynd, Wirth was a leading scholar during his time, a statesman for his discipline, and an articulate advocate of an action-oriented approach to social science, one that could produce insights not only about the state of social affairs but also about needed social changes and steps for realizing them. His election in 1947 as president of the American Sociological Society also made Wirth the first Jew to hold this position. Three years later, he became the first president of the International Sociological Association as well. As a scholar, he remains best known for his work in urban sociology, including an influential book called *The Ghetto*.[79]

Wirth's interest in linking social research to progressive reform can be seen in his 1946 essay "The Unfinished Business of American Democracy." Here, he commended the New Deal for helping to overcome widespread discrimination and for prodding his fellow citizens to live up to "America's democratic creed … of equality of opportunity for all, irrespective of race, creed, or origin." But the nation still had a long way to go in order to realize the "ideals of a democratic order." Wirth thus called upon his Americans to continue the hard work needed to achieve full employment, establish national health care, provide comprehensive social security, and minimize unacceptably large differences in educational and cultural opportunities.[80]

Wirth understood as well that social science patrons, both private and public, could have a large impact in shaping the contours and purposes of scholarship. In an extensive though unpublished 1937 report about the SSRC, Wirth criticized the council's heavy reliance on philanthropic support, especially from the Rockefeller philanthropies, because this made the council's work dependent on their decisions and interests, which, in turn, compromised its independence of judgment and undermined its ability to promote social science scholarship that ran contrary to philanthropic interests.[81] Now, regarding the NSF proposals, Wirth feared that the course of debate offered at most a sharply circumscribed role for the social sciences, which would weaken lines of inquiry needed to realize progressive goals.

Following the elimination of the social science division from the 1946 compromise bill, Wirth observed in private correspondence that this move left the social sciences confined "to the relatively narrow field of predicting and analyzing the impact of technology on social life."[82] Moreover, in a 1947 article published in the *Annals of the American Academy of Political and Social Science*, he explained that this danger was not merely a contingent fact about the course of the NSF debate; rather, the danger stemmed from widespread efforts by entrenched interests to contain scholarly threats to the status quo. Social researchers who scrutinized traditional "institutions, practices, and values" often provoked "the displeasure, if not the outright hostility, of the dominant elements in society, especially those who have a vested interest in things as they are." In response, social scientists seeking safe ground often became champions of value neutrality and nonpartisanship, which rendered social research incapable of direct engagement with pressing problems, if not indeed altogether irrelevant. At a time when the natural sciences were widely understood to be value neutral and objective, Wirth argued that "aping" their "methods" and trying "to bask" in their "reflected prestige" would simply not do.[83]

A couple years later, Wirth received $500 from the SSRC to support an interdisciplinary seminar at the University of Chicago about the manifold relationships between values and the social sciences. A series of meetings held in the first half of 1949 examined "what the social sciences have to contribute to ethics and social policy and how in turn ethical and policy considerations affect the social sciences." Uncertainties about the future of social science funding and mounting international conflict gave such issues special urgency, prompting Wirth to ask if the "source of funds" would "determine the problems to be investigated, the methods of investigation, and the nature of the results?" In light of the "ideological conflict between democracy and totalitarianism and between capitalism and communism," Wirth asked: could and should "the social sciences be 'neutral'?"[84]

After the seminar ended, Wirth reported to the council that although the discussions had revealed little consensus on key issues, they had demonstrated the need to move beyond the simplistic view that values only shape the selection of topics but not the process or interpretation of research. It now seemed clear that "the value problem impinges upon all phases of social science, including empirical research."[85]

However, Wirth's efforts and the broader left-liberal critique of a value-neutral, apolitical investigative approach had no impact on the NSF debate. The SSRC had not asked Wirth to testify at the initial set of legislative hearings in the fall of 1945. Nor did any of the social scientists chosen to testify call for the sort of value-oriented, socially engaged approach to social inquiry advocated by such prominent scholars as Dewey, Lynd, Beard, and Wirth.

Moreover, shortly after the Hart amendment vote in July 1946 to exclude the social sciences, the SSRC decided to engage in "salesmanship," to enhance the "prestige of social science" by promoting awareness of and appreciation for the alleged unity of the sciences. Paying no heed to the anti-scientistic position of left-liberals like Wirth, SSRC's Problems and Policy Committee would support projects designed to strengthen "operational ties with natural scientists at all levels" and to improve "public relations … through publicity directed both to laymen and specialists in other fields."[86] One such project involved the purchase and free distribution of 5,000 copies of a 1946 pamphlet called "Should the Government Support Science?" Written by the science writer Waldemar Kaempffert, this pamphlet supported Kilgore's liberal science policy plans and called for a "closer union of the natural and social scientists."[87]

In a more ambitious project, the SSRC commissioned the sociologist Talcott Parsons to produce a national statement on the social sciences that would be "analogous" to Bush's *SEF*.[88] Parsons, who received $10,000 for this project, had recently been appointed head of Harvard's newly created Department of Social Relations. His research on social systems theory and the structural-functional approach to social analysis would make him enormously influential as well as deeply controversial. It must be noted that Parsons's writings rendered his advocacy of the unity-of-science viewpoint complex and, in the context of the NSF debate, potentially problematic. On one hand, he claimed that in key respects, the social and natural sciences had much in common: natural and social scientists alike aimed to be systematic and logically consistent, both groups sought to establish a close integration of empirical evidence and theoretical structures, and both aimed to produce reliable knowledge that was, at least when properly done, value free. On the other hand, he argued, much like the great German social scientist Max Weber had, that there are some basic differences in the subject matters of the natural and social sciences. The latter also requires a special, interpretive methodology to make sense of people's feelings, thoughts, and values.[89]

Regardless of such complexities in his thinking, in his SSRC-commissioned essay and in some scholarly articles about the NSF debate, Parsons followed the strategy of presenting the social sciences as part of a unified scientific enterprise. Failure to understand the true nature of modern social science, he claimed, rested on "naive, popular misunderstandings," which wrongly supposed that this branch of study provided a "haven for crack-brained reformers," represented "a glorified form of social work," and was "primarily concerned with promoting sexual libertarianism." Such views, along with the suspicion that these sciences were not objective and thus could not "rise above partisan politics," contributed to the "problem of obtaining political support."[90] So did "antilonghair sentiment in Congress." Against such views, Parsons argued that "in general it is not possible to draw the line and say that the scientific method shall be applied to nature but not to man in society."[91] Science is "a fundamental unity," not only in "methods and outlook" but also in the resulting "knowledge and its application."[92]

As it turned out, Parsons's essay had no impact on the NSF debate. By the time he had a full draft for the council's consideration in 1948, the social sciences had already been relegated to the sidelines in the ongoing legislative deliberations. Furthermore, his draft received serious criticism from many scholarly reviewers, among other things for its turgid prose. Certain passages would also have alarmed conservative critics, including one that claimed the social sciences could provide the "basis of rational, 'engineering' control" in modern society.[93] In the end, SSRC leaders decided that his manuscript was not ready for publication.[94] (Parsons's essay was eventually published in the mid-1980s, when it appeared in a book coedited by the sociologists Samuel Klausner and Victor Lidz.)

In sum, the liberal-left critique of scientism advocated by Wirth and others failed to gain much traction. By the middle of 1946, conservative interests had already managed to place the social sciences on the sidelines for good, although the NSF debate itself continued on. Meanwhile, developments associated with the early Cold War and the postwar Red Scare led, as the historian of science Jessica Wang has shown, "to the exclusion of the progressive-left" from the national science policy arena.[95] Against this background, the viewpoint that said the social sciences should embrace a critical, socially engaged approach fell on deaf ears, a point reinforced by the SSRC's ongoing efforts in salesmanship dedicated to the unity of the sciences.

CONCLUSION

During the immediate postwar and early Cold War years, the social sciences struggled intensely with the problems of scientific identify, social relevance, and public support. Although these were longstanding problems, they appeared in distinctive ways due to their connections to pivotal developments and debates in American political culture and national science policy. The need to work out those problems came into particularly sharp view during the postwar NSF debate.

The events in this episode raised basic questions about whether, to what extent, and under what conditions the federal government should fund the social sciences in the nuclear age. Since the initial NSF proposals called for a new comprehensive science agency as the centerpiece of a greatly expanded postwar national science system, the arguments for and against including the social sciences were burdened with unusual significance. As we will see in the next chapter, even though that expansive vision for the agency receded during the drawn-out NSF debate, the quest for inclusion retained considerable importance. To have been left out of this agency altogether would have been a substantial blow. If that had happened, the social sciences would have been excluded from the only science agency devoted to the advancement of basic science, which, following Vannevar Bush's discussion in *SEF*, was widely seen as the basis not only for scientific advance but also for technological progress deemed essential to the nation's defense, health, and economic strength.

Moreover, obtaining support for basic social research was difficult, a point underscored in 1948 by the psychologist Hadley Cantril when he observed that "funds to carry out really basic research in the social sciences are practically unheard of." Yet, Cantril continued, "Industries and government spend millions, even billions, of dollars financing research in the physical sciences for the quite understandable reason that such research in the past has been shown to have some concrete pay-off."[96]

In addition, the NSF's legislative origins brought into sharp relief substantial difficulties facing the social sciences as a group of powerful and mainly conservative figures in the nation's political and scientific communities questioned their worth. Not too long before, during the New Deal era, social scientists had enjoyed relatively strong support and influence in federal science and politics. However, by the early postwar era, they stood in the shadows of natural scientists and especially physical scientists. Furthermore, influential members of a conservatively oriented scientific elite, including Vannevar Bush and his close associates in the NSF debate, raised significant doubts about the social sciences—that they were not real sciences, that their work was not objective, and that they were involved with crackpot schemes to alter government. That group was joined by conservatively minded politicians who advanced similar charges, as seen in Republican Representative Brown's warning about short-haired women and long-haired men and in Republican Senator Smith's comment about the social sciences being involved with such things as the theory of life rather than the discovery of truth. By mid-1946, conservative critics had succeeded in placing the social sciences on the sidelines. Social scientists could hardly afford to ignore their criticisms.

Equally significant, in crafting a strategy for inclusion, the SSRC and its scholars embraced a scientistic position. Trying to make the most of a difficult situation, an interdisciplinary group led by Wesley Mitchell presented Congress with an argument based on the alleged unity of the sciences and that took the natural sciences as the gold standard. A wider circle of scholars associated with the council, including the sociologist Talcott Parsons, came on board as well. To be sure, all of these scholars supported the unity of the sciences in other settings as well—although Parsons's advocacy for an interpretive methodology could have been used to underscore the unique character of social inquiry. However, in the context of the NSF debate, their claims about scientific objectivity and value neutrality in modern social science, about its technical and nonideological character, and about the distinction between the social scientist and the social reformer also had a significant strategic dimension, as this set of claims promised to allay the damning worries expressed by conservative critics.

The position put forth by council scholars has additional historical importance because of what it left out, namely, a well-known left-liberal critique of scientism and value-neutral social inquiry advanced by prominent American figures—Lynd, Beard, Dewey, and Wirth—and also by well-known Europeans—Mannheim and Myrdal. During the NSF debate, it was the Chicago sociologist Wirth who intervened by raising objections to the scientistic strategy. Wirth's attempt to redirect the council's efforts and the wider discussion came to naught, however. Although he received modest funds from the council to examine relevant questions about values and social science in the setting of a university seminar, the council and its scholars never acknowledged the left-liberal critique in their congressional testimony or in other public relations efforts during those years, including the report commissioned from Talcott Parsons.

Following passage of the 1950 NSF enabling legislation, the opportunity to revisit basic questions about the standing and worth of the social sciences would arise again. Whatever understanding of these sciences one may have preferred, their marginalization during the agency's legislative origins was striking and could not be ignored in the coming years. Thus, the new agency had to reexamine questions about whether and how to include the social sciences with great care. Similarly, social scientists and their advocates had to reconsider the value of scientism as a strategy for gaining support in the context of the new agency's character and priorities and in light of broader developments in politics, science, and society in midcentury America.

2

STAKING OUT THE HARD-CORE, FROM THE MCCARTHY ERA TO SPUTNIK

As a matter of fact, we do very little in this area [social science] and only in those subjects which border closely on our research program in the physical and in the life sciences.
—Alan Waterman, physicist and first NSF director, 1955[1]

Invidious hierarchical distinctions among specialists of the various disciplines must disappear. An institutional environment must be created to encourage all scientists in intimate association to cooperate effectively toward achievement of the common goal of better understanding of the ways of nature, man, and society.
—Harry Alpert, sociologist and the NSF's first social science policy architect, 1958[2]

The establishment of the NSF is a major landmark in the history of American science and federal science policy. The agency's enabling legislation (the National Science Foundation Act of 1950) created a governing structure consisting of a full-time director and a twenty-four-member board (the National Science Board or NSB). The director and board members were to be appointed by the president, upon the advice and consent of the Senate, normally for a six-year term and eligible for a one-time reappointment. (In 2012, legislation eliminated the requirement of Senate advice and consent for board members.)

In the early years, the agency's main responsibilities and activities included awarding grants for basic scientific research and providing fellowships for advanced scientific training at the predoctoral and postdoctoral levels. Programs for adjudicating and awarding research grants resided in two research divisions: one for the mathematical, physical, and engineering sciences and the other for the biological and medical sciences. From early on, however, it was understood that major responsibility for supporting medical research rested with the National Institutes of Health, not the NSF. A third division for scientific personnel and education handled NSF fellowship and curriculum development programs.[3]

Of special importance were two central commitments at the agency: to support basic science and to promote the best science. As the first NSF annual report observed, its legislative mandate to promote basic science, also sometimes called pure or fundamental science, "broke new ground." "The American people, through their elected representatives, [had] decided that fundamental, scientific research" had such great importance that federal funds should support it. The NSF was thus created to "ensure the wise expenditure" of those funds.[4] During the early Cold War years, the great preponderance of federal science funding—90 percent as of 1954—supported not basic research but applied science, further underscoring the special nature of the NSF's mission to advance the frontiers of scientific knowledge without much if any attention to practical applications.[5]

In addition, the agency had a firm commitment to the promotion of first-rate science and scientific excellence. As the historian of science Daniel Kevles has put it, NSF leaders were "strongly disposed to a best-science approach." Agency grants were thus awarded based on judgments of scientific merit rendered by the scientific community itself.[6]

As for the social sciences, their status had been left vague by the enabling legislation, which explicitly mentioned NSF support for basic research and advanced training fellowships in the mathematical, physical, medical, biological, and engineering sciences but also permitted support for "other sciences."[7] The latter phrase could be used as a basis for including the social sciences. In light of the serious charges leveled against them by an alliance of conservative critics from the scientific and political communities during the postwar NSF debate and pervasive anti-communist attacks against them during the McCarthy era, NSF leaders approached this matter with some trepidation. But when other federal agencies, including the military, raised questions about the new agency's intentions, the need to gather additional information and develop an informed policy became evident.

Thus, in 1953, the NSF hired the sociologist Harry Alpert (1912–1977) to carry out a background study and develop policy recommendations regarding the social sciences. In the coming years, Alpert became head of the agency's earliest programs in this area as well. In effect, Alpert became its first social science policy architect.

This chapter examines the mix of factors—political, science policy, institutional, intellectual, and biographical—that shaped the early development and increasing importance of NSF social science during Alpert's tenure at the agency, from 1953 to 1957.[8] The first section analyzes Alpert's role in developing policy recommendations that focused on providing support for the so-called hard-core end of the social research continuum. The second section explores how Alpert's policy recommendations became the basis for the agency's early and cautious efforts to fund basic studies in the social sciences. Meanwhile, the project of establishing a viable space for these "other" sciences faced substantial resistance inside the agency from skeptical representatives in the physical sciences and in conservative quarters, as will be discussed in the third section. This chapter's fourth and last section takes a closer look at Alpert's views, revealing his qualms about the narrow scientistic approach that the NSF embraced and that he had helped to craft under difficult circumstances.

Here, we will see how a tightly circumscribed strategy for promoting the social sciences became established at the levels of policy, programs, and practices during the NSF's formative years. This enabled the agency to become a valuable source of federal funding and scientific legitimacy for this branch of scholarship. Equally importantly, when placed in a broad historical perspective, the scientistic framework is as important for what it left out as for what it included.

HARRY ALPERT AND THE UNDERDOG'S STRATEGY

Although he could not have foreseen it, Alpert's intellectual training and early professional background prepared him well for his work at the NSF. In 1938, Alpert completed a PhD in sociology at Columbia University. A serious researcher with wide-ranging interests, he developed an outlook on political and social affairs characteristic of scholars inspired by the New Deal and a moderately progressive strand of liberalism. As a scholar, he became best known for his studies of the seminal French thinker Emile Durkheim.[9] Alpert's broader scholarly passions included the sociology of science, scientific methodology including statistical analysis, the interrelations of the social sciences and natural sciences, public opinion research, and the social sciences in international perspective. Before his NSF years, Alpert held a number of academic positions, including posts at City College of New York

(1933–1947), Columbia University's Bureau of Applied Social Research (1946–1948), Queens College (1948–1950), and Cornell University Medical College (1951–1956). He also acquired substantial government experience, through research and consultant positions at the Office of War Information (1943–1944), the Office of Price Administration (1944–1945), the Bureau of the Budget (1945–1948, 1950–1953), and the Air Force Research and Development Board (1948–1950).

Upon Alpert's arrival at the young NSF, various factors encouraged him to pursue a prudent approach. First, the NSF's general counsel determined that its legislative origins and charter permitted the support of research and fellowships in the social sciences. But it was also clear that Congress expected the agency to "exercise a fair amount of restraint in the use of this authority."[10]

Second, Alpert found the NSF budget to be "pitifully small."[11] The initial proposals for a new science agency had envisioned it as a comprehensive centerpiece of a dramatically expanded postwar federal science establishment. But during the drawn-out NSF debate, other agencies, including the Department of Defense and its main branches (Air Force, Army, and Navy), had launched their own ambitious science programs. In this context, congressional appropriations for the NSF were rather modest. Although the agency requested $14 million for its first full year of operations in 1952, Congress appropriated only $3.5 million. A few years later, the level of appropriations surpassed that initial request. But in 1955, the agency's budget was still a meager $14.25 million.[12] Furthermore, with such small budgets, the potential for political controversy that could damage the NSF's reputation, which any amount of funding for the social sciences might inspire, had to be considered.

Third, as Alpert understood, the agency had an overwhelming natural science bent, reflected in its leadership and its priorities and wholly in line with broader developments in the post–World War II federal science establishment that curtailed the influence of social scientists while giving natural scientists, especially physical scientists, a dominant presence. The general tone was set by the first NSF director, Alan Waterman. Before taking up this position in 1951, Waterman had been a physics professor at Yale University, a member of the wartime Office of Scientific Research and Development headed by Vannevar Bush, and then the chief scientist at the Office of Naval Research, arguably the single most important federal science

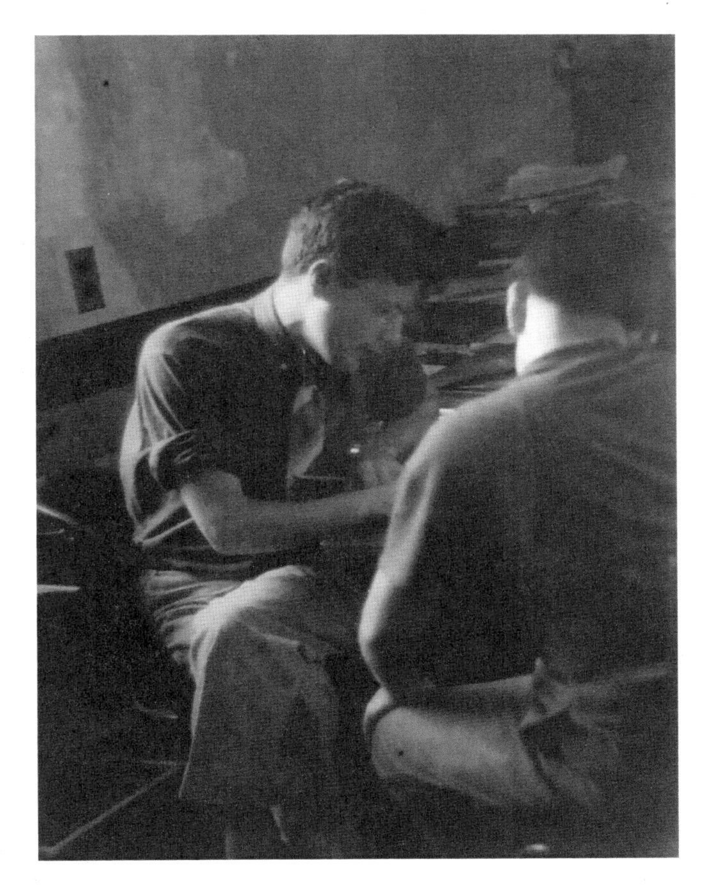

Figure 2.1
Harry Alpert, instructor and PhD candidate at Columbia University, meeting with a student, ca. 1937. Courtesy of the Alpert family.

patron during the late 1940s and early 1950s. Waterman was also a strong supporter of Bush's vision for the NSF as an agency dedicated to supporting basic research in the natural sciences and insulated, as much as possible, from political pressures.[13] The natural science orientation of the agency's governing board was also unmistakable, starting with its first two chairmen, James Conant, a prominent chemist, and Detlev Bronk, a biophysicist.

In contrast, the social sciences had minimal representation in the agency's governing structure. Furthermore, NSB members with social science interests were appointed to the board for other reasons. This group included

Chester Barnard and Charles Dollard, who had been the presidents of the Rockefeller and Carnegie foundations, respectively. It also included Sophie Aberle, one of only two women in the first group of twenty-four NSB members. Her main qualifications for this position derived from her medical expertise, although she also published on the economy, society, and culture of the Pueblo Indians.[14]

Among the first group of NSB members, only Frederick Middlebush had advanced training in the social sciences. After completing a doctorate in political science at the University of Michigan, he became a professor at the University of Missouri and coauthored a 1940 introductory textbook on international relations.[15] But his NSB appointment rested on his leadership in higher education: as president of the University of Missouri, president of the Association of State Universities, and president of the Association of American Universities. Moreover, Middlebush's home discipline, political science, was not explicitly eligible for NSF funding during the 1950s, for reasons discussed later in this chapter and the next.

The inhospitable climate during the McCarthy era provided Alpert with a fourth reason to proceed with care. Intense anti-communist pressures could not be ignored or avoided. Even the appointment of Waterman as the first NSF director appeared problematic when it became known that his wife had gone to the Soviet Embassy for tea twice. National security fears also led to a loyalty oath required of all NSF fellowship applicants.[16] But most relevant to the social sciences were a series of congressional investigations in the early 1950s that placed them and their major philanthropic patrons—including the Rockefeller, Carnegie, and Ford foundations—under suspicion for supporting a variety of threatening "isms." In the midst of McCarthyite attacks, the sociologist Louis Wirth observed, "Since the problems with which the social scientist deals are characteristically controversial issues among the rival groups comprising society, social science runs the risk of being considered 'dangerous thought,' and the social scientist a 'subversive' character."[17] Those were prescient words—from what turned out to be Wirth's very last publication before his untimely death in 1953.

Deeply concerned himself, Alpert sent his assistant Bertha Rubinstein to the 1954 hearings of a McCarthyite investigation led by Tennessee Republican Representative B. Carroll Reece. According to Reece, his staff members, and a few choice scholarly witnesses, the social sciences had contributed to a dangerous leftist movement within the U.S. The investigation's majority final report alleged that numerous social scientists, research projects, and patrons had contributed to a range of troubling trends, including the growth of atheism, socialism, and moral relativism as well as the expansion and centralization of power in the federal government. In addition, the Reece Committee roundly attacked the scientistic strategy in American social science, claiming it undermined sound values and morality and also supported the rise of centralized planning and social engineering.[18]

Fifth and last, knowledgeable figures explicitly advised Alpert to proceed cautiously. Through extensive consultation with members of the NSF governing board, private foundation leaders, personnel from social science organizations, and other scholars, Alpert heard that he needed to keep the social sciences and the agency safe from political attack, at least as much as possible. In this context, focusing on their "hard" as opposed to their "soft" aspects seemed like a promising approach—despite the hostility to scientism during the Reece Committee investigation.[19]

With the above matters in view, Alpert proposed a carefully restricted strategy. Specifically, he recommended to his superiors that the NSF should develop a program based on an "effective integration and partnership between the natural and social sciences." This quotation, not incidentally, came directly from Vannevar Bush's postwar congressional testimony.[20] In explaining what integration and partnership would mean in practice, Alpert said that the NSF should support investigations at the "hard-core" end of the social research continuum. This particular formulation came from NSB member Chester Barnard, who had stressed the importance of drawing "a sharp distinction between the application of scientific methods in the social sciences, and the essentially political, ethical and welfare activities which frequently pass for social science."[21]

In a surprising twist, we know from Barnard's other writings that he actually had serious doubts about the supposed unity of the sciences. In a 1952 essay, he pointed out that reading the literature in different fields of science suggested "variations so great in the actual research procedures ... that these fields would have to be considered as different species, if not classes, of science." In addition, he did not agree that progress in the social sciences required them to take the natural sciences as their investigative model.[22] Nevertheless, to create a viable space for the social sciences that could attract support from the agency's natural science–oriented leaders, Barnard and then

Alpert recommended that the NSF should confine its support to hard-core research.

To ward off criticisms, Alpert also recommended that the NSF leave a variety of controversial topics and approaches to other funding sources, especially philanthropic organizations. These organizations were presumably less vulnerable than a public agency to political pressures and thus seemed better positioned to fund risky ventures. Thus, despite the wave of anti-communist investigations of private foundations, Alpert proposed that these organizations should be the "major source of … 'risk capital' in the social sciences." Not the new and small NSF but the well-established and large private foundations should shoulder the "major responsibility for supporting the unorthodox, the unusual, and the 'big gamble', as well as areas, like sex and politics, which can easily lead to public controversy." "Private foundations," Alpert added, mainly funded "problem-oriented" studies that focused on "social and political action," giving the NSF yet another reason to limit its support to "research-oriented" studies aimed at advancing knowledge per se.[23]

Furthermore, Alpert recommended that the NSF limit its support, at least initially, to convergent studies, that is, to social research that converged or overlapped with the natural sciences. Convergent criteria would restrict the agency's support to "basic research" that meets "the highest standards of scientific inquiry and fulfills the basic conditions of objectivity, verifiability, and generality."[24] In the years to come, these last three criteria—objectivity, verifiability, and generality—would appear over and over again in NSF documents. Regarding fellowships, Alpert similarly proposed that support for scholars who applied "concepts and methods of the mathematical, physical, medical, biological, and engineering sciences." As examples of convergent work, Alpert mentioned studies in "experimental social psychology, mathematical sociology, econometrics, operations research, information theory, communications theory, and similar fields in which there is direct application of natural science techniques, methods, and concepts to social behavior."[25]

"Underdogging." That's how Henry Riecken, Alpert's successor at the NSF, described Alpert's prudent approach. Its essence involved "allying one's cause with stronger others, in this case the physical and biological sciences."[26] Indeed, due to a combination of political, institutional, and intellectual considerations, Alpert, following advice from Chester Barnard and others, had proposed that the agency should concentrate on hard-core studies and convergent research. This move placed the social sciences within a larger arena of scientific activity where the natural sciences were the acknowledged leaders. Revealingly, in internal documents, Alpert himself never suggested that social research beyond these parameters was in any sense less worthy as scholarship or less valuable to society. Neither had Barnard. Nevertheless, that restrictive approach soon became official policy.

A LIMITED PROGRAM FOR CONVERGENT RESEARCH

In August 1954, the NSF approved Alpert's recommendation to establish a limited program in the social sciences. This program would be "exploratory" and "experimental." And it would be developed "cautiously," with the main goal, anticipated by Vannevar Bush, of establishing an "integration and partnership between the natural and social sciences."[27]

Although the new program remained rather small in size and scope, we will see that it soon became an important source of federal funding for the social sciences as well as a crucial source of scientific legitimacy. The NSF itself lost no time in highlighting the importance of this last issue, by identifying the agency's interest in the longstanding problem of determining the interrelations between the social and natural sciences. As the 1955 NSF annual report noted, the great nineteenth-century French theorist of science, August Comte, had posited a scientific hierarchy, with mathematics at the top followed by physics, biology, and then the social, psychological, and ethical sciences. Furthermore, "all" of these fields "were viewed as integral parts of a unitary [scientific] scheme"—just as the NSF itself now claimed, except without any consideration of ethics as a science.[28]

After providing useful recommendations, Alpert was put in charge of the new convergent program, actually two separate programs. One for Anthropological and Related Sciences, located in the division of biological and medical sciences, supported work in human ecology, anthropology, archaeology, psycholinguistics, demography, and also quantitative and experimental social psychology. The second program for SocioPhysical Sciences, located in the division of mathematical, physical, and engineering sciences, funded studies in mathematical social science, economic geography, econometrics, and the history, philosophy, and sociology of science.

Under Alpert's guidance, the convergent programs developed as anticipated, that is, cautiously. Project proposals that received funding focused on topics such as the mathematical theory of economic models, research in multiple factor analysis, the biometrical study of evolution,[29] statistical variation in suicide rates,[30] social psychological experiments on cognitive dissonance, and experimental and mathematical research on choice behavior.[31] Noteworthy as well, Alpert emphasized that to evaluate research proposals submitted by scholars, the convergent programs relied on "the same techniques and procedures as those employed in the evaluation of natural sciences proposals." In "most instances," the evaluation process entailed review by both natural and social scientists.[32] Although Alpert did not mention it, social scientists were not involved in reviewing proposals handled by other programs in the physical science or biological science divisions.

Alpert's comments already reveal that social science research was expected to live up to standards associated with the natural sciences and that the evaluation process was designed to ensure that they did. However, to fully appreciate the importance of the evaluation process as it was applied to the

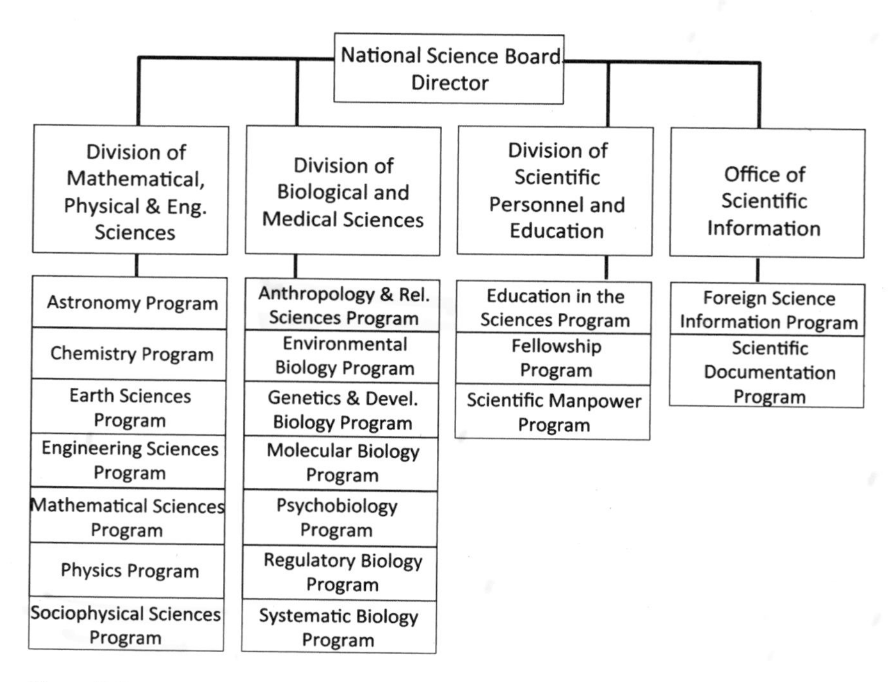

Figure 2.2
National Science Foundation organizational chart as of 1955. Nondivisional administrative offices have been omitted.

social sciences, one needs to understand how that process worked more generally.

With the goal of identifying and promoting the best science, the agency set up a system of scientific peer review. In a typical case, a research proposal submitted to the agency landed in the hands of the leader of the relevant organizational unit, often called a program. If the proposal looked appropriate, the program leader sent it out for review to a few scientists in the general field of study. Their reports and recommendations then informed a multilayered evaluation process inside the agency, involving the program's scientific advisory board, the program's leader and other staff members, the leader of a higher-level unit called a division, and then, at the final stage of approval, the agency's governing board and director. This review process was supposed to guarantee that funding decisions depended heavily on judgments concerning scientific excellence made by those best qualified to make them, namely, other practicing scientists as well as agency personnel at various levels who had extensive experience in science, science administration, and higher education.[33]

Thus, a social science proposal that was awarded funding also received a stamp of scientific excellence, issued by an agency committed to a stringent understanding of such excellence.

Although small and circumscribed by intent, the convergent programs gave NSF social science visibility on the national funding landscape. By 1955, the NSF already ranked fourth among all federal agencies that supported basic social research, behind the Department of Agriculture, the Air Force, and the Department of Health, Education, and Welfare—the last one held the top spot mainly because it included the National Institute of Mental Health, a big patron of research in social psychiatry and social medicine during those years.[34]

By August 1956, the convergent programs had processed about 100 proposals. Eighty-three of them focused on basic research, with a combined funding request of $1.8 million—about $16.6 million in 2018 dollars. Thirty-three of those proposals received support, for a total of $400,000, about $3.7 million in 2018 dollars.[35] If one included the agency's psychobiology program, located in the biological sciences division, the figures would be higher.[36] Starting in 1956, the NSF also opened its predoctoral and postdoctoral fellowship programs to applicants in convergent areas.[37]

Meanwhile, the political climate was becoming friendlier to the social sciences and their patrons. The darkest days of McCarthyite investigations passed after the Wisconsin senator disgraced himself during the nationally televised 1954 Army-McCarthy hearings. In addition, a small group of liberal Democrats and moderate Republicans in Congress began to advocate for greater federal social science support, based on the argument that the country needed more social research, including, but not exclusively, research that could help address pressing national problems, such as juvenile delinquency. This group of legislators included the Democrats Hubert Humphrey from Minnesota, Charles Porter from Oregon, Wayne Morse also from Oregon, and Estes Kefauver from Tennessee, as well as the Republican Jacob Javits from New York.[38]

Under these more favorable conditions, Alpert proposed a modest expansion in the agency's efforts along lines approved by a small social science advisory committee, which included scholars from various disciplines who were not on the NSF payroll and could therefore provide a measure of independent advice. Starting in 1956, Alpert recommended the creation of a separate and unified social science unit plus elimination of the convergent criteria, so that scholars working on a broader range of social research would be eligible for funding.[39] According to Alpert, the cautiously developed convergent programs had succeeded, demonstrating "that it is possible to identify significant and scientifically meritorious projects in selected social science areas which meet the criteria of basic research, science, convergence, and the national interest." With the anticipated three-year exploratory period now complete, Alpert advocated "a limited broadening ... for support of basic social science areas which do not necessarily meet the criterion of convergence," although still keeping in mind the congressional mandate to "proceed cautiously." In addition, the agency should continue to steer clear of "applied and mission-related social science research such as that pertaining to delinquency, social welfare, health, agriculture, and labor," because such studies were more properly supported by other agencies.[40]

In August 1957, NSF leaders once again followed Alpert's advice, by ending the two convergent programs, by eliminating the convergent criteria for research proposals (although convergent criteria still applied to predoctoral and postdoctoral fellowship proposals), and by establishing a unified social science program. Unlike the emphasis on interdisciplinary work during the convergent period, this new program had a disciplinary structure, with separate programs for anthropology, economics, and sociology, and another one for history and philosophy of science.[41] It should be noted that apart from the agency's support for studies of science grounded in historical and philosophical inquiry, it did not fund research in history or philosophy more generally, because these were considered humanistic rather than scientific disciplines. In addition, the agency did not have a program for political science, due to concerns that funding research on controversial topics in this discipline would land the agency in hot water. Nevertheless, as of 1957, no other federal agency provided regular support for basic research in a number of areas now funded by the NSF, including anthropology, economics, demography, and also the history and philosophy of science.[42]

NSF social science gained additional stature due to the emphasis among private patrons on practically oriented research, a point Alpert had highlighted before and that by the late 1950s was even more evident. The Ford, Rockefeller, and Carnegie foundations all provided extensive support for social science carried out in universities as well as in private and public research institutes or centers. But these foundations concentrated mainly on addressing practical issues, such as stimulating development in the so-called Third World, countering communist expansion, and strengthening the national economy. In light of these conditions, Alpert told Director Waterman that scholars seeking support for basic social research increasingly relied on the NSF. Appreciation of this point grew stronger after the Ford Foundation announced in 1957 that it was going to close its behavioral sciences program, effectively eliminating the nation's single largest source of private funding for basic social research since the early 1950s.[43]

Social scientists also valued funding from the NSF because of its growing reputation as a major federal patron of first-class scientific research and an ardent proponent of what Roger Geiger, a leading historian of the American research university, has called "an ideology of basic research."[44] As early as 1952, "scientific merit" stood at the top of the NSF's list of criteria for evaluating research proposals, followed by other closely related criteria, including the anticipated contributions to scientific knowledge, the scientific ability of the investigator, and the scientific resources of the host institution.[45] Two years later, an executive order signed by President Eisenhower clarified the NSF's role by specifying that, while other agencies could support basic research deemed relevant to their missions, the NSF would be "increasingly responsible" for providing support for "general-purpose basic research," meaning that the agency had to look after the general well-being of the nation's basic science effort.[46] Drawing on the viewpoint articulated by Vannevar Bush in *SEF*, the NSF, together with the National Academy of Sciences and the American Association for the Advancement of Science, promoted a basic research ideology with vigor throughout the 1950s.[47]

For social scientists, establishing a foothold in an agency known as a place where "good science gets funded" had great symbolic value.[48] In a 1960 editorial published in *Science*, the psychologist Dael Wolfle, a former executive secretary of the American Psychological Association and a leading expert on relations between science and government, drew attention to this aspect of the agency's significance. Its social science program only supported "basic research that meets high standards of conceptual and methodological rigor," explained Wolfle. This approach, which was entirely consistent with the approach in the agency's natural science programs, helped allay "any remaining fears that the term *social science*" was "merely a cloak for action on important but sometimes controversial social issues." Wolfle—with more than a dash of optimism—reasoned that this strategy would hasten the day when people would "no longer" have "any doubts concerning the appropriateness of the word *science* in *social science*."[49]

In a series of scholarly articles about NSF social science, Alpert emphasized that the agency promoted scientific progress above all.[50] In these publications, one finds numerous references to the unity-of-science stance as it had been articulated by prominent voices from the natural and social sciences. Putting to good use the views of James Conant—the accomplished chemist, first NSB chairman, and influential albeit amateur historian and philosopher of science—Alpert declared that just as the natural sciences had done, the social sciences "must progress toward firmer theoretical formulation and conceptualization." Similarly, Alpert invoked the words of Charles Dollard—NSB member and Carnegie Corporation president—in his declaration that social scientists had to "become scientists in fact, dealing in fundamental theory which can be tested" and searching for "laws and generalizations which will enable us to predict what men or groups of men will do or not do under stated conditions." Furthermore, Alpert urged social scientists to stay within the boundaries of scientific inquiry proper, by eschewing "identification with the sum total of speculation and intuition about human social behavior" and rejecting "identification with social reform movements and welfare activities, and especially, the unfortunate phonetic relationship to socialism."[51]

These writings are also revealing because Alpert omitted serious consideration of the political, institutional, and pragmatic factors that, as we have seen, strongly shaped the NSF's restricted scientistic approach. This omission was surely not simply an oversight, for Alpert was well aware of those factors. But he had good reasons to underplay their importance in publications about the NSF written while he was working there. Drawing attention to them would have undermined the general points he sought to convey about the wisdom in the NSF's outlook: namely, that the agency was committed to the unity of the sciences and hard-core research because this was the best path to scientific progress in the social sciences—and not because of other concerns that had a pragmatic or political character—and that the agency was therefore an especially valuable source of funding.

SISYPHUS HAD THE EASIER TASK

In those writings, Alpert also spent little if any time discussing dissent within the agency. Internal criticisms are an important part of this story, however. Evidence from oral interviews and archival documents indicates that criticisms from physical scientists and politically conservative figures in particular posed major obstacles and thus ensured that the social science programs remained on a tight leash.

Reminiscences from two knowledgeable figures, Bertha Rubinstein and Pendleton Herring, attest to the existence of ongoing challenges, including opposition from physical scientists at the agency. As a senior at Bryn Mawr in the late 1940s, Rubinstein worked with the psychologist John T. Wilson on a project funded by the Office of Naval Research. When Wilson became an administrator in the NSF's biology division in the early 1950s, he helped to get her a job at the agency. Soon she became Alpert's assistant. For the next thirty years, until her retirement in 1981, Rubinstein worked closely with the social science programs, giving her ample opportunity to witness the struggles for legitimacy behind the scenes. When I interviewed her many years later, Rubinstein emphasized that some physical scientists on the governing board had been critical of the agency's social science efforts.[52] Herring, a political scientist who produced influential studies of the political process, pressure groups, and public administration, served as president of the Social Science Research Council for twenty years starting in 1948.

He also served as chairman of the NSF social science advisory panel in the late 1950s and early 1960s. Along the same lines as Rubinstein, he remembered that the agency's physical scientists distrusted the social sciences and feared that controversy over funding for them would inspire political attacks and budgetary cutbacks, thereby weakening the agency's efforts in the natural sciences.[53]

In addition, Kevin McCann, not a scientist but a conservatively minded NSB member and close friend of President Eisenhower, had serious reservations. McCann had been writing speeches for Eisenhower since 1946, published a flattering biography of his friend just in time for Eisenhower's successful 1952 presidential campaign and from 1955 to 1957 served as the president's special assistant. It seems no mere coincidence that Eisenhower himself had little admiration for the social sciences. In any case, McCann, whose bond with the president ensured that his views were taken seriously at the NSF, stated that "except for a few extremely limited areas," this field of study was worse than "anything released by Pandora."[54]

Such challenges caused splitting headaches. As Henry Riecken observed in a retrospective essay, "There seemed to be an endless need for justification, a continuous procession of questioners and critics." After one especially "wearying" session with NSF leaders, Rubinstein—at that point Riecken's assistant—"captured our mood when she declared that Sisyphus must have had the easier task."[55] Moreover, the long line of doubting toms meant that any effort to loosen the constraints on the social science program would meet stiff resistance.

Such resistance became manifest in an episode involving the liberal Democratic senator Estes Kefauver, who encouraged the agency to fund research on juvenile delinquency. Kefauver served as chairman of the Senate Subcommittee on Juvenile Delinquency during the mid-1950s, a time when national concern about American youth culture was widespread, as represented in Hollywood movies such as *Rebel without a Cause* starring James Dean and in the Broadway musical *West Side Story*. Prominent social scientists tackled the issue as well, including Harvard's Talcott Parsons; two other Harvard scholars, Sheldon and Eleanor Glueck; and Robert Merton and Paul Lazarsfeld from Columbia University.[56] Despite the availability of some studies of juvenile delinquency already undertaken or in progress, Kefauver believed that a satisfactory understanding of the problem, as well as the development of effective measures to bring it under control, required much further study. Accordingly, in 1956, Kefauver's subcommittee recommended that the NSF provide funding for research about the impact of the mass media (i.e., television, radio, and published materials, including comic books) on American youth and about the relationship between school dropouts and delinquency. Democratic Senators Wayne Morse and Hubert Humphrey supported this recommendation.[57]

But the NSF responded with hesitance. After receiving a letter from the senator in early 1956, the governing board found reason to be skeptical—presumably, based on the notion that the agency was dedicated strictly to scientific studies and thus should not get involved with research closely related to social reform or public policy.[58] The matter received additional attention through NSF exchanges with Kefauver. Moreover, a 1957 report from Kefauver's subcommittee emphasized the need to support basic or fundamental research—the agency's forte: "the insistent demands for immediate action programs to prevent and treat juvenile delinquency should not obscure the equally important need for long-range, fundamental research," to establish the "basic knowledge of human social behavior upon which rational, effective, action programs must be based."[59] Nevertheless, Director Waterman maintained that while the agency remained receptive to proposals for "good basic research in the field," it considered studies of juvenile delinquency to be "applied research" and thus something that an action-oriented agency should support, most likely Health, Education, and Welfare, but not the NSF.[60]

Constraints on NSF social science came into clearer view in another episode involving Father Theodore Hesburgh, an NSB member and arguably the nation's most influential priest. Hesburgh served as an adviser to four popes and six presidents, became a prominent civil rights advocate, received the Presidential Medal of Freedom in 1964, and was a towering figure in Catholic higher education. As a long-term president of the University of Notre Dame (1952–1986), he sought to rebuild the institution's social science programs, personnel, and resources, as part of his broader goal to make Notre Dame the greatest Catholic university in the world. Based on an understanding of education and knowledge grounded in a theological perspective, Hesburgh wrote that among all the branches of secular knowledge, including the natural sciences and social sciences, philosophy was the "highest of the sciences." But "in the right order of things," philosophy, along with these other fields, was, in turn, subordinate to theology.[61]

Figure 2.3
Father Hesburgh at Illinois Rally for Civil Rights, Chicago's Soldier Field, June 21, 1964. Left to right: Rev. Edgar Chandler, Rev. Theodore M. Hesburgh, Rev. Martin Luther King, Jr., and Msgr. Robert J. Hagarty of Chicago. Copyright Unknown. Courtesy of Notre Dame Archives.

In early 1958, the NSB put this larger-than-life Catholic leader in charge of a special panel on the social sciences. He was joined by three other board members: the geologist Donald H. McLaughlin, a full professor at Harvard and the president of a gold-mining company; the biologist Douglas M. Whitaker, a former provost at Stanford University, the chief biologist for the atom bomb testing program at the Bikini Atoll, and the vice-president for administration at the Rockefeller Institute for Medical Research (later renamed Rockefeller University); and the political scientist Frederick Middlebush, still the only NSB member with advanced social science training. The four-member panel managed to strengthen the social science program by recommending an upgrade in its organizational status, which received board approval and was implemented shortly after Alpert left the agency in the summer of 1958.

However, Hesburgh's panel had little success when it tried to broaden the scope of that program beyond the hard-core. As a first step in this direction, the panel proposed a reconfiguration of the social science advisory panel, such that "all of the various points of view" would be represented.[62] But this call for diversity alarmed the skeptics, including Julius A. Stratton, an electrical engineer with extensive military ties and a high-level administrator at MIT. Stratton summed up his worries and those of other board members in dire terms:

> Extreme caution should be exercised in expanding this program since such a step would introduce a totally new problem. It would change the character of the foundation and might necessitate a change in the composition of the Board.

Stratton also underscored "the controversial nature of the social sciences in the Congress and possible repercussions on the regular research budget of the Foundation."[63]

In agreement with Stratton, NSB member Warren Weaver, an accomplished mathematician and renowned leader of the Rockefeller Foundation's natural science division (1932–1952), emphasized that the agency should not move too far or too fast. Later, in his 1970 autobiography, Weaver would write that when he spoke about science, he meant "of course, real science, basic science, such as physics, chemistry, biology, mathematics, and astronomy." In response to the proposal from Hesburgh's panel to expand the agency's social science efforts, Weaver asked if his fellow NSB members really wanted to move beyond convergent fields—though the board had, in fact, already taken this step, as Director Waterman pointed out.[64]

With no resolution in sight, the geologist McLaughlin suggested that Hesburgh's panel should reconsider its position in light of the fundamental worries conveyed by board members.[65] Under pressure, Hesburgh's panel withdrew its recommendation to broaden the range of viewpoints represented on the social science advisory board.[66] Years later, Hesburgh remembered that many of the board's natural scientists "looked down their noses" at the social sciences because they believed this was "not hard science … not 'real' science."[67]

Before leaving the agency, Alpert wrote an unpublished background paper for Hesburgh's panel that discussed how the agency should proceed given the persistent difficulties. No other document from these early years did a better job of specifying the wide array of issues in play and how they had shaped social science policies and programs. Among other salient points, Alpert observed that the NSF had defined "its social science research program so as to omit ... controversial areas. ... We have identified such research areas as sex, religion, race, and politics as fields which might be more appropriately supported by private foundations or by government agencies with applied social science missions." He also anticipated that the agency would stick to a restricted approach, by refusing to "support research in sensitive, controversial fields no matter how significant such research might be."[68]

Similarly, Alpert anticipated that the NSF would continue to restrict its support to the "hard-core scientific end of the continuum." This meant steering clear of social research that concerned "ethics, welfare, and philosophical interpretations of man's destiny." It also meant insistence that the social scientist, like the natural scientist, produced value-neutral knowledge and did not engage in normative judgments: "Whether the atom is used for peace or destruction ... whether knowledge of human motivations is used to provide happiness or sell soap, these are alternatives which the scientist as seeker of knowledge and truth cannot determine."[69]

Under these conditions, although NSF social science grew during the mid-1950s, constraints on such growth were palpable. The number of social science proposals received increased yearly, from 8 in 1954 to 29 in 1955, 44 in 1956, 53 in 1957, and 106 in 1958. The number of NSF awards rose as well, from a low of 3 in 1954 to a high of 49 in 1958. Yet the level of social science support compared to the amount of funds requested never reached 40 percent. Moreover, as the agency received more and more proposals, the dollars granted compared to the dollars requested actually declined, and significantly so, from a high of 39 percent in 1956 to merely 19 percent in 1958.[70]

NSF funding levels were marked by glaring disparities across the sciences as well. From the total 1958 NSF research budget of roughly $16.3 million, only about $600,000 went to the social sciences, or about 3.7 percent, a point noted with dismay by Oregon Representative Porter, one of the most consistent legislative advocates of the social sciences.[71] Henry Riecken added that his predecessor, Alpert, had "found it necessary to pare all requests to the very minimum." Consequently, the average social science grant was "appreciably smaller" than the average grant in the two natural science divisions. The average duration of a social science grant was also "shorter."[72]

Similar disparities marked NSF support for advanced training. From a total of 756 predoctoral graduate fellowships awarded in 1958, only 21 went to psychology, 9 to anthropology, and 10 to convergent areas, making for a meager total of 40 for all of the social sciences put together. This figure was far lower than the number of graduate fellowships awarded in many individual disciplines from the natural sciences: physics occupied the top spot with 178. That same year, the social science share of NSF postdoctoral fellowships was even worse. Only 3 of 85 awards went to convergent research, and none went to any other field in the social sciences. In contrast, 32 went to the life sciences, the leader in this category.[73]

At a time when the natural sciences and especially the physical sciences dominated the federal science system, the struggles facing NSF social science reflected and reinforced broader developments. It is true that massive expansion in the federal science system provided significant opportunities for psychologists, economists, anthropologists, and other scholars working on an array of topics, especially those related to national security and foreign policy concerns, many of them linked more or less directly to American Cold War objectives.[74] Yet, it is also true that the social science share of the total federal commitment to scientific research and development (R&D) in the mid-1950s was only between 1 and 2 percent. Meanwhile, the physical sciences received the lion's share at 87 percent, while the life sciences followed with 11 percent.[75]

On top of this, many federal programs that supported social research had a rocky existence throughout the 1950s, mainly because of conservative political pressures. "Appropriations storms" threatened and sometimes put a halt to federal funding for studies on a wide variety of topics, including "child-rearing practices, mother-love among lambs, population dynamics, message diffusion, and other projects," observed a dismayed Alpert.[76] As noted earlier, anti-communist investigations led by conservative politicians had also targeted social science funding from the large private foundations.

In sum, although NSF publications and Alpert's writings in scholarly journals during his tenure at the agency said little if anything about skepticism toward the social sciences inside the agency, the archival record shows that, behind the scenes, influential board members continued to express sharp doubts. Moreover, the agency's cold responses to Senator Kefauver's request regarding research on juvenile delinquency and to the Hesburgh panel's request about including all viewpoints on its social science advisory committee reveal how such doubts reinforced limitations on the scope and extent of NSF social science. Those limitations, in turn, reflected and contributed to glaring disparities between the social and natural sciences at many levels.

Nevertheless, the early development of NSF policies, programs, and practices provided the social sciences with an important measure of scientific legitimacy and funding. In light of the sharp questions raised during the postwar NSF debate, the trials and tribulations during the McCarthy years, and the endless struggles inside the agency, one should not underestimate the significance of that accomplishment.

ALPERT'S LAMENT

In the summer of 1958, Alpert began a new job at the University of Oregon, as a sociology professor and graduate school dean; his resignation at the NSF went into effect August 1. Rising through the university's administrative ranks, he became provost, vice-president, and then acting president in the mid-1970s. During those years, Alpert continued to publish regularly and remained active in other professional arenas. He was vice-president of the American Sociological Association (1958) and president of four scholarly societies: the American Association of Public Opinion Research (1955–1956), Pacific Sociological Association (1962–1963), Sociological Research Association (1961), and Washington Statistical Society (1954–1955). He was also the editor of the *American Sociological Review* (1960–1962), a fellow at the Center for Advanced Study in the Behavioral Sciences at Stanford University (1963–1964), and the director of UNESCO's Department of Social Sciences (1967–1970).

Scholarly discussions of Alpert have typically been cursory and portray him as a consistent advocate of the unity-of-science viewpoint, objective research, and quantitative studies. Yet closer consideration of his views before, during, and after his NSF years reveals that he felt uneasy about that viewpoint and worried about its implications for his home discipline (sociology) and the social sciences more generally. Examination of Alpert's views therefore offers a special vantage point for appreciating not only why the unity-of-science viewpoint can appear rather restrictive in retrospect but also why it could seem unduly narrow even at the time to some of its main proponents.[77]

We can start with Alpert's extensive writings on Emile Durkheim, for these bring into focus underappreciated aspects of Alpert's views about the nature of social science inquiry, its place within the wider scientific community, and its potential to advance human welfare. During the 1930s and

1940s, the writings of Harry Alpert and Talcott Parsons positioned Durkheim as a key figure in American sociology's ancestral past.[78] Alpert spoke favorably about the French thinker's view that sociology could and should be rigorously scientific. In this context, Alpert also invoked the triumvirate of scientific criteria—verifiability, objectivity, and generalization—that he later deployed in NSF documents. However, Alpert also wrote respectfully about Durkheim's view that sociology should address the challenges of social reconstruction and the problems of ethics, so it could serve as "an instrument of social action" and "play a role in the social reconstitution of France."[79]

In addition, in other writings, Alpert championed social science inquiry that confronted social problems such as anti-Semitism, a position at odds with the NSF's refusal to support action-oriented and policy-relevant research, including work on juvenile delinquency. Alpert added that American scholars had failed to undertake reform-oriented research that could help reduce prejudice against Jews: "there is little in the existing literature that offers scientific guidance on the question of how discrimination and prejudice may be combated and controlled, and scientifically designed studies aimed at evaluation of the effectiveness of specific action programs in this field seem to be few in number."[80] Note as well that Alpert issued this charge in 1949, only a few years before he began working for the NSF, which supports the point that it was for strategic purposes, and not any heartfelt commitment, that Alpert proposed a slew of measures designed to separate the agency's social science program from such action-oriented research.

Furthermore, Alpert believed that social research, at least a good portion of it, required interpretative analysis of a sort that had no place in the natural sciences. On this issue he found Durkheim wanting, for the French thinker had overlooked the central role of interpretation in social inquiry (i.e., *verstehen* analysis along the lines emphasized by Max Weber).[81] Here, Alpert agreed with many other influential American scholars who wrote about the foundations of scientific social inquiry, including, as noted before, fellow sociologists Talcott Parsons and Louis Wirth. This position implied some fundamental disunities between the natural and social sciences at the ontological, methodological, and epistemological levels.

In this way, Alpert qualified his advocacy for the unity-of-science position, especially if this meant that the natural sciences were to be taken as the gold standard. His position comes across clearly in a 1954 essay he wrote about the Columbia sociologist Robert M. MacIver—who, not incidentally, had been a member of Alpert's dissertation committee two decades before. Alpert highlighted MacIver's "humanistic" approach, evident in his presentation of "man as an idealistic, goal-seeking, value-creating being who is especially distinguished by his capacity to impute meanings to events, objects, and people." Thus, deep study of human beings and society (as opposed to more superficial examination of observable behavior) required interpretation of man's inner life, which had no equivalent in the physical sciences, for, as Alpert put it, "there is no inside story of why a meteor falls or why a liquid freezes." Such considerations pointed to the need for "verstehen analysis," which raised the challenge of "developing objective methods for dealing with data of inner experience."[82]

While Alpert favored the development of objective and quantitative modes of analysis where applicable, he also emphasized that they had significant limitations. For example, in his essay on MacIver, Alpert applauded his former professor for recognizing that quantitative methods and statistical studies were valuable but should not be considered as the final points of analysis. Rather, scholars should use them as "guideposts to understanding."[83] Elsewhere, Alpert observed that Vannevar Bush had pointed out that many cases of "great science" in the past "involved very little in the way of measurement and mathematical formulation"—a striking point in light of the fact that NSF leaders regularly invoked Bush's views for legitimization purposes.[84]

Even though Alpert strongly supported NSF social science during his five years there, he also criticized what he believed were unhealthy constraints on federal social science funding. To begin with, he observed that an "on again, off again" pattern of political support made it impossible for many agencies to develop sustained research programs.[85] He took aim at the reigning scientific hierarchy as well, which placed the social sciences in a precarious position. In a 1958 article entitled "The Knowledge We Need Most," published just as he was leaving the NSF and thus at a moment when he no longer had an obligation to avoid publicly criticizing the agency, Alpert—as seen in this chapter's opening quote from him—argued that

> invidious distinctions among specialists of the various disciplines must disappear. An institutional environment must be created to encourage all scientists in intimate association to cooperate effectively toward achievement of the common goal of better understanding of the ways of nature, man, and society.

Alpert's article ended with a cartoon about the sorry state of social science funding.

By questioning the value of the value-neutral stance, Alpert challenged yet another cornerstone of the hard-core approach at the NSF. In a 1963 assessment of the state of sociology, he proposed that the discipline's health required the presence of social critics—something the NSF had not supported during his years there and still did not support.[86] A few years later, Alpert published an essay about the sociologist George Lundberg, a vigorous defender of the unity-of-science position and an outspoken advocate of value neutrality as the only legitimate investigative stance in scientific inquiry. Here, Alpert worried out loud that this stance tended to give scholarship a conservative direction, albeit perhaps unwittingly in the case of some scholars who advocated it for other reasons. (Lundberg was on Alpert's dissertation committee in the 1930s, along with MacIver and Robert Lynd. The clash of perspectives offered by the three of them surely stimulated Alpert's abiding interest in the foundations and purposes of social inquiry.)

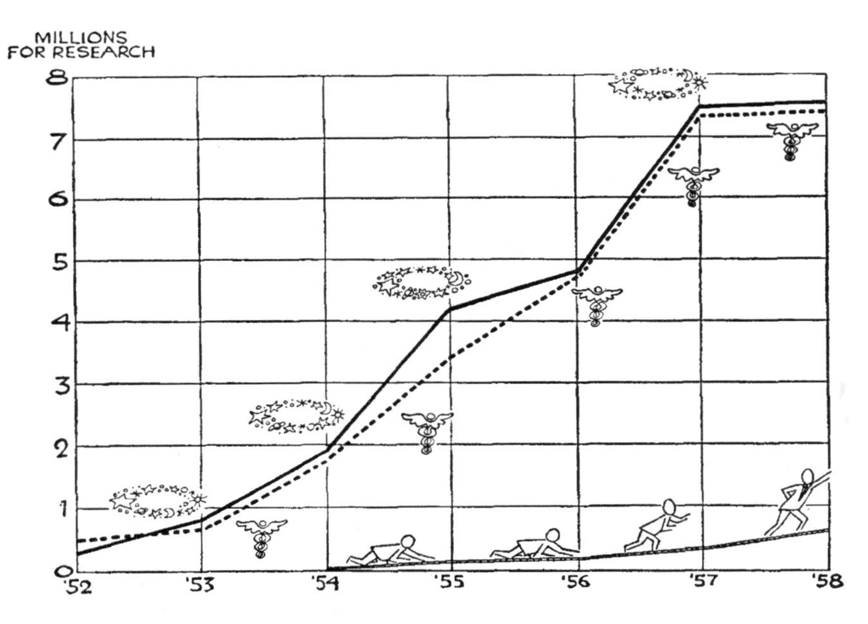

Figure 2.4
Social Science Struggles for Recognition and Funding at the NSF. Published with the article "The Knowledge We Need Most" by Harry Alpert in *Saturday Review*, February 1, 1958, pp. 36–38.

The danger of Lundberg's approach, Alpert wrote, was that although it might be very good at describing and analyzing society, this knowledge by itself could not offer an explicit assessment of the existing social order or challenge the status quo.[87] Such a critique had been advanced by other left-liberal scholars mentioned before (i.e., Lynd, Wirth, Dewey, and Beard) and was passionately taken up during the 1950s and 1960s by the sociologist C. Wright Mills and other scholars associated with the New Left.

The more one learns about Alpert, the less one is surprised to see him raising so many doubts about the scientistic project in writings where he was not in the position of explaining and promoting the NSF's approach. Alpert was a learned man. He spoke French fluently and had broad and varied intellectual interests, including not only the social sciences and the history and philosophy of science but also literature and the arts. He had a generous, caring, and compassionate character as well, which inspired the following doggerel:

Loving, too, is an Art
In which all should take part
Harry Alpert has ... taught us
To think with our heart.

No ode to a man with narrow scientistic commitments, these words were engraved on a plaque Alpert received as a departing present circa 1970 from his UNESCO staff. For the rest of his life, Alpert kept this plaque in his university office, along with another one he received years before from his NSF associates. This second plaque showed Alpert cutting through bureaucratic red tape in the name of the people.[88]

Alpert's essay on Lundberg appeared in 1968, by which time the value-neutral stance had come under heavy fire within academic, intellectual, and political spheres, while pressure for federal agencies to support social research directed at social problems had grown dramatically—developments explored in the next two chapters. A couple of years later, Alpert received the plaque about thinking with one's heart. However, it would be a mistake to conclude that Alpert developed his criticisms of the scientistic approach only during this latter period of his career. As we have seen, in his earlier writings on Durkheim and in other places, he had already put forth a complex view about the nature and relevance of social science inquiry. In addition, his years at the NSF gave him firsthand experience with disconcerting implications of the scientistic approach at the levels of funding policies, program development, organizational standing, and institutionalized relations with the natural sciences. In the final analysis, I propose that we need to see his role in promoting funding for hard-core social research at the NSF as the work of a savvy scholar-bureaucrat. To accomplish what seemed possible under difficult conditions, Alpert had put aside his more catholic vision of the social sciences, one that recognized certain ontological, methodological, and epistemological principles with no close equivalents in the natural sciences and one that appreciated the importance of value-laden inquiry and social criticism.

CONCLUSION

The early development of NSF social science from 1953 to 1957 was marked by a cautious approach and a commitment to scientism. As the agency's first social science policy architect, Harry Alpert played a key role in these developments. On the basis of his background studies, policy statements, and carefully crafted recommendations, the agency began funding studies on the hard-core end of the social research continuum. Inside the agency, such work was defined by its basic science orientation, by supposedly universal scientific criteria (i.e., objectivity, verifiability, and generalizability), and initially by "convergent" criteria as well. During his years at the agency, Alpert also advocated successfully for significant changes and upgrades. These included the elimination of convergent criteria, the establishment of new disciplinary-based programs, the consolidation and elevation of those efforts with the creation of a unified social science office, and modest increases in funding. But the NSF would still not support various types of social research, from applied studies (as the agency didn't have a mandate to support applied work in the natural or social sciences at that point) to action-oriented research, normative investigations, humanistic studies, and anything else deemed to lie beyond the hard-core. During this formative period, the agency also steered clear of work on certain topics considered too hot to handle. Alpert explained that the task of funding work on sensitive subjects, including sex, religion, race, and politics, should be left to the large private foundations.

Although the scientistic approach remained dominant during that time, this was not inevitable. Nor was the agency's commitment to it simply the obvious choice if one was aiming to promote social science progress.

Instead, a set of historically specific and contextual factors came together in ways that made that carefully circumscribed approach appealing, even for those such as NSB member Chester Barnard, who otherwise had serious qualms about the wisdom of its assumptions regarding the nature and value of the social sciences. That set of factors included all of the following: the conflicted position of the social sciences in the agency's legislative origins and the prior SSRC-led effort to advance a scientistic platform as the basis for social science inclusion; the vague reference to "other sciences" in the NSF charter and the dominant position of natural scientists at the new agency, including its first and long-time director Alan Waterman; the power of a basic science ideology and the closely associated value-neutral ideal in scientific inquiry; the persistence of conservative concerns in Cold War America, reinforced by McCarthyite attacks that threatened to make NSF support for social research a political target and thus a budgetary liability; and, last, the presence of other funding sources that, presumably, were in a better position to support social research on sensitive issues.

Equally important, under those conditions, the social sciences had no chance of being considered equals with the natural sciences when it came to representation, influence, and funding at the young agency. Indeed, by any measure, their position was weak and their standing low. Back in 1945 this situation had been anticipated by the sociologist William Ogburn during his congressional testimony: "If you put 1 social scientist in with, say, 9 or 10 natural scientists, and if you have a director who is a natural scientist, they will forget all about the social scientist being there and they will give him very scant attention."[89] In 1958, following the establishment of a consolidated program for the social sciences, Father Hesburgh noted that they had gotten a "second toe in the door."[90] While this image conveyed a real improvement, it also highlighted how marginal and vulnerable they remained.

Furthermore, the strategy of limiting NSF support to the hard-core made it vulnerable to criticism from those who had alternative views. Some conservative figures continued to charge the social sciences with being intimately involved with troublesome endeavors, as seen in NSB member Kevin McCann's comment about these sciences being worse than Pandora's box. Other doubts, raised most often by natural scientists, suggested that due to some distinctive feature of social inquiry (i.e., insufficient rigor, lack of lawlike generalizations, involvement with social agendas, susceptibility to ideological and political bias), the social sciences were not—and perhaps could never be—part of a unified scientific enterprise. Meanwhile, and most surprisingly, Alpert himself challenged the scientistic project. Looking beyond his writings where he was explaining and promoting the NSF's programs and approach, we have seen that Alpert, much like some left-liberal scholars, saw that adherence to a value-neutral and apolitical investigative outlook would compromise intellectual progress in the social sciences and inhibit their ability to contribute to progressive change.

Nevertheless, the NSF still became an important patron during the Alpert years. By the late 1950s, it was one of the leading federal agencies in the area of funding for basic social science research. In addition, in light of its commitments to first-class scientific research and hard-core studies, the NSF managed to bestow an important measure of scientific legitimacy on the social sciences.

In the coming years, the United States underwent considerable changes marked by dramatic political, social, cultural, and intellectual conflicts. Far from the common ivory tower image of them, scientists, scientific agencies, and research universities were swept up in the events and issues associated with Sputnik, the space race, the Vietnam War, the War on Poverty, the civil rights movement, and the environmental movement. Neither the NSF nor the social sciences would be exceptions. Still, when Alpert left the agency in the summer of 1958, how the mix of opportunities, challenges, and turmoil ahead would alter NSF social science and its broader significance remained to be seen.

3

HELP FROM ABOVE: A MODEST FLOURISHING DURING THE LIBERAL HIGH TIDE, 1957–1968

> [Social scientists focus on] explaining and/or predicting and/or controlling both regularities and changes [in the phenomena they study, giving them] the same general epistemological character as ... the physical and biological sciences.
> —Henry W. Riecken, First NSF Social Sciences Division Director (1961–1965), 1969[1]

> All schools of political science [should avoid] internecine struggles in favor of an open-minded eclecticism.
> —Evron Kirkpatrick, long-time executive director of the American Political Science Association, 1962[2]

Compared to the previous years, the post-Sputnik era was more supportive of NSF social science, due to four main reasons. First, this period witnessed an enormous expansion in the federal science establishment. Amid a pervasive panic following the Soviet launch of the first artificial earth satellite in October 1957, the American political and academic communities pushed hard to beef up federal responsibilities and support for scientific endeavors. The coming months and years saw the creation of many important agencies, including the National Aeronautics and Space Administration (NASA); the passage of landmark science-related legislation, including the National Defense Education Act; the establishment of the President's Science Advisory Committee (PSAC); and a dramatic uptick in science funding. In just four years, from 1957 to 1961, federal R&D support doubled, while federal support for basic research tripled.[3] The federal contribution to total national spending on university research increased as well, rising from 56 percent to 75 percent between 1957 and 1964. Skyrocketing federal funding, in turn, supported a dizzying explosion of scientific projects, buildings, research facilities, and training programs.[4] Just as a rising tide lifts all boats, soaring federal science budgets made increases in social science funding likely.

Second, although puny at first, the NSF now grew in leaps and bounds. In the wake of Sputnik, a "national self-appraisal that questioned American education, scientific, technical and industrial strength" led to rising appropriations, as former NSF historian George Mazuzan noted. Indeed, the agency's budget soared, from $40 million in 1958 to $134 million in 1959, and then to nearly $500 million in 1968.[5] On top of this, the agency acquired greater prominence within the federal science establishment and national funding system. At a time of frenzied growth in basic science, the NSF's unique mandate to promote such work stood out in particular. The agency underscored the special value of such work in a 1957 report called *Basic Research—A National Resource*, which was transmitted to President Eisenhower while the shock of Sputnik was still fresh in the air.[6]

Third, the NSF gained greater stature due to its special role as the so-called balance wheel in the national funding system. This meant that the agency was responsible for looking after all fields of basic scientific study, including ones that, for one reason or another, did not receive adequate support from other public or private patrons. The agency, "as a matter of deliberate policy," sought "to support all the fields of science in a comprehensive way," explained NSF director Alan Waterman in 1961.[7] Similarly, the agency's next director, Leland Haworth, observed that only the NSF was "dedicated to the health and welfare of science as a whole."[8] With its budgets increasing, with its stature as a patron of basic science rising, and with its responsibility for supporting basic science regardless of field, growth in the size and importance of its social science efforts became all but inevitable.

Fourth, relations between the federal government and the social sciences became tighter due to a reinvigoration of American liberal thought and reform. With the end of the McCarthy era, the nation's political culture became more receptive to liberal thinkers and politicians that promised, as John F. Kennedy did during his 1961 presidential campaign, to get the country moving again. During the New Frontier days of the Kennedy administration and the Great Society days of the Johnson administration, the impulse to use the federal government to tackle the nation's social, economic, and moral problems became stronger than at any time since the New Deal. Under these conditions, the social sciences moved into the national spotlight not because they seemed threatening to American ideals but because they promised to make government efforts to achieve those ideals more effective and rational. On the home front, social scientists stepped forth to address a broad array of troubling issues from juvenile delinquency to poverty, racial tensions, and urban decay.[9] Meanwhile, the expansion of Cold War–inflected conflicts around the world stimulated work in a variety of fields, including modernization theory, development studies, social systems analysis, psychological warfare, and counterinsurgency.[10]

This chapter examines the evolution of NSF social science in the post-Sputnik era, from the late 1950s to the late 1960s. The first section highlights significant changes, upgrades, and growth that marked this as an expansionary period for the social sciences. The agency did not treat all of the major disciplines equally, however, with political science lying at the bottom of the pack. To understand this differential treatment of the disciplines and the broader science policy implications at stake during these years, including the persistent power of scientism, the second section explores the struggle by political scientists to overcome the agency's reluctance to support their work. By focusing on the development of an ambitious NSF-supported social science–based grade school curriculum project called Man, A Course of Study (MACOS), the third and final section examines how the social sciences became part of NSF Cold War–inflected efforts to reform American science education.

The central argument is that the agency's social science efforts underwent notable expansion at many levels during these years, yet that expansion remained rather modest in size and scope, even during the liberal high tide associated with the Kennedy and Johnson administrations. NSF leaders recognized that more favorable conditions in the wider society and federal science establishment provided opportunities to push forward. But they also retained a strong preference for taking cautious steps along a path whose contours were already defined by earlier developments, from the postwar NSF debate up through the agency's founding and formative years when Harry Alpert served at its first social science policy architect.

RIDING THE POST-SPUTNIK/NEW FRONTIER/GREAT SOCIETY WAVE

Less than a year after Sputnik first circled the Earth and shortly after Alpert's departure from the NSF in the summer of 1958, the agency hired a new and highly regarded social science leader, Henry Riecken. Riecken also became the associate director of the scientific personnel and education division. When the agency created a social sciences division a few years later, it placed Riecken in charge.

Riecken arrived with outstanding professional credentials. Before pursuing an academic career, he had acquired significant government experience, first during the New Deal, through a job in the U.S. Department of Agriculture where he learned techniques of statistical survey sampling and open-ended interviews, and then during World War II, when he worked in the Army Air Corps on aviation personnel selection. Following the war, Riecken pursued graduate studies at Harvard's Department of Social Relations (DSR), where he studied with notable figures including Gordon Allport, Jerome Bruner, and Samuel Stouffer. After completing his PhD in 1949, Riecken became a lecturer at the DSR, before taking an academic post at the University of Minnesota's Laboratory for Social Relations. There, he worked with social psychologists Leon Festinger and Stanley Schacter on research that was informed by Festinger's theory of cognitive dissonance and led to their classic 1957 book *When Prophecy Fails*—about how a group of people who had predicted the end of the world on a particular date responded when that prophecy failed.[11]

In addition, Riecken had an abiding interest in the social relevance of social research along with special expertise in evaluation research, a nascent field of study inspired by growing political and scholarly interest in creating effective means for assessing the impact of social policies and programs. Before his NSF job, Riecken had already published a book called *The Volunteer Work: A Psychological Evaluation*, which examined the impact on participants of a domestic work camp program supported by the American Friends Service Committee, a Quaker organization. That book included a more general method for evaluating the psychological effects of "social action programs." Riecken situated his analysis in relation to percolating interest, especially among liberal thinkers, in determining why programs to reduce prejudice toward minority groups succeeded or failed.[12] Two decades later, Riecken's interest in practical relevance was manifest in *Social Experimentation: A Method for Planning and Evaluating Social Intervention*, an edited book on the use of randomized experimental trials in a wide array of social policy arenas, including crime, education, and social services.[13] It seems fair to say that these studies by Riecken were good examples of the sort of action-oriented social research that Alpert found important as well.

Figure 3.1
Henry W. Riecken during his early NSF days, ca. 1958. Courtesy of the Riecken family.

However, in contrast to Alpert, who complained about the underdeveloped state of research on prejudice reduction and who harbored serious doubts about the NSF's approach to the social sciences, Riecken's writings suggest that he was satisfied with its scientistic orientation, including the principle of separating social science from social action and public policy. As Riecken later recalled, his own "professional identification with experimental social psychology fitted well with the generally quantitative and empirical orientation of the NSF program."[14]

This orientation also fit well with Riecken's perspective on the scientific trajectory of the social sciences. "The social and behavioral sciences," he wrote, had "become differentiated from ethical and moral philosophy" not that long ago. In fact, "only in the last two or three decades" had "any substantial number of scholars" begun to practice "social *science* rather than social opinionating." The difference was profound, "as great as that between chemistry and alchemy."[15] As this chapter's opening quotation from him indicates, Riecken further asserted that contemporary social science work, with its emphasis on prediction and control, had "the same general epistemological character" as studies in the natural sciences. Given his professional background and his understanding that the social sciences were part of a unified scientific enterprise, Riecken himself fit easily into the NSF's "gentleman's club," as the agency's long-time social science assistant Bertha Rubinstein put it.[16]

During his NSF years, Riecken thus sought to strengthen social science funding within the basic framework established during the Alpert years. We will see how he did this at a number of levels in the rest of this chapter, until he left the agency in 1965 to take a job at the Social Science Research Council. (Riecken served as the council's vice-president for one year and then its president for five years, from 1966 to 1971. Later, he moved to the University of Pennsylvania as the Francis Boyer Professor of Behavioral Sciences, until his retirement in 1985.)

But, first, we must note that midway through the decade, the social science division acquired a new leader named Howard Hines, who would seek to stay the course. A native of Iowa City, Hines served in the U.S. Army and worked during World War II in the Office of Strategic Services in Europe. His studies included an undergraduate degree from the University of Iowa, followed by a PhD in economics from Harvard. Subsequently, he became an economics professor at Iowa State University and, according to Otto Larsen, had a teaching position at the University of California–Berkeley before going to the NSF. Unfortunately, not much more is known about Hines, either as a scholar or as an NSF staff member. Larsen simply says that Hines "did not possess his predecessor's persuasive, diplomatic style," although he had "staying power" and served ten full years as the head of the social science division. In 1975, Hines was "eased out" of that position and given other minor roles, until his retirement from the agency in 1984.[17]

From fragments of evidence, we can also surmise that Hines was fully behind the scientistic strategy. For one, there is no indication that he raised any questions about its soundness, though there were ten good years when he could have done so. More revealing, though, is an unpublished 1971 review of the agency's social science program. Here, Hines wrote that the program's "most fundamental goal," which includes "each and every part" of it, is "to strengthen the scientific character of the social sciences, to make them more scientific." Above all, this entailed an emphasis on rigorous scientific methodology. For Hines, this was the program's bedrock and appropriately so. It had been laid down at the beginning, and it would serve as a solid foundation well into the future. In his own words: "Consequently, from the origin of the NSF program until as far into the future as any of us can now foresee we are going to be concerned with developing more highly scientific methods of doing research." As an example, he pointed to advances in quantification in his home discipline, economics: "in no area of social science has there been greater progress in the lifetime of the NSF than in the field of quantified economics or econometrics." Moreover, the NSF was "the principal source of support of this kind of research in the United States today and has been for some years."[18]

In the post-Sputnik era, however, it was during Riecken's years that NSF social science gained considerable strength, beginning with major upgrades in organizational standing. In December 1959, following recommendations from the agency's social science advisory committee and from Riecken himself, the agency replaced its social science "program" with a higher-status "office." And in December 1960, the agency gave the social sciences their own division, which included the disciplinary-based programs for anthropology, economics, sociology, and the history and philosophy of science.[19]

The division's establishment marked a milestone, for the social sciences now enjoyed the same organizational status as the natural sciences—whose divisions were created when the agency first became operational and thus had already been in place for a full decade. According to a *New York Times* editorial, with this new division, the social sciences had "received a long-sought recognition ... by being elevated to the status of the physical and biological sciences in the Government's program of basic research." The editorial also credited Director Waterman with emphasizing "the symbolic importance" of placing the "social sciences on an equal footing with the more traditional fields."[20]

Greater organizational standing went hand in hand with expanding financial commitments. NSF funding for social science research rose from $2.2 million in 1960 to $10 million in 1965 and then to $15.3 million in 1969. As a percentage of the total research budget, the social science's share also rose substantially. The most impressive growth occurred in the decade's early

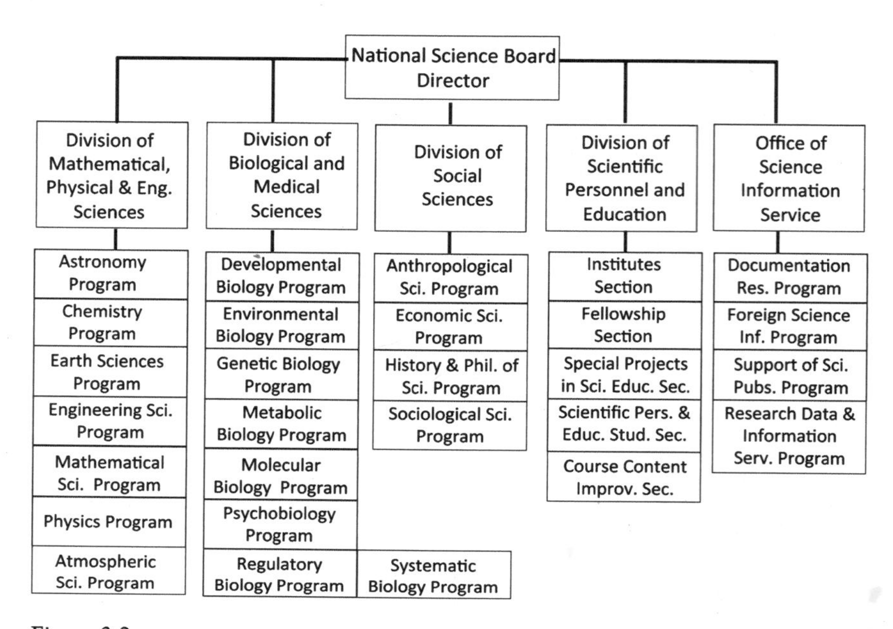

Figure 3.2
National Science Foundation organizational chart as of June 1961. Nondivisional administrative offices have been omitted.

years, rising from 3.2 percent in 1960 to 5.8 percent in 1965 and a tad higher to 5.9 percent in 1969.[21] In addition, funding increased for social science–related activities handled by other agency units dedicated to science education, institutional resources, science planning and policy studies, and science information. So much so, in fact, that by 1968, total funding from these other units exceeded funding from the social science division.[22]

The social sciences's position also improved as national interest in their practical relevance grew. The NSF had placed such work off limits before out of respect for its basic science mandate and due to the worry that engaging with practically oriented social research could stir up political controversy. But in the late 1950s and 1960s, growing interest in social science that could help solve social problems pushed the agency to rethink its position. In 1959, Director Waterman noted an "increasing need for special attention to areas of science which may be critical in the national interest."[23] In 1963, Henry Riecken pointed out that "scientific advance" had to some extent "been bought at the price of reducing the relevance" of social science findings to large-scale problems and thus did not adequately meet the "needs of the social order." The new division could address this deficiency by focusing on "problem-oriented basic research."[24]

In fact, the division had already taken steps in this direction. For example, it supported the social psychologist Stanley Milgram's famous and controversial experiments on obedience to authority, in which volunteer subjects were told to deliver electric shocks to another individual whenever that individual gave the wrong answer on a memory test. The results revealed shockingly high levels of obedience among ordinary Americans.[25] The division also supported experiments conducted by another social psychologist named Leonard Berkowitz that aimed to determine if certain factors prevalent in American society led to violence, including "exposure to violence on films and the physical presence or absence of weapons."[26]

The NSF funded social research relevant to American foreign policy and military operations as well. Its 1963 annual report mentioned a "cross-national study" about the "social ramification of modernization" in Chile, Nigeria, and Pakistan.[27] Five years later, when the social science division awarded 426 grants, 171 of them—more than 40 percent—involved research on foreign areas. The amount of funding, $5.2 million, accounted for one-third of the division's total financial commitments, $15.3 million. Furthermore, by this point, the NSF was the third largest federal sponsor of foreign area research, behind the Department of Defense and the Department of Health, Education, and Welfare, in first and second place, respectively.[28]

Yet the changes noted above—in leadership, organizational status, funding, and program scope—hardly made a difference at many other levels, thus ensuring the continued dominance of the natural sciences. Although, as seen in the funding chart below, the social sciences' share of the NSF research budget rose to just about 6 percent by the end of the 1960s, this still left the natural sciences with 94 percent. So, for every dollar channeled to the social sciences, almost nineteen went to the natural sciences.

Support for particular programs provides additional perspective on the glaring disparities. For example, in 1966, when total funding for research in the social science programs stood at nearly $13 million, funding for the oceanography research program, just one of many programs in the physical sciences division, received $29.1 million.[29]

NSF leaders continued to come mainly from the natural sciences as well, especially the physical sciences. By the time the physicist Alan Waterman retired in 1963, he had served as director for a dozen years. Similar to

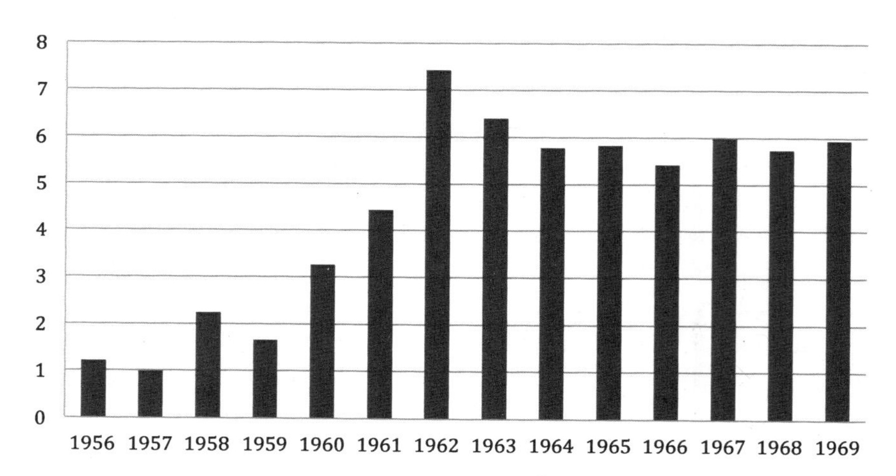

Figure 3.3
NSF Social Science Research Obligations as Percentage of NSF Total Research Obligations, 1956–1969. Data compiled from Tables 3.2 and 4.1 of Otto N. Larsen, *Milestones and Millstones: Social Science at the National Science Foundation, 1945–1991* (New Brunswick, NJ: Transaction Publishers, 1992).

Waterman, his replacement, Leland Haworth, was a physicist with strong ties to military science programs. An expert in particle physics, accelerators, and nuclear energy, Haworth had worked at the University of Illinois, MIT's Radiation Laboratory, and Brookhaven National Laboratory, a leading center for nuclear research that he directed from 1949 to 1961. During the Kennedy administration, Haworth served on the Atomic Energy Commission and had close relations with the president's chief science adviser Jerome Wiesner, the President's Science Advisory Committee, and the National Academy of Sciences. Haworth's tenure at the NSF lasted from 1963 until 1969, when he returned to Brookhaven.[30]

Social science representation on the twenty-four-member governing board remained minimal as well. Although not trained in the social sciences, Father Hesburgh had been among their strong supporters and pushed, although without success, for inclusion of a much broader range of viewpoints at the agency. But his NSB term ended in 1966. In the period from Sputnik to the late 1960s, only a half-dozen board members had advanced training in this area: Charles Hardin, an agricultural economist and chancellor of the University of Nebraska; Roger Heyns, a social psychologist and chancellor of the University of California–Berkeley; James March, a political scientist at Stanford University; Katherine McBride, a psychologist and president of Bryn Mawr College; Fred Thieme, an anthropologist who was provost of the University of Washington and then president of the University of Colorado; and Ralph Tyler, an educational psychologist, a renowned leader in educational evaluation, and a long-time director (1953–1967) of the Center for Advanced Study in the Behavioral Sciences at Stanford. Remember that the standard NSB appointment lasted six years, with the possibility of a one-time renewal. So, at any given moment, the number of board members with backgrounds in the social or psychological sciences was in the low single digits. At the tail end of this period, in 1969, there were exactly three such members—the social psychologist Heyns, the political scientist March, and the anthropologist Thieme.[31]

Concurrently, the agency's approach continued to be defined by a unity-of-science viewpoint and hard-core emphasis. After the social science division's establishment in 1961, Director Waterman announced that this upgrade did not entail any shift in viewpoint. Reiterating what had been asserted on many occasions, he explained that the agency "pledged to assist in the study of basic problems in the social sciences and the development of research techniques, rather than applied craftsmanship in social affairs or in contemporary social problems."[32] We have seen that a gap was emerging between this sort of rhetoric and the agency's granting practices, for studies relevant to contemporary social problems did receive funding. Still, NSF documents continued to underscore its basic research commitment, as seen in the 1963 annual report's claim that it provided support only for "basic research," not for any "studies of public policy, social issues, or other applied problems." This meant that NSF-funded research projects employed "objective methods," typically accompanied by a strong quantitative component, and that such studies aimed to produce "independently verifiable results ... which are general in nature rather than specific to a particular time, place, or event."[33]

To sum up, developments in program leadership, funding levels, organizational status, scope of support, and symbolic affirmation reveal that NSF social science expanded and deepened in the dozen years following Sputnik, a time when the agency itself acquired greater importance within the federal science establishment and the nation's scientific enterprise more broadly. At the same time, however, NSF leaders and documents continued to promote hard-core social research as its exclusive concern. Objectivity, verifiability, and generalizability: these funding criteria, originally articulated by Harry

Alpert, had become firmly embedded in policy. Furthermore, the distribution of NSF funding and the composition of its leadership show that the status of the social sciences remained rather modest in ways that reflected and reinforced their subordinate status vis-à-vis the natural sciences.

The social sciences' second-class status at the NSF contributed to their persistent marginalization within the federal science establishment as well. Consider this fact: as of 1966, the Defense Science Board, the central science policy unit within the U.S. military, did not include even a single social scientist. In addition, from the President's Science Advisory Committee's founding in 1957 up through 1967, 77 percent of its members came from the physical sciences and the rest from the life sciences.[34] As the science reporter John Walsh remarked in 1968, "Tokenism is the prevailing policy in the integration of behavioral sciences into the science advisory structure."[35]

The matters discussed above also shaped NSF social science in other ways. To bring the science policy issues at stake into clearer view, let's start with the case for political science, a discipline whose epistemological character and political salience made it especially problematic.

WITH ALL DELIBERATE SPEED: FINDING AND FUNDING THE SCIENCE IN POLITICAL SCIENCE

Indeed, political science had PROBLEMATIC written all over it from the start. During the mid-1950s, the agency had determined that the discipline's subject matter—politics—was off limits. Race, religion, and sex were the other main subjects in this category. I have found no evidence suggesting that the decision to relax those restrictions took place at any particular moment in time. But in the post-Sputnik era, NSF discussions no longer invoked those restrictions on a regular basis. Still, as we will see below, the underlying worry—that funding for research on certain topics and for certain types of inquiry would lead to unwanted public criticism and political repercussions—did not disappear. Moreover, the story of the agency's special caution vis-à-vis political science, a discipline struggling at the time with profound changes associated with the so-called behavioralist movement, reveals how scientistic tenets continued to define opportunities and constraints in NSF funding.

In the middle of the twentieth century, proponents of behavioralism recognized that their disciplinary predecessors had studied some worthwhile things, including the development of political ideas, the structures of political institutions, and the proper functions of political parties. But, according to an up-and-coming generation of scholars, such as the University of Chicago's David Easton, as well as more established scholars who provided support, including SSRC president Pendleton Herring and Harvard's V. O. Key, earlier political scientists had not studied political behavior and political processes adequately.[36]

According to the disciplinary historians Robert Adcock and Mark Bevir, the most striking thing about the behavioralists was their aggressive insistence on making political science "more rigorously 'scientific.'" This typically entailed a "cumulative interplay between theoretical innovation and empirical research." It also included a commitment to the scientific explanation and prediction of political behavior that went beyond particularistic and historical case studies. The behavioralist movement has often been remembered for its emphasis on greater rigor through statistical analysis and other quantitative techniques. Furthermore, many behavioralists aimed to establish their work on a value-free and ethically neutral basis.[37]

On the flipside, behavioralists sought to marginalize other types of work that did not fit the bill required by their understanding of scientific rigor. This included qualitative investigations, critical analysis, and philosophically oriented studies advanced by prominent European émigrés such as Eric Voegelin and Leo Strauss. These scholars contributed to a strong current in political theory that found modernity and the Western world to be mired in political decay and moral crisis. Moreover, they argued, against the behavioralists, that the intertwined rise of liberalism and a value-neutral, ethically bereft scientism threatened Western civilization.[38]

Equally important for understanding the characteristic outlook and momentum of the behavioralist movement, its leading scholars and lines of research rose to prominence with the help of generous support from major private organizations and foundations. Of special importance were the Social Science Research Council, the Carnegie Corporation, and the Ford Foundation.[39]

As for the NSF, the growth of behavioralist scholarship and the support from private patrons made the agency's reluctance to fund political science problematic. After all, NSF policy stated that it funded hard-core research in any discipline. So, why didn't it welcome proposals from scholars working within a behavioralist framework? In 1959, Director Waterman reported to the NSB that individuals in Congress and elsewhere—although he did not identify any by name—wanted the agency to fund research in this discipline.[40] As its social science effort grew in size, status, and visibility, its caution in this particular case became more and more worrisome for political scientists and their advocates.[41]

With the establishment of a social science division in 1960, the question of why political science was missing would become impossible to avoid. As it turns out, this same year, the agency awarded its first grant to a political scientist, Duncan MacRae from the University of Chicago.

Unsurprisingly, MacRae's work fit well with the NSF's scientistic framework. In 1950, MacRae had completed a Harvard PhD with a dissertation based on an empirical test of the famous Swiss psychologist Jean Piaget's theory of moral development.[42] After switching research topics but retaining a commitment to rigorous empirical testing of theories, MacRae published a 1958 book that presented a statistical study of congressional voting behavior. In step with the behavioralist movement, MacRae wrote that "the method of science ... includes the formulation of hypotheses and theories." At least "in the natural sciences ... these have taken symbolic and mathematical form." While he acknowledged that in the social sciences, "the same sort of mathematics as is used in physics, for example, may not always be applicable," he added that "the aim of concise symbolic representation of findings and predictions is striven for in any event." MacRae's book aimed to do exactly that, by developing "a mathematical model" of legislative behavior.[43] In his 1961 annual social science report to the director, Henry Riecken explained that MacRae's NSF-funded project built on that previous study, as he proposed to "develop a rigorous statement of the relationship between popular and legislative votes which is, in effect, a mathematical statement of the process of representation."[44]

Yet support for MacRae remained the exception, for political science was still not recognized as a field of study eligible for NSF support. This meant that the agency had to handle proposals from political scientists on an ad hoc basis (i.e., MacRae's proposal was reviewed by the sociology program).

An anonymous editorial in the *American Behavioral Scientist* characterized the broader implications in stark terms: only if you wore "the proper scientific garb," avoided "political science and all controversial topics," and were "properly respectful and grateful" might you receive an invitation to "an NSF cocktail party." Thus, despite recent upgrades in their status, social scientists, in the editorial's words, continued "to resemble the 'official Negroes' of NSF grants policy."[45]

Such treatment irked the discipline's leaders, including Evron Kirkpatrick, who would lead the battle to overcome the NSF's caution. A firm Democrat and anti-communist, Kirkpatrick has been described as "something of a pooh-bah in political science ... an agile academic and political operative."[46] After teaching political science at the University of Minnesota during the late 1930s and early 1940s, Kirkpatrick worked as a government research and intelligence analyst, first in the Office of Strategic Services during World War II and afterward in the State Department. Starting in 1954, he served for nearly three decades as executive director of the American Political Science Association (APSA). In this role, Kirkpatrick oversaw enormous growth in the association's size and influence.[47] According to the political theorist and disciplinary historian John Gunnell, Kirkpatrick strongly supported the behavioralist movement.[48] Similarly, Kirkpatrick's *New York Times* obituary mentions that Thomas E. Mann, a distinguished political scientist and Brookings Institution scholar, remembered that Kirkpatrick "wanted to treat the arguments and evidence of human political behavior scientifically."[49]

Yet, despite these claims about Kirkpatrick's outlook from respected sources, closer analysis shows that Kirkpatrick's views were more complex, and he was certainly no fan of the scientistic outlook. In his contribution to a 1961 edited book on the behavioralist study of politics, Kirkpatrick stated that he would "ignore the slang usage which equates science with natural science." Instead, he would use what he called the "correct meaning of science," namely, "a body of systematic and orderly thinking about a determinate subject matter." Thus, he did not endorse the behavioralist movement, as it reflected "the impact of the natural sciences on the social sciences," and its leaders typically assumed "a qualitative continuity of knowledge between the natural and social sciences." He also doubted that this movement would "preempt" the discipline in the foreseeable future. If it did, he would be unhappy, because he valued alternative approaches. Furthermore, he pointed out that this movement within the discipline, just like the behavioral sciences project more generally, had failed to discover any "laws of human behavior" that could provide a reliable "basis for accurate prediction and control." As seen in this chapter's opening quote by him, Kirkpatrick recommended that "all schools of political science" should avoid "internecine struggles in favor of an open-minded eclecticism."[50]

A decade later, Kirkpatrick also wrote against the value-neutral, disinterested investigative ideal: "Many of our concepts are inevitably value- and affect-laden. Concepts such as 'efficiency,' 'function,' 'success,' 'effectiveness,' 'democracy,' and 'progress,' [have] involved us in value judgments even when they appear to be simply 'descriptive.'"[51] These are hardly the views of a committed behavioralist.

Nevertheless, Kirkpatrick did not ask the NSF to support all schools of political science or even any of the alternatives to behavioralism that he found worthy. Instead, much as Alpert had done from his insider's position during the 1950s, Kirkpatrick, from his outsider's position, pursued a less ambitious albeit more realistic goal; he simply insisted that the agency support political science in the same manner that it supported other social sciences.

In September 1963, Riecken wrote to Leland Haworth—the new director—about the controversy brewing over political science. According to Riecken, Kirkpatrick had first begun to challenge the agency's policies and practices four years earlier, shortly before the social science division was created. Since the division had separate programs for sociology, anthropology, economics, and history and philosophy of science but not one for political science, Kirkpatrick presented this as evidence of unfair treatment. He also found it unfair that political science was not included in NSF documents that identified specific social science fields as being eligible for research grants and fellowships.[52]

Following some discussion within the social science division, Riecken had responded to Kirkpatrick by noting that much research by political scientists did not meet the agency's scientific criteria. A great deal of it was "applied, normative or policy-oriented." As such, it could not pass "the normal test of scientific investigation (objectivity, verifiability and generality)." Riecken told Kirkpatrick that the agency could not in good faith identify political science as a field eligible for support. But that response hardly satisfied Kirkpatrick, who argued that the agency should not "discriminate in advance" against his discipline.[53]

The NSF's peculiar treatment of political science rested not simply on the matter of fulfilling scientific criteria or not but also on the fear of political fallout. As had been true during the Alpert years, agency leaders continued to believe that limiting support to scientistic studies on the hard-core end of the social research continuum would help minimize public scrutiny and political attacks. In January 1963, while Waterman was still director, the social science divisional committee reaffirmed its support for this strategy. Regarding research proposals in any "sensitive" area, the committee advocated "considerable caution" and noted that political science seemed to be "more centrally concerned with questions of public policy and partisan controversy" than other social sciences. Furthermore, although Congress had expressed some interest in providing funding for political science, the divisional committee found "reasonable grounds" to believe that politicians would question "government-supported research at least on American political questions." The strategy of invoking scientific criteria for public relations purposes thus remained compelling: "the danger of a negative Congressional reaction is minimized by holding to a stringent definition of eligibility in terms of basic nature and scientific (rather than policy) orientation."[54]

Unhappy with the NSF's handling of the matter, Kirkpatrick decided to go over Riecken's head by taking his complaint to the new director. In November 1963, Kirkpatrick told Haworth that neither in writing nor in conversation had he received "a rational explanation or justification." In fact, the agency's defense of its hard-nosed stance vis-à-vis political science could hardly "survive an objective analysis." Furthermore, whatever criteria the agency used to "exclude" his discipline should be "just as relevant" to other social sciences. Similar to the case he put before Riecken, Kirkpatrick told Haworth that the agency's "unfortunate and arbitrary singling out" of political science amounted to a "discriminatory policy."[55]

In an even more significant shift in strategy, Kirkpatrick turned to national legislators for support. Kirkpatrick had extensive government experience and understood the growing public policy importance of the social sciences. He also knew some legislators personally, including Hubert Humphrey, the liberal Democratic senator from Minnesota who, after President Kennedy's assassination in November 1963, became vice-president in the Johnson administration. During the 1930s, Humphrey had been a star undergraduate student in Kirkpatrick's political science classes at the University of Minnesota. After graduating in 1939, Humphrey completed a master's degree in political science at Louisiana State University, then returned to Minnesota's political science doctoral program with the thought of becoming a professor himself someday. But with Kirkpatrick's support, Humphrey chose a life in politics instead.[56] In late November 1963—just a couple weeks after writing to Haworth and only days after Kennedy's assassination—Kirkpatrick sent letters to Humphrey and other senators asking them to tell the NSF to stop discriminating against political science.

In these letters, Kirkpatrick outlined the problem and proposed a solution consistent with NSF policies and practices in other fields. He started with numerical indicators, showing that in 1961 and 1962, the agency had given four political science grants, for a total of $115,000. In contrast, the agency had given its four established social science programs 334 grants, for a total of $10.3 million. So, political science received only about 1 percent of the grants and 1 percent of the dollars. These facts, reasoned Kirkpatrick, suggested that political science had not even obtained "token integration." As for the solution, Kirkpatrick said the agency should establish a political science program and place a political scientist in charge. Only then would political science applications "receive full consideration on a basis of equality with the other social sciences."[57]

Although it is not clear how many letters Kirkpatrick sent or how many responses he received, we do know that his effort generated considerable interest in the Senate. Among those who sided with the push to include political science was vice-president Humphrey. Other supportive senators included the Democrats Frank Church, Paul Douglas, Philip Hart, Eugene McCarthy, Patrick McNamara, and George Smathers, as well as a few Republicans, Everett Dirksen, Tom Kuchel, and John Tower.[58]

Moreover, Kirkpatrick soon told Haworth that he was seeking additional supporters. Kirkpatrick planned to send letters to all 11,000 APSA members and to all members of Congress to explain that political science was "completely" excluded from the NSF fellowship program and "virtually" excluded from its research grants program. In addition, the agency had still not provided a "rational justification" for such unequal treatment. Last but not least, every APSA member and every legislator would hear how this state of affairs inflicted "undue, unfair, and discriminatory hardship on political science as a discipline and a profession."[59] However, if the NSF agreed to rectify its ways, Kirkpatrick promised Haworth that he would not proceed with these letters.[60]

Not wanting the NSF caught in an escalating confrontation, and especially not with Congress, Riecken told Kirkpatrick that he should discourage "further public debate" because his request to include political science in the fellowship program had already been approved. He even suggested that

Kirkpatrick would be "foolish" to continue the debate in public, as doing so could make it more difficult to keep "friends" inside and outside the agency.[61]

Although equal treatment for political science was still hardly a reality, Kirkpatrick welcomed Riecken's pledge to be more inclusive. So, he agreed to stop the letter writing campaign. As for Riecken, he made a note to himself that the NSF should "proceed with reasonable expeditiousness in this matter."[62] Moreover, political pressure to do so mounted. According to the March 1965 issue of the *American Behavioral Scientist*, "some 150 congressmen" had questioned the agency about its approach to political science.[63]

The process of including political science moved forward but always slowly and cautiously. In August 1964, Director Haworth specified, in an internal policy statement, that various lines of study in the discipline remained ineligible for funding, including "efforts to evaluate the merits of competing ideologies, political parties, interest groups, and other organizations." The agency was also "not interested in polemics, doctrinaire pronouncements, or statements about public policy or controversial issues."[64] The following year, the agency created a separate program for political science. Yet another year passed before the agency appointed a political science advisory panel, equivalent to the advisory panels for other social science programs Moreover, in his 1968 internal annual report, Social Science Division leader Howard Hines noted that the agency had not yet hired a full-time director for its political science program[65]—the agency would finally do so in 1970, when it gave the job to the political scientist William A. Lucas, from the State University of New York at Buffalo.[66]

No small wonder that compared to the other social science programs, the newcomer got off to a slow start. Consider the year 1966, when the political science program and its advisory board were in place. Funding for political science research amounted to $335,000. Meanwhile, funding for anthropology amounted to $3.98 million or more than ten times as much, and funding for history and philosophy of science amounted to $1.02 million or nearly three times as much.[67] In the coming years, the gap between political science and other programs became smaller, as one would anticipate. For example, two years later, in 1968, funding for political science had risen to $788,000 and thus had more than doubled since 1966. Meanwhile, funding for anthropology had fallen slightly to $3.60 million. Nevertheless, as these figures also show, anthropology still received more than four times as much as political science.[68]

Strict boundaries on the type of research approved for funding remained in place as well. Successful project proposals that met the hard-core research criteria appeared in NSF annual reports, highlighting the sort of work that it welcomed. One project, led by Yale's James Barber, involved the "transfer of sophisticated measurement and observational techniques of small group research to the study of the behavior of governmental committees."[69] The agency broadcast its support for the University of Rochester's William H. Riker as well, emphasizing that his investigation of the "behavior of subjects in game experiments" could help determine whether human beings in society act "as predicted in the mathematical theory of games."[70]

While the discipline's scientistic practitioners had at least a modest chance of securing funding by the late 1960s, other scholars continued to find the NSF's approach too limited. In 1967, James A. Robinson, the director of the Mershon Center for Education in National Security at Ohio State University, complained during congressional testimony that the NSF remained "timid" in its support for political science. This created a serious problem because not all of his political science colleagues carried "the banner of science in the narrow sense" required by the agency. Unfortunately, large and important fields of study that engaged with normative matters still did not qualify as real science in the agency's judgment, thereby excluding political theorists and philosophers who addressed fundamental questions about the "good society," the "just state," and the "good life." Equally troubling, the discipline's policy researchers, who examined "goals and values in relation to scientific analysis of alternatives for achieving specified ends," could not get funding. Moreover, these were not merely the speculations of a disgruntled scholar, for Robinson was a member of the political science program's first advisory panel. His views rested on a detailed knowledge of the processes involved in awarding grants at the agency.[71]

These limitations also meant that the agency kept its distance from the rise of politically engaged scholarship. In 1967, the same year as Robinson's complaint, hundreds of APSA members organized the left-leaning Caucus for a New Political Science. As two of its leaders, Marvin Surkin and Alan Wolfe, explained, this new group charged the discipline with being a "handmaiden to the counterrevolutionary reflex of America in its policies at home and abroad." From this perspective, political scientists had, with few exceptions, failed to carry out research that would "serve the interests of the poor and oppressed around the world."[72] Similar to other groups that formed around the same time (such as the Sociology Liberation Movement, the Union for Radical Political Economics, and the Caucus for Black Sociologists), the left-leaning critics in political science found the discipline's center to be elitist, conservative, and antidemocratic. Despite the many assertions about scientific maturity and objective methodology, the discipline's dominant investigative framework, often identified as behavioralism, seemed to be value laden and ideological in ways generally "supportive of the political status quo in the United States."[73]

While the NSF now welcomed studies in a behavioralist vein, the array of charges mentioned above contributed to the rise of postbehavioralism or, as David Easton's 1969 APSA presidential address called it, the "postbehavioral revolution."[74] Drawing on the historian and philosopher of science Thomas Kuhn's landmark work about scientific revolutions grounded in paradigm shifts, political science caucus leaders Surkin and Wolfe argued for a radical alternative to the behavioralist paradigm. The urgent task involved not revising the current paradigm but "shattering it," to bring "an end to political science." In its place, socially engaged scholars would erect a new "*political* social science."[75]

To sum up, Kirkpatrick's multiyear campaign to stop the agency from discriminating in advance against his discipline brings into sharp focus the combination of expanding opportunities and persistent constraints at the NSF. At a time of increasing social science funding, Kirkpatrick seized the opportunity to push for his discipline's inclusion. Success came with the help of political pressure, applied through Kirkpatrick's enlistment of Senate allies, including his former student and long-time friend Hubert Humphrey. During the second half of the decade, the NSF extended its fellowship program to political scientists, created a political science program within the social science division, and increased funding for political science. In addition, by appealing to well-established agency criteria such as objectivity and value neutrality, along with a preference for quantitative methods and empirical testing of hypotheses, the agency lent its support to work aligned with the behavioralist movement.

At the same time, the process of including political science took place slowly, with agency leaders exercising caution at every step along the way to ensure that the research it supported remained confined to the hard-core. Initially, this caution took the form of outright resistance to including political science on more than an ad hoc basis. Later, after the creation of apolitical science program, finding the "science" in political science remained a paramount concern. All of the following voices expressed concern that much of the research carried out by political scientists was not really scientific and thus didn't deserve funding: NSF social science division leader Henry Riecken, the social science divisional committee, and the agency's top leaders, including directors Alan Waterman and Leland Haworth. Equally important, the fear that funding for political science would make the agency vulnerable to damaging charges that it was advocating controversial viewpoints on sensitive matters remained strong. Hence, the agency's growing support for political science was not simply a matter of deciding where the most promising lines of scientific advance lay but also of insisting that its funding policies and practices should help ward off unwanted political scrutiny.

Meanwhile, if behavioralists could be happy with growing NSF support, its detractors argued that the movement's conception of science left out much that was truly important in the discipline. Most surprisingly, we have seen that Kirkpatrick, contrary to the standard view, was among the detractors. In his dealings with the agency, however, Kirkpatrick withheld his own pointed criticisms of behavioralism and its scientistic tenets. Had he argued for supporting political science research outside the scientistic realm, the case for including his discipline would have been much harder to win. However, as behavioralism came under increasing attack in scholarly and political circles by the late 1960s and a postbehavioral era seemed imminent, other critics such as James A. Robinson openly expressed their dismay with the agency's "timid" approach.

In charting the course of NSF social science in the post-Sputnik years, the overall growth of research support stands out, as does the establishment of the social science division in 1961 and the creation of a political science program in 1966. Another important thread in this story involves a social science educational project called Man, A Course of Study, or simply MACOS. An exploration of MACOS's origins shows how core ideas about the nature, current standing, and future promise of the social or behavioral sciences became important beyond its social science division.

WHAT THE KIDS NEED NOW: SOCIAL SCIENCE, NOT SOCIAL STUDIES

MACOS was part of a powerful movement to reshape American science education that first gained vigor in the late 1950s. Before that point, federal support of American grade school education had remained limited due to a strong tradition of local control. But after the Soviet launch of Sputnik in October 1957, national concern about the state of American education created an opening for a larger federal role. In the coming years, the government, with encouragement from a group of elite scientists and educators as well as support from various private organizations and foundations, assumed greater responsibility for strengthening the educational system, especially in science and other fields deemed central to national defense. The 1958 National Defense Education Act during the Eisenhower administration, together with new federal programs during the Kennedy and Johnson administrations, provided massive increases in funding for science, math, and foreign language studies.

In this context, the NSF became a star patron. The agency's mandate to support science education, established by its 1950 charter, rested on the notion that the nation's educational institutions were responsible for ensuring an adequate supply of scientific manpower. At first, the agency concentrated on advanced training, mostly through graduate fellowships. But its programs soon supported predoctoral and postdoctoral fellowships as well as grade school science education. After Sputnik, NSF budgets for these programs soared.[76]

With Cold War competition as a central motivating factor, the agency channeled substantial support for precollegiate science education to projects associated with the MIT physicist Jerrold Zacharias. In 1956, Zacharias established the Physical Sciences Study Committee (PSSC), which produced textbooks, movies, and laboratory materials as part of a more general quest to transform high school physics by teaching students to think more like physicists. Two years later, Zacharias founded Educational Services, Inc. (ESI), an organization dedicated to the advance of educational research and programs. NSF-funded projects focused on the development of new curriculum and course materials, including science equipment and audiovisual support, as well as summer training programs. The NSF also provided subsidies to help schools purchase the new materials.[77]

Crucial to the social science piece of the story, the trajectory of NSF precollegiate educational efforts reflected the prevailing scientific hierarchy. As noted by the historian Scott Montgomery, common views about the relative "hardness" of the sciences and their relevance to the nation's scientific, technological, and security interests influenced the allocation of attention and resources. Funding first went to study groups in the physical sciences (such as the PSSC) and mathematics, followed by support for comparable groups in chemistry, biology, and earth sciences.[78]

By the early 1960s, the social sciences were lagging behind the natural sciences in this realm, just as the agency considered them to be lagging at the levels of empirical research findings, theoretical achievements, and investigative methods. Minutes from a May 1962 board meeting reported that NSF leaders reaffirmed their strong commitment to progress in the teaching of the natural sciences and mathematics. Yet the board's social science committee warned that an imbalance favoring the natural sciences could have serious consequences: the American educational system, without "some degree of parallel accomplishment in the behavioral sciences," would become "warped."[79]

This issue also received attention in a landmark report from the President's Science Advisory Committee. Published in *Science* just one month before the May board meeting, "Strengthening the Behavioral Sciences" called for increased support of these sciences. Regarding science education, it asserted that "students should be exposed earlier and more effectively to the possibility of investigating behavioral phenomena by scientific techniques." This goal could be accomplished by "inserting material on behavior in existing courses in biology and by emphasizing newer, empirical approaches in existing courses in social studies."[80] As the NSF itself noted, the PSAC report was "a most significant event" in drawing attention to the national importance and needs of the behavioral sciences. The report also had "clear implications for the future conduct of the Foundation's programs in support of behavioral science."[81]

Piggybacking on the statements from the PSAC and the NSF board, social science division leader Henry Riecken advanced his own ideas about educational reforms. Riecken was already a participant in this discussion by virtue of his appointment as associate director in the division for scientific personnel and education along with his role as a consultant to the panel that produced "Strengthening the Behavioral Sciences." In his 1962 annual report to the NSF director, Riecken discussed four different approaches to what was often called social studies education. The most common approach focused on "the transmission of the cultural heritage" and aimed to "make the student acquainted with the history of western civilization." A second approach concentrated on "preparation for citizenship," with an emphasis on teaching students about the "problems of democracy." The third approach centered on "life adjustment," including activities designed to improve the student's social relationships and emotional life. However, none of these approaches was firmly rooted in modern social science, Riecken emphasized. Certainly, they did not draw on the sort of work promoted by the NSF.[82]

Riecken proposed that the agency only had a "legitimate interest" in a fourth approach, which would introduce the student "to the study of human behavior and society as phenomena, as objects of scientific inquiry just as natural and physical worlds are." Although this approach was not well developed or well represented in the schools, it was supported by the NSF social science and educational divisions. Before this approach could be implemented, however, substantial work would be needed, a point Riecken used to bolster his recommendation that the agency should assume a leading role.[83] Following his recommendation, the agency's precollegiate science education program welcomed the social sciences and began supporting project MACOS.[84]

Although many individuals had a hand in shaping MACOS's framework, Jerome Bruner was, arguably, the single most influential figure. Bruner had completed a PhD at Harvard in 1941, served during World War II on psychological warfare projects, and returned to Harvard in 1945 as a psychology professor. A prolific scholar, Bruner criticized key tenets of behaviorism, including its rejection of mind and the study of thinking, and became prominent in the rise of cognitive psychology. He also gained renown as an institution builder. In 1960, he and his colleague George Miller cofounded Harvard's influential Center for Cognitive Studies. Support for Bruner's work came from a cluster of prominent patrons, including the NSF, the National Institute of Mental Health, the U.S. Office of Education, the Carnegie and Ford foundations, and Jerrold Zacharias's ESI.[85]

In the post-Sputnik years, Bruner was among the movers and shakers in the drive to reform American science education. He participated in conferences supported by the PSAC, the National Academy of Sciences (NAS), and the U.S. Office of Education. He also made valuable intellectual contributions. Among his early publications were two books, *A Study of Thinking* (1956) and *Process of Education* (1960), the latter based on a major NAS-sponsored conference about developing new scientific curricula. *Process* had enormous success, with translations in some twenty languages and 400,000 copies sold within its first four years.[86] In another influential book

Figure 3.4
Jerome Bruner and a young grade school student working with MACOS materials, ca. 1969. Courtesy of Educational Development Center, Inc.

called *Toward a Theory of Instruction* (1966), Bruner presented an outline for a new social science–based educational program—Man, A Course of Study.[87] The idea for such a course had emerged a few years earlier, when Zacharias created an elementary science education project at ESI that was initially directed by the anthropologist Douglas Oliver. But when Oliver left the project in 1964, Bruner took over.

Bruner's pedagogical vision rested on certain ideas about how to study the human condition scientifically, which included a combination of naturalism and evolutionism alongside an implicit secularism. MACOS, as Bruner described it, would focus on "man," especially "his nature as a species" and the "forces that shaped and continue to shape his humanity." These ideas followed from the generally accepted scientific understanding that human beings have evolved through natural processes, just as all other living beings. Bruner made the centrality of this approach impossible to miss, as he wrote that the course would combine insights from the "natural history of the human species, evolutionary biology, and cultural anthropology."[88]

Rather than relying on more traditional historical and social studies approaches, Bruner emphasized that the course would also draw on the "newer" behavioral sciences. Naturally, he included his specialty, cognitive psychology and its applications to education. But he also included ethology and cultural anthropology. He had found the latter field exciting ever since his college years at Duke University in the mid-1930s, when he read the work of Bronisław Malinowski and attended a talk by Margaret Mead.[89]

To provide MACOS with a more specific focus, Bruner identified three central questions. "What is human about human beings?" "How did they get that way?" And "How can they be made more so?" He wanted students to examine discernible continuities between humans and their "animal forebears." But he also wanted them to consider how "man is distinctive in his adaptation to the world."[90]

By engaging with these issues, students, Bruner anticipated, would become familiar with certain fundamental truths about human societies. These were not truths about some supposedly universal human nature or fixed social order; rather, these were truths commonly found in anthropological studies that emphasized how changeable, malleable, and variable human beings and society are. The specific truths identified by the Harvard psychologist included the following: that there is a "structure in a society and that this structure is not fixed once and for all"; that each society has a certain "world view, a way of defining what is 'real,' what is 'good,' what is 'possible'"; and that "men everywhere are humans, however advanced or 'primitive' their civilization." Thus, "the difference" among people and societies, Bruner continued, "is not one of being more or less human, but of how particular human societies express their human capacities."[91]

As these statements imply, the vision underlying MACOS opposed any national, racial, religious, or other sense of superiority about one's particular group. So, in addition to its secular, naturalistic, and evolutionary outlook, MACOS would promote consideration of cultural relativism. Children would come to understand, as Bruner put it, that "all cultures are created equal."[92] In short, MACOS would encourage children to develop a deep respect for and appreciation of all cultures, by revealing how each culture, in its own particular way, addressed universal human needs and concerns.

Last but not least, Bruner wanted students to be active learners, in science and in life. Rather than treating students as passive recipients of information that they would have to regurgitate, MACOS would encourage them to discover things for themselves. Rather than being taught scientific facts, they would participate in the exciting process of inquiry and investigation. As shown by historian of science Jamie Cohen-Cole, Bruner and many others working along this line believed that "children should be understood not only as active, autonomous agents, but also as creative, inventive scientists."[93] As Bruner himself put it, students should develop a "respect for" and a "confidence in the powers of their own mind[s]." This confidence should extend to their conclusions about the "human condition, man's plight, and his social life." If all went well, students would finish the course "with a sense of the unfinished business of man's evolution."[94]

One should note that neither Riecken, nor the NSF, nor Bruner identified MACOS as part of a partisan or ideological project. But as Cohen-Cole's work has also shown, MACOS was part of an energetic movement led by liberal elites and intellectuals who placed great value on open as opposed to closed minds. As they saw it, education had a crucial role to play in the development of tolerant, creative, and reasonable American citizens who could protect liberal democracy against its enemies inside and outside the nation's borders.[95]

In sum, MACOS drew strength from growing national interest in the social and behavioral sciences during the early to mid-1960s. It was also rooted more specifically in the educational reform movement inspired by Cold War competition and fears about failing to keep up with the Soviets in education, science, and military might. NSF support for MACOS followed recommendations about incorporating the behavioral sciences into this reform movement put forth by the PSAC, by Riecken, and by other supporters within the NSF. Besides being consistent with the agency's general approach to the social sciences, those recommendations also reflected the more particular notion that new pedagogical ideas and programs for the natural sciences—associated with the movement led by Jerrold Zacharias—could be fruitfully extended to the social sciences. Jerome Bruner fleshed out a more detailed vision for MACOS that drew on notions from evolutionary biology and the newer behavioral sciences.

As the development of MACOS proceeded in the late 1960s and early 1970s, schools around the country took up the course. When that happened, MACOS's partisan and ideological implications provoked widespread uproar from conservatives who feared that it threatened dearly held principles of morality and social order. We will return to this story in chapter 7. But when the 1960s came to an end and controversy over MACOS had not yet erupted, this project had all the trappings of a landmark development in the annals of American educational reform and the social sciences.

CONCLUSION

From the late 1950s to the late 1960s, NSF social science benefited from a more favorable environment compared to previous years. The impressive post-Sputnik expansion of the federal science establishment, the agency's own considerable expansion together with its special role as the balance wheel in the national science funding system, and the gathering force of liberal ideas, social movements, and public policies during the Kennedy and Johnson administrations all helped to energize the agency's social science efforts. This happened at many levels, as seen in the social sciences' rising share of total NSF research expenditures from 2 to nearly 6 percent, in a major upgrade in their organizational standing with the establishment in 1960 of a social science division, and in the growing interest in practically relevant social research. Expansion also took place through new programs and initiatives. These included the opening up of research and fellowship opportunities to political science, followed by the creation of a political science program, and the move to include the social sciences in the agency's renowned program for precollegiate science education through the development of MACOS.

However, the post-Sputnik expansion and deepening of NSF social science remained modest in many ways. During the final years of Alan Waterman's directorship, and continuing through Leland Haworth's tenure as director, the agency retained a strong natural science orientation, reinforced by the dominant position of natural scientists in top decision-making positions. In addition, while NSF directors and board members now expressed greater interest in the social sciences, they continued to insist on an approach that placed the social sciences within a unified scientific enterprise led by the natural sciences. What is more, even though previous restrictions on eligible subject matters, such as race and sex, were no longer in place and thus the agency could support a wider range of research, NSF leaders continued to worry about funding for studies that might attract controversy. As revealed by the events surrounding the campaign led by Evron Kirkpatrick to have political science included on a basis equal to other social science disciplines, that worry among NSF leaders was also shared by Henry Riecken in his role as social science division director.

The agency's continued caution and the persistence of its scientistic approach also had a powerful impact on new programs and initiatives. In the case of political science, NSF leaders continued to invoke principles associated with hard-core research in order to keep clear of research on contentious matters (i.e., public policy, ideology, political philosophy) that might provoke public criticism, compromise the agency's reputation, and hurt its budget. Even though in his scholarly writings Kirkpatrick expressed serious reservations about the behavioralist movement, the struggle he led on behalf of political science never challenged it. As the agency's leaders considered his demands for equal treatment and eventually made a series of changes in response to them, they never saw any reason to reconsider its exclusive focus on hard-core studies. In the case of MACOS, the idea to include the social sciences in the science educational reform movement led by trailblazers from the physical sciences first attracted attention when it appeared in the 1962 PSAC report on strengthening the behavioral sciences. Riecken pursued this idea by arguing that the NSF should support an educational curriculum grounded in modern social science as opposed to other available approaches deemed unscientific. Subsequently, the agency provided generous funding for MACOS's development under the influence of Jerome Bruner who put forth a pedagogical vision grounded in evolutionary biology and the behavioral sciences, including cognitive psychology, ethology, and cultural anthropology.

In short, funding practices and new initiatives during these expansionary years adhered closely to NSF policy principles governing the type of research eligible for funding. While clichés may never do justice to the messiness of historical complexities, the saying "the more things change, the more they stay the same" applies reasonably well here.

During the mid-to-late 1960s, the groundswell of interest in the social sciences also stimulated wide-ranging discussion about how federal funding for them could be strengthened. Given the NSF's increasing prominence in this area, its work became a major focal point of debate among scholars and politicians. Two legislative initiatives in particular would have far-reaching implications for the agency's future.

4

TWO CHALLENGES, TWO VISIONS: THE DADDARIO AND HARRIS PROPOSALS

As it has taken us many billions of dollars to solve some of the health problems that face us through medical research and application, so we are going to need comparable sums in social science. We are going to need many billions of dollars.
—Fred Harvey Harrington, historian and University of Wisconsin president, 1967[1]

It is a regrettable consequence of the criteria that the National Science Foundation and the Congress have established for the support of social science research that, in principle, NSF must constantly determine if proposals in these fields are or are not "scientific," whereas in the natural sciences, NSF need only determine if they are good.
—Harold Orlans, sociologist, 1967[2]

During the mid-to-late 1960s, the social sciences had considerable influence on major governmental policies and programs, from the War on Poverty to the War in Vietnam. "More than anything else, the vision of the Great Society has called for the insights of the behavioral sciences," declared a 1966 *Newsweek* article on "The Proper Study of Mankind." Further elaboration came from Joseph Kershaw, an economist who directed the Office of Economic Opportunity's planning division and was thus responsible for coordinating and administering federal programs supporting the War on Poverty: "We want to make 32 million people better, richer, happier and more productive, and that is a behavioral-sciences problem."[3] In this context, the question of whether federal policies and programs for the social sciences were sufficient to meet the great challenges of the day acquired a special urgency in political, scientific, and academic circles.

At the same time, however, leftist critics of the so-called Establishment argued that closer ties between the federal government and the social sciences undermined the freedom of scholars and compromised their ability to speak truth to power. Joy Rohde and other historians have examined the questionable involvement of social scientists with military and intelligence agencies as well as associated efforts to reform the politics-patronage-social science nexus. During the 1960s, the notion that "knowledge was influenced by the matrix of power relations in which it was produced" became widespread, observes Rohde.[4] While this notion was certainly not new, the claim that the substance of modern social science was and should be free of ethical judgments, social ideology, and political agendas now faced an avalanche of leftist criticism both inside and outside the academy. As shown in the previous chapter's discussion of behavioralism in political science, such criticism included sharp challenges to the scientistic approach because it seemed to undermine the social scientist's ability to critique the unjust and antidemocratic status quo and promote progressive goals based on substantive inquiry into the good life and the good society.

Notwithstanding the important contributions made by historical studies on military and intelligence agencies during the 1960s, this literature has tended to obscure the increasing importance of NSF social science. Although this was far from the only agency that came under scrutiny, its special role vis-à-vis these sciences made it one of the main focal points of concern and debate. This chapter seeks to better understand the trajectory of NSF social science as well as its larger significance in the context of shifting political and intellectual currents associated with 1960s liberalism and leftist criticism. To do this, we will focus on two legislative initiatives that had far-reaching implications. The first section begins by examining the legislative history of Representative Emilio Daddario's proposal to amend the NSF charter by making its mandate regarding the social sciences more explicit and by giving it authority to support applied research. The rest of this section explores the position of the social sciences in the first NSF venture in applied research as it took shape in the late 1960s. The second section examines the history of another legislative proposal from Senator Fred Harris that sought to give the social sciences an agency of their own.

By placing these two stories side by side, it is clear that the two politicians and their respective proposals advanced rather different views about the social sciences. While Representative Daddario accepted the NSF's view of them as immature and needing to catch up to the natural sciences, Senator Harris questioned the very premises of that view. This basic difference informed other significant differences in their proposals as well, concerning the standing of social sciences at the NSF and in the federal science system more broadly, their social relevance, and their relations with the natural sciences, on one hand, and the humanities, on the other. Moreover, the fates of these proposals—Daddario's succeeded while Harris's failed—would have deep implications for the future of NSF social science and for the place of the social sciences in the federal science system and wider society.

THE PROMISE AND PERILS OF APPLIED RESEARCH: FROM DADDARIO TO IRRPOS

Following the vision laid out in *Science—The Endless Frontier*, common wisdom in the early postwar and early Cold War years said that progress in basic science led to practical applications of great value. But any effort to establish a timeline for this process would resist precision, as would any effort to specify the nature of the practical applications themselves. Following this line of thinking, the NSF charter gave the agency a mandate to support basic science, but no responsibility for supporting applied studies. This gave the agency a strong academic orientation, enabling it to concentrate on scientific advances by providing financial support for research projects, advanced training, and material resources such as computer facilities, equipment for laboratory and field research, and buildings. Up through the mid-1960s, NSF leaders repeatedly upheld, and mainly without quarrel, its special basic science mandate.

Basic science ideology helped to insulate the agency from political meddling, giving elite scientists considerable control over its internal affairs and priorities. Furthermore, an oft-cited passage from an appendix to *SEF* had warned that "under the pressure for immediate results ... *applied research invariably drives out pure*. ... It is pure research which deserves and requires special protection and specially assured support."[5] This provided another reason for the NSF to encourage the pursuit of knowledge for its own sake.[6] Speaking to this point in 1966, director Leland Haworth noted that the country needed to exercise "great caution" because "attempts to mold basic science in the direction of immediate usefulness" might "not only harm science itself but also, at least in the long run, thwart its every purpose."[7]

Regarding the social sciences, NSF leaders had additional reasons for maintaining a distinction between agency-supported research and any practical benefits associated with such research. As seen in previous chapters, this distinction was a key part of the strategy—originally established during the Harry Alpert years—for separating social science from the contentious spheres of ideological conflict, social reform, partisan politics, and public policy. By the early 1960s, however, the agency was, in fact, supporting various research projects that addressed topics of practical concern, such as the social psychological studies by Stanley Milgram on obedience to authority and by Leonard Berkowitz on whether watching violence leads to increases in violent behavior. Still, within the agency, interest in providing funding for social research that promised practical applications remained muted. And NSF leaders invoked its lack of authority as a reason to keep political science funding within strict limits.

But by the mid-1960s, national science policy figures were questioning the wisdom of those limitations on the agency's responsibilities and activities. Nobody pushed harder than Emilio Q. Daddario. A Democrat and lawyer from Connecticut, Daddario had served in the armed forces during World War II and the Korean War. He first won election to the House of Representatives in 1958. The following year, he obtained a position on the recently created House Committee on Science and Astronautics. And in 1963, he became chairman of that committee's new Subcommittee on Science, Research, and Development. In the coming years, Daddario became one of the nation's leading science policy authorities. Years later, he attributed his "great interest" in this area to his wartime years when "leadership in science and the applications of technology" had been "so important." Daddario's position as subcommittee chair also made him responsible for overseeing annual hearings on NSF appropriations.[8]

In 1966, Daddario introduced "A Bill to Amend the National Science Foundation Act of 1950," which called for a number of changes. Two of them involved the social sciences directly. First, Daddario proposed changing the charter so that it explicitly mentioned them—up to this point, the charter's original wording regarding "other sciences" was still seen as the basis for the agency's authority to support them. Second, he called for expanding the agency's mandate to include support for applied science. In the 1966 NSF annual report, Director Haworth drew attention to these proposed changes by noting that Daddario's subcommittee recommended "greater emphasis" on the social sciences and the "important role they can play in the development of solutions for the problems of society."[9]

Figure 4.1
Emilio Daddario (Center), U.S. Democratic Representative from Connecticut and Chairman of the House Subcommittee of Science, Research and Development, with U.S. Astronauts Neil Armstrong, Edwin Aldrin, and Michael Collins, September 16, 1969. Courtesy of Wesleyan University Library, Special Collections & Archives.

Hearings held by Daddario's subcommittee before and after the introduction of his bill revealed a range of views regarding NSF social science. Some of the strongest advocates of expanding its mandate focused not only on supporting work that would yield practical benefits but also on understanding the particular character of the social sciences in ways that the natural science–oriented NSF had not yet done. California Representative George E. Brown Jr. complained that "a certain 'attitude' had taken hold" within the federal science establishment, which recognized the natural sciences as "the sciences" while denying "other fields of science ... full [scientific] status or stature." Yet the nation desperately needed these other fields to address "the whole problem of pathology in our society today," such as criminology. To Brown, more robust support for social research seemed just as important as more funding for "high-energy physics."[10]

Much as the politician Brown found the reigning scientific hierarchy problematic, the political scientist Pendleton Herring, though he supported the Daddario amendment, suggested that a stronger commitment to the social sciences would be more likely if people stopped trying to force them into an inappropriate natural science mold. Herring, as noted before, was the SSRC president and the chairman of the NSF social sciences advisory committee. Because agency leaders were "predominantly concerned with the natural sciences," "very often," observed Herring, "those in charge of the social sciences have to think: 'Now, what would be an equivalent?'" Thus, "you start off in the language, let's say, of physics, and try to find some counterpart to that in the field of sociology."[11]

Director Haworth also favored Daddario's proposal, although without criticizing the established scientific hierarchy as Brown and Herring did. According to Haworth, the time had come to give the agency an explicit mandate to support the social sciences. He also favored giving the agency the authority to encourage these sciences to "play an increasingly important role in coping with some of the major problems facing society today."[12] However, Haworth parted ways with Brown and Herring when it came to the relationships between the social and natural sciences. He was particularly concerned about the possibility that alleged differences between the two branches of science would undermine their ability to work together on research relevant to practical problems. Thus, he warned that increased federal support, which he favored, should not be used to encourage the social sciences to work separately from the natural sciences. Working in isolation would yield piecemeal and fragmented approaches. To avoid this danger, he proposed that researchers concerned with practical applications should adopt a "broad systems-type approach," one that would "draw upon and unite in one common effort" the natural and social sciences.[13] It seemed to him "particularly appropriate" for the NSF to fund applied social research "in coordination with physical sciences and engineering."[14]

Beyond revealing conflicting views about the social sciences and their relationship with the natural sciences, discussion of the Daddario proposal exposed worries within the NSF and the federal science establishment about giving the social sciences a bigger role. The agency's governing board told Congress that "expanded efforts" in this field had to be informed by a "clear awareness of the many difficulties inherent in such studies."[15] In addition, Guy Stever, a physicist who served on a science advisory panel to Daddario's subcommittee and subsequently became the NSF director, recalled years later that while he supported the effort to strengthen support for the social sciences, some other panel members disapproved, saying, basically, that "the social sciences aren't really sciences at all, and worse, they'll take money away from the physical and biological sciences which are really important."[16]

Sharp criticism came from the chemist and influential science policy figure Donald Hornig as well. A member of the Manhattan Project generation and one of the youngest scientists elected to membership in the National Academy of Sciences (NAS), he was arguably the nation's most powerful science adviser during the mid-1960s. The number of high-level positions held by him simultaneously is remarkable: president's chief science advisor, chairman of the President's Science Advisory Committee, director of the Office of Science and Technology, and chairman of the Federal Council for Science and Technology. Commenting on the notion that the social sciences had grown into adulthood and were finally ready to take their place alongside the natural sciences, Hornig argued that the former were, in fact, woefully immature. They had "not yet evolved the kind of basic laws and principles" found in the natural sciences. Furthermore, social researchers were "only learning to become quantitative and predictive." Hornig therefore proposed, much as Haworth did, that federal science policy should encourage the social sciences to "develop more productively through close interaction with the physical and biological sciences."[17]

Tempered enthusiasm characterized Daddario's own views as well. This is not surprising considering that the Connecticut representative worked closely with the NAS, NASA, and other natural science–oriented nodes of power inside the federal science establishment.[18] In agreement with some of the skeptical views, Daddario emphasized that any additional federal support for the social sciences needed to be kept in check. Lest he be misunderstood, Daddario specified that "the intent" of his amendment was "by no means to direct a disproportionate amount of total NSF support for the social sciences." For good measure, he added that these sciences, although "extremely important to human welfare," were "still relatively primitive."[19]

Daddario also did not seek to rock any boats regarding research on explosive political and social issues. At one point, Pennsylvania Republican Representative James Fulton asked if proposed changes would lead to NSF supporting social research on sensitive topics such as segregation, civil rights, transportation, urban renewal, or housing. "Just where does it stop?" Fulton queried. In response, Daddario explained that, no, the intent of his legislation did not "lend credence" to such fears.[20]

On July 8, 1968, after a few years of consideration and debate, Congress passed the Daddario amendment,[21] also known as the Daddario-Kennedy bill, in recognition of Massachusetts Democrat Edward Kennedy's role in sponsoring a companion Senate bill. The standard account of the NSF's efforts to support the biological sciences, by the historian of science Toby Appel, reveals that this legislation marked a key inflection point for the agency.[22] More recently, science studies scholar Janet Abbate explains that the Daddario amendment contributed to an emerging understanding during the 1960s that "categorized applied research as a form of science rather than technology."[23] Regarding the social sciences, the agency's new responsibility to support relevant research could hardly have been anticipated when its original charter was approved back in 1950 or at any point during the agency's first dozen years or so. Note as well that although the Daddario amendment succeeded in the sense that it was approved and became law, that success had not put to rest conflicting views raised during the legislative hearings regarding the social sciences' scientific nature, their practical promise, and their position at the NSF and in the federal science system more generally.

★★★

With the Daddario amendment's passage, uncharted waters lay ahead for NSF social science. Seeking guidance, the governing board set up a Special Commission on the Social Sciences, tasked with producing a report about increasing "the useful application of the social sciences in the solution of contemporary social problems." The commission, led by Orville G. Brim Jr., a sociologist and the president of the Russell Sage Foundation, published its final report under the apt title *Knowledge into Action* (1969). At all "levels in the federal government where major policy is made, the social sciences should be deeply involved," stated the Brim Report. In addition, since "disciplinary structures" often compromised the "attack on social problems," greater support for interdisciplinary work was needed to address "complex social problems."[24] Following that line of thinking, as well as suggestions from an advisory committee on engineering, in 1969, the NSF created a new program called Interdisciplinary Research Relevant to Problems of Our Society (IRRPOS).

The agency's first "catalyst effort" in this area, IRRPOS provided support for "projects involving several disciplines but with a central engineering component and closer to 'action' than had been customary during the agency's first two decades."[25] The program's leader was Joel A. Snow, a physicist with a PhD from Washington University in St. Louis who first joined the NSF as program director in theoretical physics. (After nearly a decade with the NSF, Snow would move on to the White House Office of Science and Technology and then the Department of Energy.)[26]

From the outset, the role of the social sciences in IRRPOS seemed problematic, for a few reasons. For one, the new program's strong physical science and engineering orientation, evident at the levels of vision and personnel, meant that the social sciences would probably not be well represented or well integrated.

In addition, as the bold social science–informed policy initiatives launched under the Kennedy and Johnson administrations turned into battlegrounds marked by political, social, and ideological conflicts, the practical value of the social sciences came under increasing scrutiny, even by many figures who had recently looked to them for guidance. To take just one example from what soon became a large body of scholarly and more popular literature, consider the shifting position of Daniel Moynihan, a prominent figure when it came to thinking about the policy-making value of the social sciences. A sociologist, national policy adviser, and in later years New York Democratic senator, Moynihan was a well-known but also controversial figure due to his role in crafting the 1965 governmental study *The Negro Family: The Case for National Action*, commonly known as the Moynihan Report.[27] Here, Moynihan relied heavily on social science statistics, research, and analysis as a basis for understanding and addressing the problems of family life in black communities. He sought to deploy the social sciences on behalf of the "general assumptions of optimistic, 'can-do' liberalism," notes his biographer Godfrey Hodgson.[28]

But a few years later, in 1969—just as IRRPOS was getting off the ground—Moynihan published *Maximum Feasible Misunderstanding*, which included a scathing assessment about social science expertise in the War on Poverty. He wrote,

> Social science is at its weakest, at its worst, when it offers theories of individual or collective behavior which raise the possibility, by controlling certain inputs, of bringing about mass behavioral change. No such knowledge now exists. Evidence is fragmented, contradictory, incomplete. Enough snake oil has been sold in this Republic to warrant the expectation that public officials will begin reading the labels.

The truth might seem harsh, Moynihan argued, but "forward vision" generated by the social sciences remained "rather blurred."[29]

Furthermore, the NSF's own social science advisory committee warned that failure to meet overblown expectations could hamper the agency's efforts. In his 1968 annual report to Director Haworth, the advisory committee chairman, M. Brewster Smith, pointed out that "some of the same spokesmen who call upon social science to produce near-magical solutions to highly intractable problems" will probably be "vehement in their criticism when no magical solutions are forthcoming." Throughout a long and successful career, Smith, a social psychologist, was steadfast in his pursuit of publicly oriented scholarship in such areas as race relations, social justice, and international conflict. He also served as president of the Society for the Psychological Study of Social Issues (1958–1959), helped create the field of peace psychology, and received the American Psychological Association's 1988 Award for Distinguished Contributions to Psychology in the Public Interest. Nevertheless, in 1968, at a time of growing attacks on federal initiatives informed by social science expertise, Smith warned that "inappropriate expectations" for the NSF's nascent applied research activities could pave the way for "disillusionment and rejection." (In his academic writings, Smith also criticized the effort by social scientists to adopt a natural science model of inquiry, with its strong emphasis on quantitative methods and a value-neutral, objective approach. It is not clear, however, if Smith critiqued the NSF's scientistic approach specifically.)[30]

Also worrisome was the possibility that the new mandate to support applied science would compromise the NSF's longstanding commitment to basic research. This threat, which had already aroused some concern during the Daddario subcommittee hearings, informed further worries toward the decade's end due to mounting budgetary pressures. As part of the broader expansion in federal science funding since Sputnik, national support for applied research, including applied social research, had grown dramatically—as had national support for basic research. But expansion ended in the late 1960s. In fact, when measured in constant dollars, federal science support began to decline in 1967. Although nobody could have known so, another seventeen years would pass before the federal science budget once again reached the level achieved in 1967.[31]

The onset of the decline, due to a downturn in the national economy but also to increasing tensions in government-science-university affairs—

marked by the antiwar movement, campus protests over military funding, and frustrations in the Congress and White House with the nation's scientific elite—naturally worried NSF leaders. In the 1968 annual report, Director Haworth explained that due to budget cuts, the agency had been "compelled to limit its support for new projects and to reduce the funds for its programs, especially graduate facilities, specialized equipment ... fellowships and traineeships." At the same time, the costs of scientific research, education, and instrumentation were soaring. Although Haworth himself did not make the inference, these "converging pressures" made it all but inevitable that any new social science ventures requiring significant resources would receive even closer scrutiny than they would have had budgetary conditions been more favorable.[32]

Still, whether those concerns and trends would have a lasting impact on national science policy and federal funding priorities could not be discerned in the late 1960s. Similarly, how the contradictory forces of enthusiasm for relevant social science, on one hand, and criticism of the practical effectiveness of social science, on the other hand, would shape the overall balance of NSF social science activities in the coming years remained unclear.

It was also not clear that the changes based on the 1968 Daddario amendment were sufficient in light of other concurrent challenges regarding the social sciences, their standing within the federal science system, and their broader social relevance. Perhaps the time had come, as SSRC president Pendleton Herring proposed, to stop viewing social research as an underdeveloped version of the type of inquiry associated with a hard science such as physics. Perhaps as well the federal science establishment should stop treating social scientists as second-class citizens, as Harry Alpert had proposed in the wake of Sputnik and as Representative Brown told Daddario's subcommittee in the mid-1960s. As it turns out, while discussion of the Daddario amendment was still under way, these challenges gained a powerful advocate in the figure of Senator Fred Harris.

RETHINKING THE UNEASY PARTNERSHIP: SENATOR HARRIS'S NSSF PROPOSAL

During discussion of the Daddario amendment, NSF leaders expressed interest in supporting social science with practical as well as scholarly aims in view, but this did not challenge the scientistic approach. Nor did passage of that amendment provide any such challenge. However, during the mid-to-late 1960s, more and more voices inside and outside the academy suggested that to address matters of human welfare and social policy effectively, social scientists had to abandon various tenets associated with the scientistic vision, including the disinterested investigative stance and the commitment to value neutrality. In addition, social movements focused attention on the nature and uses of the natural sciences in ways that made their status as a purportedly objective and apolitical model of inquiry worth emulating more problematic than before. From the antiwar and environmental movements to those for racial and gender equality, critics charged that the natural sciences had an elitist and antidemocratic character; that they supported unjust social hierarchies, rapacious capitalism, and murderous militarism; and that they preferred mechanistic explanations and technical solutions at the expense of humanistic values and democratic reforms.[33]

Against this complex background, Oklahoma Democratic Senator Fred Harris developed a proposal to create a National Social Science Foundation (NSSF). As we will see below, the case advanced by Harris in support of this proposal presented a powerful challenge to the Cold War federal science system's scientistic commitments, its treatment of the social sciences as second-class citizens, and the NSF's approach to them in particular.[34]

Born in 1927 into a poor Oklahoma sharecropper family, Harris developed a thirst for knowledge and a passion for helping individuals, communities, and regions that faced special hardships. After completing high school during the Great Depression, Harris did his undergraduate studies with a major in government at the University of Oklahoma, followed by legal studies at the same institution. The year 1954 proved to be big, as he graduated from law school, got married, established a law firm, and entered politics as a Democrat in the Oklahoma State Senate. A decade later—in 1964—he moved into national politics as a U.S. senator. Bright, hardworking, and ambitious, Harris became close friends with prominent liberal Democrats, including Senators Robert Kennedy, Eugene McCarthy, and Walter Mondale. He had good relations with President Johnson and Vice-President Hubert Humphrey as well.[35]

At mid-decade, Harris established himself as an expert on science and government affairs. When, upon Harris's urging, the Senate Government Committee on Operations created a Subcommittee on Government Research, he was appointed chairman. Besides attending to a wide range of policy issues, the so-called Senator for Science revealed a willingness to challenge the "science establishment," for example, by criticizing federal research policies that dramatically favored some regions of the country over others.[36]

Harris also became involved in the raging controversy over Project Camelot, a massive, unclassified social research project designed by the Army's Special Operations Research Office. Couched in the language of social systems analysis, Camelot's grandiose goals included developing a scientific model of revolutionary movements to help the U.S. influence the course of such movements around the world, from South America to Southeast Asia. In 1965, when political communications and media exposure raised worrisome questions about Camelot's implications, the project became a key focal point in a widespread debate about the nature and impact of the growing military–social science partnership. According to many critics, Camelot and a slew of related studies showed that scholars were willing to sell their souls to the highest bidder—the military. The following year, Harris used his new subcommittee to investigate these problems, including the question of whether civilian sources of federal funding should be augmented to complement, and perhaps in some cases replace, military patronage.[37]

In October 1966, Harris introduced a Senate proposal for a new social science agency (S. 836), accompanied by a statement about the need to improve the militarized image of the U.S. abroad. A recent trip to Latin America—where criticisms of Project Camelot and other projects funded by U.S. military, intelligence, and propaganda agencies had erupted—convinced Harris that the nation desperately needed to "civilianize" its image and protect its scholars working there. With this in mind, he put forth a proposal for a new civilian agency that would provide the social sciences with the funding they needed. The suggestion to create a separate social science agency had initially arisen twenty years earlier, during the postwar NSF debate. As we saw, that suggestion had received little attention during discussions that placed an overwhelming emphasis on the natural sciences and inspired a range of anti–social science sentiments from powerful and mainly conservative natural scientists and politicians. But now, in the mid-1960s, a time of growing national interest in social science–informed solutions to domestic and foreign policy problems, Harris and his staff leader Steve Ebbin, who had a doctorate in political science from Syracuse University, anticipated that the idea of giving the social sciences an agency of their own would receive serious consideration.[38]

Figure 4.2
Fred R. Harris, U.S. Democratic senator from Oklahoma, with members of the Kerner Commission, created in 1967 by President Johnson to study the problems of racial unrest and develop solutions. Harris is seated in the center and looking at the camera. Courtesy of Fred R. Harris.

Indeed, Harris's proposal became a major item of consideration in political and scholarly discussions about national funding for the social sciences and their position in the federal science establishment. After he introduced S. 836, the Senate took no official action before the year ended. However, behind the scenes, Harris convinced many of his peers to support his bill. Thus, when he reintroduced it in February 1967, S. 836 had the approval of nineteen cosigners, among them Senators Edward Kennedy, Robert Kennedy, and Walter Mondale.[39]

Later that year, Harris's subcommittee held hearings on the proposal. About 100 witnesses from civilian and national security agencies, scientific organizations, higher education, and the social sciences testified over the course of twelve days.[40]

The following year, Harris's subcommittee approved the NSSF bill, as did the parent committee on government operations. The latter issued a supporting document called a committee print, which presented an extensive defense of the bill based on points Harris had assembled between 1966 and

1968.[41] Harris also made sure his initiative received widespread attention by giving speeches on the Senate floor and writing articles for *Science* and various scholarly journals.[42]

Harris's NSSF legislation, which went through a number of iterations in the late 1960s, drew heavily on the NSF example while also advancing an ambitious vision for stimulating progress in social science as a scholarly and socially engaged enterprise. The proposed agency's structure would be similar to the established NSF, with a twenty-four-member governing board, a director, and a deputy director, all to be appointed by the president with the advice and consent of the Senate. Regarding fields of interest, the new agency would seek to promote social science scholarship broadly, including but not necessarily limited to the following disciplines and fields of study: political science, geography, linguistics, communications, and international relations. The agency would also encourage social science contributions to democracy, based on the premise that "democracy demands knowledge and insight into the problems that man and nations face in interacting one with the other." Furthermore, the agency would promote these sciences because it seemed clear that the nation's strength depended not only on "superior power, wealth, or technology" but also on "leadership in the realm of knowledge and ideas."[43]

Harris's initiative also addressed questions of scholarly freedom and political sensitivity that had become so worrisome during the Camelot controversy. In general, the proposed agency would strive to establish "a climate encouraging freedom of thoughts, imagination, and inquiry," based on the premise that the country's "long-range interests" would be "best be served by a free and independent academic community." Although the agency could support international activities, NSSF-funded projects in foreign countries would require prior notification of and approval by their governments. The results of NSSF-funded research would be made freely available to the public as well. No security restrictions would be permitted.[44]

Regarding the agency's budget, Harris suggested that first year appropriations would be capped at $20 million, although he expected much larger budgets in the future.[45] Dankwart Rustow, a political scientist at Columbia University's School of International Development and a RAND consultant, suggested that a higher starting budget of $50 million would be more appropriate.[46] An even bolder suggestion came from the historian and University of Wisconsin president Fred Harrington, who spoke—as seen in the opening quote by him—of needing billions for the social sciences and mentioned a starting budget of $.5 billion, which, he added, would represent but "a small percentage" of the "nation's annual investment in the repair of social damage."[47]

As Harris's thinking matured, his pro-NSSF case developed into a substantial challenge to the federal science establishment. Although initially the Oklahoma senator focused on protecting the image of the U.S. and the social sciences from foreign criticism, his subcommittee's investigation of federal support for and use of these sciences led him to a more critical viewpoint that incorporated ideas associated with the New Left. This group of organizations and individuals found inspiration in various advocates of critical social inquiry, including members of the Frankfurt school and the American sociologist C. Wright Mills, who charged mainstream social science with upholding and obscuring the realities of power and exploitation in the modern world—points we encountered before in the postbehavioralist challenge within political science.[48]

Harris modified his NSSF proposal accordingly. In 1966, he had suggested that because of its status as a civilian funding agency, the NSSF would help to soften the militarized image of the U.S. and its scholars abroad. However, his initial proposal stated that the agency could carry out studies requested by other governmental units such as the military on a reimbursable basis, with total support from such transfers capped at 25 percent of the agency's annual expenditures. But in later versions of his bill, Harris eliminated this possibility, because he now understood that transfers from mission-oriented agencies and especially national security agencies would threaten scholarly freedom and intellectual integrity in the social sciences, which, in turn, would harm the agency's reputation.[49]

In addition, although Harris always supported expansion in NSF social science, he grew critical of the established agency's cautious approach. Because of the NSF's special role in looking after the general well-being of American science, Harris saw its carefully circumscribed funding policies and practices as a special problem. The NSF had helped to place the natural sciences "at the center of the stage," while the social sciences, in his words, had been "left out in the cold." Agreeing with the political scientist Rustow, the historian Harrington, and many other critics of the NSF, he charged that its support for the social sciences was paltry. He blamed the agency for relegating them to "second class citizenship." He also found its adherence to the "unity of science theme," as "interpreted" from the perspective of the natural scientist or physical scientist, stifling.[50]

As the 1968 Committee Print supporting Harris's NSSF proposal put it, the NSF had "given very little or no support at all to certain disciplines and methodologies within the social sciences."[51] More generally, its cautious approach had hampered the intellectual progress and practical contributions of the social sciences, Harris noted, thus limiting their ability to help the nation "confront its myriad social problems." The latter included unacceptably high levels of "discontent and alienation" and serious problems of "unemployment and indigency," which persisted "despite increased efforts to solve them."[52]

In making his case, Harris drew attention to limited NSF support for political science. We have already seen that the struggle for inclusion of this discipline had been rather difficult in the preceding years. In light of the perception that political science, perhaps more so than other disciplines, dealt with matters that were likely to attract unwanted political scrutiny, agency leaders appealed to established hard-core funding criteria to ensure that various lines of inquiry within the discipline would not be eligible for funding. Even after the agency created a political science program, the level of support remained low, as Harris noted. Remember that in 1966, the year when Harris first introduced his NSSF proposal, the NSF had approved funding for only seventeen applications, amounting to only $336,000, as Harris himself pointed out.[53]

Some other fields of social research had fared even worse though. For example, the miniscule level of funding for research in law seemed extreme. Neither the NSF nor any other federal agency had a program dedicated to supporting this work. Yet many figures in the legal and social science communities had recently argued that law, or at least certain types of empirically grounded legal research and practice, deserved recognition as an important field within the social science enterprise.[54] At one point during the 1967 NSSF hearings, Harry Eckstein, a Princeton University development studies expert, went so far as to declare that the existing NSF treated not just his discipline, political science, but all of the other social sciences as well "as poor and stupid stepchildren."[55]

Also worrisome, the NSF seemed unduly hesitant to support social research relevant to social problems, due to a fear that controversy over such research might jeopardize the agency's core commitments and its "natural science orientation." According to the Committee Print on Harris's bill, in order for social scientists to address the nation's "growing social problems" effectively, they needed encouragement to undertake "innovative and, perhaps, controversial thinking and research." Such work could offer "a critical assessment of our social values, our priorities, and the existing national programs and approaches for implementing them." Seen in this light, a new agency, with a "strong legislative mandate," was needed to support social researchers investigating the "root causes of social problems."[56]

The pro-NSSF case developed by Senator Harris also noted that relatively meager support for the social sciences at the NSF and elsewhere in the federal science system was the result of massive changes privileging the natural sciences since the 1940s. Not long before that, during the New Deal era, the position of the social sciences vis-à-vis the natural sciences had been much stronger. For example, in 1938, the social sciences had received 24 percent of the total federal support for scientific research. But during World War II and the Cold War years, shifts in American science policy rendered support for the social sciences "relatively superficial." The decline in the percentage of funding seemed alarming. In 1966, of the total federal support for basic scientific research, the social sciences received only 2.4 percent. And of the total federal support for applied research, they received merely 3.5 percent.[57]

With this historical perspective in mind, Harris anticipated that the proposed agency would tackle a number of vital issues. With an initial budget of $20 million, followed by substantial increases in subsequent years, the agency would provide a "quantum leap" in federal social science funding. Substantial NSSF support would, in turn, "revitalize" social research programs in other agencies. Harris explained that he hoped the existing NSF would beef up its social science efforts as well. However, it seemed unrealistic to expect it to undertake the deep reforms needed to make the social and natural sciences "coequals."[58]

Historical commentary has typically claimed that Harris's bill had little support from social scientists themselves. Consider the remarks from the early 1990s by the eminent Harvard historian I. B. Cohen and his Harvard colleague Harvey Brooks, a physicist and a former member of the NSF governing board and of the President's Science Advisory Committee. In a published interview with Brooks, Cohen asserted that social scientists who testified on the NSSF bill at congressional hearings expressed "serious misgivings." Brooks concurred, declaring that "most" social scientists "opposed" Harris's proposal. Harris's biographer, Richard Lowitt, also claimed that the response among social scientists, as well as among natural scientists, was "tepid." More recently, Joy Rohde, in her book on the militarization of the social sciences, has stated that "most social scientists were unreceptive to his proposal."[59]

Yet this viewpoint is not supported by my review of the relevant bodies of evidence. Consideration of social science testimony at congressional hearings, correspondence from social scientists to Harris's subcommittee, and academic and government publications reveal the presence of deep disagreements and a wide division of opinion within the social science community. Interestingly, this division of opinion was noted at an early date by the sociologist Harold Orlans. One of the most knowledgeable commentators on the social science–government partnership during the 1960s, Orlans had been an analyst at the NSF in the late 1950s and a senior fellow at the Brookings Institution from 1960 to 1973. After careful study of the NSSF debate, Orlans reported, in 1971, that "as a group, social scientists" were "thoroughly divided" over Senator Harris's proposal.[60]

The proposal's supporters included many accomplished scholars and scholar-administrators who understood the political, institutional, and intellectual contexts that shaped social science funding. In addition to the political scientist Rustow and the historian Harrington, as mentioned above, this pro-NSSF group included the following: the psychologists Launor Carter, Rensis Likert, and Gardner Murphy; the sociologists Kingsley Davis and Daniel Moynihan; the economists Samuel Hays and F. Max Millikan; the anthropologists Paul Bohannan and Margaret Mead; the political scientists Don Price, Warren Miller, and Evron Kirkpatrick (who, as we saw in chapter 3, led the push to include political science in the NSF in a constrained manner that did not sit well with his more eclectic vision for the discipline); the historian Arthur Schlesinger Jr., and the physicist-turned-historian of science Gerald Holton; and the legal scholars Wex Malone and Myres McDougal.[61]

The largest contingent represented the "have nots" at the NSF and within the federal science system more generally. Some were from disciplines that received little funding, such as law, political science, and history, as well as more recently established specialties, including Latin American studies, Asian studies, and African studies. Some of them, such as the political scientist James Robinson (who, as noted in the previous chapter, complained that studies about the good life and the good society were not eligible for NSF funding), saw value in investigative approaches including historical and normative inquiry that seemed suspiciously soft, if not altogether unscientific, to the natural science–oriented establishment and to social scientists who themselves were committed to the hard-core end of the social research continuum.[62]

Also noteworthy, a number of scholars who did not support the NSSF initiative were nevertheless adamant about the need for dramatic improvements in the standing of the social sciences at the NSF. Some of them also voiced deep skepticism about the NSF's scientistic approach. Unlike Harris, who thought that only a new agency could accomplish those goals, these individuals proposed that, with sufficient encouragement from the political and scholarly spheres, the existing NSF might be able to. Thus, even when they had doubts about whether this would actually happen, they recommended giving the agency a chance to change its ways.

A good example is the economist Carl Kaysen, who had served as President Kennedy's deputy special assistant for national security affairs. Starting in 1966, Kaysen replaced the physicist J. Robert Oppenheimer as director of the Institute for Advanced Study in Princeton, New Jersey. Regarding the proposed NSSF, Kaysen feared that it would be more "vulnerable to the winds of political controversy" than the established NSF, but he also rejected the latter's scientistic approach. The English usage of the word *science*, Kaysen noted, typically referred to the natural sciences. But a much broader understanding, which he strongly preferred, could be found in continental Europe, where the term *science* covered "the whole range of organized intellectual and scholarly activity." Kaysen wanted the NSF to adopt this broader understanding. In addition, regarding institutional representation and power within the NSF, Kaysen proposed that a social scientist should serve as deputy director and then director.[63]

Along these same lines, Aaron Wildavsky, chairman of the University of California–Berkeley political science department, told Congress that his first inclination was to stick with the NSF. Wildavsky added, however, that "no single method or school of thought should be allowed to prevail." He agreed that major steps needed to be taken to avoid "timidity" and, more positively, to provide adequate support for "controversial projects." He recommended changing the agency's name to the "National Physical and Social Science Foundation." And he proposed that it should consider having a social scientist as its director.[64]

Meanwhile, the strongest critics of Harris's proposal typically came from the "haves." The majority were "establishment" social scientists, as Harold

Orlans put it, including "influential advisers to government agencies and leaders in the affairs of the Social Science Research Council, the National Academy of Sciences, and the National Science Foundation."[65]

This group applauded the NSF's emphasis on the unity of the sciences, even if some of them also believed that the agency should greatly increase its level of social science funding. Take the case of Herbert Simon, originally trained as a political scientist but at the time of the 1967 NSSF hearings a professor of computer science and psychology at the Carnegie Institute of Technology and also the chairman of the NAS's Division of Behavioral Sciences. Simon testified that the unity-of-science approach was "basically sound and aimed in the right direction." However, finding the NSF's social science budget woefully inadequate, he proposed a 20 percent increase every year until it reached $1.5 billion.[66] The latter figure was ninety times greater than the NSF's 1967 social science research budget.

NSF leaders themselves defended its social sciences efforts in familiar terms. Consistent with their support for the Daddario bill, Leland Haworth noted that he agreed with Senator Harris that finding solutions to "crucial problems" required greater contributions from the social sciences. However, Haworth reasserted that success would depend on "their integration with the natural sciences and engineering." Furthermore, Haworth argued, contrary to the pro-NSSF contingent, that the NSF had in fact been doing an excellent job. Since the establishment of the social science division in 1960, the agency's financial commitments had grown between 15 and 20 percent annually, from $1.8 million in 1960 to $17.6 million in 1967. As of 1967, the agency also provided about 20 percent of the federal government's total annual commitment to basic social science research. In light of the NSF's central commitment to scientific advancement, Haworth reported that it concentrated on "the sorts of things that can be studied in a quantitative, objective way—in the same sense that one can do in the natural sciences." The agency also stood out as virtually the sole source of federal funding for certain fields, including archaeology, history of science, and philosophy of science. And it remained the "major source of basic, non-mission oriented work" for political science, economics, linguistics, and demography.[67]

Haworth also informed Congress that NSF leaders had "grave reservations" about Harris's proposal. They believed that giving the social sciences their own agency would inspire an undesirable "splintering effect" between the natural and social sciences, thereby inhibiting interactions and undermining the valuable "unity of scientific knowledge," which the NSF worked so hard to promote. Creating a separate agency would also undermine the sort of "interdisciplinary collaborative research" needed to develop effective public initiatives regarding urgent problems such as environmental pollution and urban renewal. Furthermore, if people believed a new agency was primarily responsible for federal social science funding, financial appropriations from other agencies, including the NSF, could—contrary to Harris's expectations—decline, warned Haworth.[68] In addition, although Haworth didn't mention it in his testimony, the NSF social science advisory committee anticipated that "an independent NSSF would be enormously vulnerable to political attack"[69]

From Haworth's testimony and other documents, it is also clear that NSF leaders saw Harris's proposal as a direct threat at two levels. First, regarding the crucial matter of budgets, an internal agency memorandum warned that a new social science agency "could very likely produce at best a leveling off of the NSF's social sciences program or at the worst a steady decline and eventual elimination of this part of the NSF effort."[70] Second, Harris's pro-NSSF case directly challenged the agency's longstanding claim that the social sciences would progress best within a unified scientific framework. In defense of that claim, Haworth insisted to Congress that federal science policy should fortify the effort to "integrate the social sciences with the natural sciences, not separate them."[71]

In the end, Harris's proposal failed. When the Senator for Science introduced it for the last time in January 1969, the number of cosponsors had risen to thirty-two—nearly a third of the entire Senate.[72] But a number of factors undermined his initiative. Criticism from the federal science establishment and from the NSF certainly hurt. Had Director Haworth and other prominent science policy figures from the natural sciences supported Harris's bill, its chance of success would have been much stronger. By mentioning the social sciences explicitly in the revised NSF charter and by including them in its expanded mandate to fund applied research, the 1968 Daddario amendment also played a role. In addition, by this point, the recently established National Endowment for the Humanities (f. 1965, NEH) provided some support for humanistic social research: in 1967, a total of $150,000 for grants and fellowships spread across the four fields of sociology, political science, anthropology, and government, plus $1.6 million spread across the four additional fields of archaeology, history, jurisprudence, and linguistics.

Of these eight fields, history received $1.2 million— more than two-thirds of NEH funding for areas considered to have social science and humanistic components.[73] NEH funding may have led some legislators to believe that the task of supporting humanistic social inquiry could be accomplished without the NSSF proposal. In addition, in the House of Representatives, companion bills to Harris's Senate bill never attracted much interest.

Transformations in the national political landscape and in Harris's own career ultimately removed the NSSF proposal from further consideration. In the 1968 national elections, Nixon's victory ushered in a more conservative White House that would have been unsympathetic to any proposal for an agency whose mission included supporting controversial social research and studies critical of the status quo. Although those elections left the Democrats with majorities in both congressional branches, Republicans had gained seventeen positions in the House and three in the Senate. At the same time, Harris's views on hot-button issues such as race relations, poverty, and the Vietnam War moved him away from the increasingly embattled liberal center, marked him as a leader of the "New Populism," and placed him in conflict with moderate and conservative Democrats, including John L. McClellan, a southern senator from Arkansas who chaired the Government Operations Committee. At the decade's end, with the relationship between Harris and McClellan deteriorating and budgetary concerns in the Senate mounting, McClellan eliminated Harris's government research subcommittee. As a result, Harris lost the institutional base needed to promote his NSSF bill.[74]

CONCLUSION

When compared to the difficult times for the social sciences associated with the original NSF debate, the postwar Red Scare, and the McCarthy Era, their standing within American political culture and within the federal science system had improved considerably, first during the immediate post-Sputnik years and more so during the Kennedy and Johnson administrations. Within this context, the NSF's expanding social science efforts acquired greater importance as well. In the second half of the 1960s, two legislative initiatives, put forth by liberal Democrats, focused discussion on NSF activities and their relationship to a wide range of issues regarding the social sciences: their scientific character, their relationships with the natural sciences and the humanities, their contributions to human welfare, their position within the federal science system, and the level, scope, and aims of public funding.

The legislative history and passage of the Daddario amendment in the summer of 1968 marked a major dividing point in the history of the NSF its general and its engagements with the social sciences in particular. The NSF charter now mentioned the social sciences explicitly. Here, the change was largely symbolic, as the agency had already been supporting these sciences for many years and doing so on the basis of well-established authority. Still, in light of recurring questions about its responsibilities, NSF leaders, together with social scientists and their advocates inside and outside the agency, welcomed this symbolic affirmation. More important, the agency now had the authority and responsibility to support applied research, in the natural and social sciences. This change also received strong support from the NSF as well as from a wide array of politicians, science policy leaders, and social science representatives.

In the coming years, the agency's social science policies and programs were destined to become both broader and more complicated. Hints of these complications were already evident during the early development of IRRPOS, the first NSF program to promote scientific research relevant to national needs.

The other legislative proposal put forth by Senator Harris called for a new agency dedicated to the social sciences. The "Senator for Science" clearly favored expansion of NSF work in this area. But through his subcommittee's research and hearings, Harris became convinced that it was unlikely to make the sort of substantial changes he and many supporters of his NSSF proposal advocated. The mature version of this initiative envisioned a new agency designed to bring about major reforms: by strengthening the position of the social sciences in the federal science establishment; by making these sciences coequals with the natural sciences rather than second-class citizens; by fortifying their scholarly independence and integrity so they could withstand pressures to support the Establishment coming from powerful patrons, such as the military and intelligence agencies; by encouraging studies that could challenge the status quo and promote controversial, innovative thinking in ways often associated with the New Left; and by promoting a broader vision of social science that embraced the value of humanistic social inquiry.

Placing the stories of the Daddario and Harris proposals side by side has revealed that the success of the former would not achieve the goals of the latter. In an important book on American psychology from 1940 to 1970, Ellen Herman has suggested that although Harris lost the NSSF battle, "it is arguable that the social sciences won their war with the federal government during the 1960s." As evidence, she points to the Daddario amendment, increased NSF funding during the 1960s, and the agency's initial forays into applied social research.[75] Yet we have seen that the legislative history of the Daddario amendment served to confirm rather than challenge the view that said the social sciences were rather immature compared to the natural sciences. The chemist and powerful federal science leader Donald Hornig emphasized that the social sciences had failed to establish basic scientific laws and principles and were also inferior to the natural sciences in the areas of quantitative analysis and prediction. Even Representative Daddario called the social sciences relatively primitive. Harris, in contrast, sought to undermine the reigning scientific hierarchy. But the failure of his proposal left intact the entrenched interests, structures, policies, and practices that kept the social sciences in a subordinate position.

It is thus also misleading to claim, as a couple of historical commentaries have done, that Harris's proposal failed because it was "redundant." According to this view, after the Daddario Amendment passed, Harris's NSSF initiative really had nothing more to offer.[76] The truth, however, is that Harris's initiative offered a major alternative to the Daddario Amendment and its position on the proper place of the social sciences at the NSF and within the federal science system. Furthermore, as Michael Reagan put it in his 1969 book, *Science and the Federal Patron*, with the failure of Harris's initiative, the social sciences still had "no explicit home in government (i.e., no agency primarily concerned with their development and application), and … little representation in the policy-making framework of government-science affairs."[77]

The juxtaposition of these two episodes has also brought into clearer view the growing power of the NSF in articulating, cultivating, and advocating the unity-of-science approach. In the case of the Daddario amendment, NSF discussions about the agency's new responsibilities for promoting applied science, including Director Haworth's testimony, made it clear that the social sciences would be expected to work closely with the natural sciences and engineering on projects. Haworth also mentioned that such collaborations would be facilitated by the adoption of particular methodological approaches such as systems analysis. When challenges to the unity-of-science stance arose during debate over Harris's initiative, Director Haworth further argued that separating the social sciences from the natural sciences would have deleterious consequences. This position, supported by other influential nodes in the federal science system, became an important factor in curbing support for the proposed NSSF.

As the decade came to a close, the future of NSF social science was uncertain. One could reasonably expect that the momentum acquired during the recent expansionary period would retain some power, especially in new programs for social science education and applied research. On the other hand, unless the nation's economic troubles subsided, budgetary pressures promised to dampen enthusiasm for initiatives that would require significant new resources, especially if this implied a redistribution of resources away from the natural sciences. With the Nixon administration now in power, federal science policy would surely be headed in a more conservative direction as well, which did not bode well for the social sciences. But what exactly this all might mean for NSF social science remained to be seen.

5

LOSING GROUND: MOUNTING TROUBLES DURING THE MORE CONSERVATIVE 1970S

> The results of social research are often disappointing when measured against the immediate practical demands of current and recurring social problems.
> —1975 NSF Annual Report[1]

> Mr. President: Will you please investigate this [MACOS]? In Texas we're scared.
> —John D. Plummer, U.S. citizen, circa 1975[2]

In the coming years, "the behavioral sciences will be called upon … to contribute to the solution of many complicated and pressing social problems." So wrote the sociologist David Sills in a *New York Times* article published January 20, 1970. Sills was a former fellow at Stanford University's Center for Advanced Study in the Behavioral Sciences. He and Columbia University sociologist Robert K. Merton had recently finished coediting the massive seventeen-volume *International Encyclopedia of the Social Sciences* (1968). In 1989, Sills would receive the American Sociological Association's distinguished career award for the practice of sociology. Writing as a leading authority, Sills anticipated that the United States would need the social and behavioral sciences more than ever before. Fortunately, to do their work more effectively, they had recently acquired "new powerful tools," including "the sample survey, mathematical models, and the high-speed computer."[3]

Yet, despite Sills's optimistic forecast, four developments made the political and intellectual climates increasingly problematic for the social sciences and their policy contributions. First, the trajectory of federal funding took a worrisome turn. After rising dramatically from the 1940s through the mid-1960s, the federal R&D budget hit a wall in the late 1960s and then fell. When adjusted for inflation, federal science spending during the mid-1970s was fully 20 percent below its highpoint in 1967.[4] Whereas in the post-Sputnik era, a rising tide had helped to lift all boats, budgetary woes now threatened to halt any further expansion in NSF social science.

Second, discussions in academic, intellectual, and policy circles registered a growing skepticism about the social sciences and their practical relevance. Building on charges advanced during the turbulent 1960s, left-leaning critics claimed that their scholarship and policy influence was tainted by a conservative bias supporting such evils as patriarchy, racism, militarism, and imperialism. Equally important were vocal critics on the right, backed by an increasingly powerful conservative movement, who charged the social sciences with supporting a different set of evils, from the erosion of American power and influence in international affairs to the spread of welfare dependency, the hobbling of capitalism, the decline of the traditional family, and the subversion of Christian culture and morality.[5]

Third, across the political spectrum, distrust in government and its experts deepened following an array of unnerving developments. These included continuing protests over the Vietnam War and the eventual withdrawal of American troops without honor and without victory; the Watergate scandal, capped by President Nixon's resignation; soaring oil prices and alarming gas shortages; the advent of "stagflation" (i.e., low economic growth coupled with rising inflation and high unemployment); the partial nuclear meltdown at Three Mile Island; and the overthrow of the American-supported shah in Iran by revolutionary religious forces, followed by the seemingly interminable Iranian hostage crisis. Against this background, the notion that social science research and advice could be of much help in addressing such problems lost considerable support in the scholarly and political communities.[6]

Fourth, the demands for political responsiveness and public accountability became more intense within the federal science system. Starting in the late 1960s, science policy became "much more a creature of the political process," as Daniel Kevles and other historians have pointed out. "Appointments to advisory and administrative posts" in science agencies now "took into account" a candidate's political views to a greater extent than before.[7] At the NSF, this trend was reinforced by a passage in the 1968 Daddario amendment that required the positions of deputy director and assistant directors to be appointed by the president and confirmed by the Senate.[8] Previously, the process of political approval only applied to the director and board members. In addition, the White House Office of Management and Budget acquired greater control over the development of budget requests from the NSF and other executive branch agencies, before the submission of their requests to Congress.[9]

The NSF faced a new layer of congressional scrutiny as well. Ever since its founding, congressional appropriations committees had been responsible for approving its budget and reviewing its work. Due to another provision in the Daddario amendment, additional rounds of hearings before science committees in the House and Senate became obligatory. Thus these committees were responsible for reviewing and authorizing NSF annual appropriations requests before they went to the appropriations committees. By making the NSF more responsive to political pressures from the White House and Congress, these changes exposed the agency to partisan conflict more so than ever before. That, in turn, meant greater vulnerability for NSF social science, especially in the mid-to-late 1970s as the influence of conservative critics rose.

These four developments are crucial for understanding the trajectory and broader importance of NSF social science during the 1970s. Budgetary pressures, growing skepticism about the social sciences and their policy relevance, a conservative resurgence that brought mounting attacks on left-leaning influences, and heightened demands for political responsiveness along with greater vulnerability to partisan conflict all posed problems.

This chapter focuses on three episodes that generated extensive controversy about NSF-funded projects in ways that put the agency and its social science advocates on the defensive. The first section focuses on the rise and fall of Research Applied to National Needs (RANN), a program created in the early 1970s as a successor to IRRPOS. RANN's story enables us to explore mounting concerns about the place of applied social science at the agency and growing criticisms about the nature and uses of applied social research more generally. The second section centers on Democratic Senator William Proxmire's widely publicized claims that some NSF-funded social science projects had no practical value and thus did not deserve taxpayer support. The third section considers stinging and mainly conservative attacks on MACOS, the social science–based grade school curriculum whose origins we considered in the previous chapter.

If trends during the 1960s and early 1970s looked auspicious, these three episodes reveal how changes in American political culture, federal science policy, and intellectual life during the new decade informed serious challenges to the social sciences and the NSF's involvement with them. Moreover, these challenges intersected in ways that made their overall impact powerful and damaging.[10]

PROBLEMS ON RANN'S MANAGED FRONTIER

Although the sociologist David Sills only commented on the social and behavioral sciences in his 1970 *New York Times* article, his remarks reflected a broader interest in the practical benefits of scientific research. As the historian of higher education Roger Geiger has observed about these years, "in research, findings applicable to immediate problems were demanded, while current rhetoric disparaged traditional scholarship in academic disciplines."[11] In step with this trend and seeking to capitalize on its new applied science mandate, the NSF had created IRRPOS. As a sign that social science would be of some importance, the 1970 NSF annual report used the case of economics to illustrate the potential for tremendous benefits: if "improvement in public and private policies resulting from research in economics" increased "the gross national product by only one-tenth of 1 percent," this would "add $1 billion yearly to our nation's economy."[12]

The turn toward applied science and specifically applied social research also posed challenges for the agency, as we saw in chapter 4. A large gap between high expectations and modest, perhaps even meager, results could lead to disenchantment and negative repercussions. Greater support for applied research could also result in diminished enthusiasm and inadequate support for basic research. And the leveling off of the NSF budget could make it difficult to sustain the momentum needed to place its new applied research initiative on a stable footing.

Regardless of those anticipated difficulties, however, political pressures and financial incentives initially encouraged expansion in NSF applied research in the early 1970s. The Nixon White House, through its Office of Management and Budget (OMB) and its Office of Science and Technology, encouraged rapid growth, as did Congress. With Nixon's support, OMB leaders encouraged the NSF to become more active in guiding research to maximize its relevance to specific social and economic problems. This became evident through OMB dealings with the new NSF director William D. McElroy, who had replaced Haworth in 1969 and was the first biologist to lead the agency.[13] After McElroy submitted a proposed budget for fiscal year (FY) 1971 that included $10 million for IRRPOS, the OMB suggested that if the agency made certain desirable changes, it would support a much higher overall budget request to Congress, as much as $100 million more than McElroy had requested and with half—$50 million—designated for applied research.[14]

The NSF responded favorably, as one would expect. The agency agreed to replace IRRPOS, which had been led by the physicist Joel Snow, with a more ambitious program called RANN and placed Alfred John Eggers Jr. in charge. An aerospace scientist-engineer with a Stanford PhD, Eggers specialized in supersonic flight research and human spaceflight, had more than two decades of high-level administrative experience at NASA, and had been a professor and administrator at MIT shortly before taking up his NSF post. One of many science-administrators who moved to the NSF in the midst of budgetary woes and layoffs at NASA due to Project Apollo's termination, Eggers was well equipped to reinforce RANN's engineering emphasis.[15]

The NSF also placed RANN in a new organizational unit called the Research Applications Directorate, which gave RANN greater visibility and independence than its predecessor. These changes underscored the NSF's expanding commitment to what McElroy called "problem-oriented research." With the general aim of stimulating the use of science and technology to solve national problems, the agency engaged in a number of activities: it set out to identify specific national problems, sponsored research to address such problems, and took measures to decrease the time lag between the conduct of research and the application of results.[16]

Moreover, with the new RANN program, the NSF departed from its traditional management practices of waiting for the scientific community to submit research proposals and expecting only minimal involvement of agency leaders in shaping those proposals. RANN program managers, by contrast, were instructed to be proactive. Accordingly, they helped to establish problem-oriented research priorities. They sought out promising lines of interdisciplinary research that might otherwise not receive adequate attention from disciplinary-oriented scholars. They engaged potential users in the formulation of research projects and in the assessment of results. They also required RANN projects to include specific plans for the dissemination and use of results.[17]

Within a short period of time, RANN became a major program. Although the agency did not get all it requested, the 1971 budget approved by Congress included $54 million for RANN, a rather large sum for any new program. It was also substantially larger than the initial figure of $10 million requested for IRRPOS.[18] As of 1971, RANN already accounted for about 10 percent of the total NSF budget. A few years later, that figure had risen to nearly 25 percent, an impressive amount in light of the agency's original and still-dominant basic science commitment.[19]

RANN supported projects involving a wide array of work in the physical, biological, environmental, and engineering sciences. RANN also became important for the social sciences, for many reasons. RANN provided encouragement and validation to social scientists, especially scholars working on research relevant to national problems. RANN reflected wider hopes that such research could make demonstrable contributions to human welfare. Last but not least, RANN provided social scientists with extensive funding. Between 1972 and 1975, NSF commitments to applied social research channeled through RANN increased from about $7 million to $20 million.[20] The latter figure represented 18 percent of RANN's annual budget.[21]

Initially, most of RANN's social science funding was channeled through its Social Systems and Human Resources Division (SSHRD). This division had three main areas of concern: municipal systems, operations, and services; social data and community structure; and social program evaluation methodology. RANN also had three other divisions, for environmental systems and resources, for advanced technology applications, and for exploratory research and problem assessment. Each of these three supported some social research as well.[22]

In 1975, the SSHRD was closed as part of a broader reorganization undertaken so that RANN could "concentrate more directly on major problem areas." Henceforth, support for social research was concentrated in a new Advanced Productivity Research and Technology Division, which had a strong focus on economic matters and "improving the productivity of the public sector." Three programs within this division focused on more specific lines of interest: public-sector technology, public policy and economic productivity, and public policy and human resources. After the reorganization, RANN had four other divisions as well, for the environment, for exploratory research and technology assessment, for intergovernmental science and public technology, and for resources, which supported the development of "strategies and new technologies" to make "more effective use of renewable and nonrenewable resources in the national interest." Much as before, each of these four supported some social research. For example, investigations about the social impact of natural hazards and disasters received funding under environmental research.[23]

RANN's importance to the social sciences becomes more fully apparent when one considers its substantial financial contribution to the overall NSF social science effort. In 1975, RANN's $20 million for applied social research was equivalent to nearly 75 percent of the social science division's budget of $27 million. In addition, whereas up until 1968, the agency officially supported basic research exclusively, the percentage of total social science funding allocated to basic research fell dramatically during RANN's expansionary period, reaching a low of 63 percent in 1976.[24]

According to a 1977 analysis from the House Subcommittee on Science, Research, and Technology, some applied social research projects sponsored by RANN were successful. These included a public opinion survey on the energy crisis, carried out by the National Opinion Research Center and designed to help public policy makers "develop more effective energy conservation and price control measures." Another project, carried out by the Hastings Institute of Society, Ethics, and Life Sciences, led to a useful book about "the ethical, social and legal issues in the field of applied behavioral technology," which "included proposed guidelines for psychosurgery being considered by presidential commissions looking at biomedical experimentation." A third study deemed successful involved an investigation of the "economics and social impact of alternative work schedules." According to the subcommittee, RANN's section on economic productivity "made several other notable awards."[25]

Nevertheless, RANN social science also generated serious criticisms within the NSF. Some of these appeared as early as 1971, when an internal assessment from the social science advisory committee noted that the array of projects receiving support was too diverse. This criticism, in turn, fueled a more general worry that RANN was "so diffuse in the number of problems considered" that it seemed destined to "fall short of achieving significant results with respect to any one of the activities selected for support."[26] In addition, although RANN provided substantial funding for social research, the social science advisory committee noted that this area was "not really very heavily represented" within the program's overall structure. Nor did RANN's top leaders from aerospace engineering have much specific knowledge about the social sciences. Many applied social research projects were of uncertain quality as well. In light of these many problems, the advisory committee suggested that RANN should at least make sure that its projects were of "high scientific quality."[27] This was hardly a resounding endorsement. After all, the goal of promoting scientific excellence was assumed from the outset, regardless of the particular NSF program in question.

Furthermore, any successes like those mentioned above provided little comfort as a barrage of negative assessments from the nation's political and scientific communities found RANN and NSF applied social science activities problematic. Critics in government included the powerful liberal Democratic senator from Massachusetts, Edward M. Kennedy. The "Lion of the Senate," Kennedy advocated for an activist government that promoted social justice and economic opportunities for all. In the late 1960s, he joined Representative Daddario in marshaling legislative support for the successful amendment to the NSF's charter. Now in the mid-1970s, Kennedy chaired the Senate appropriations committee responsible for determining its budget. Although the Massachusetts senator supported its applied research mission in general, he found RANN disappointing.

In October 1974, Kennedy expressed his concerns to NSF director Guy Stever, who two years earlier had replaced the biologist McElroy. A physicist with expertise in guided missiles and space flight, Stever had been chief scientist at the U.S. Air Force in the mid-1950s and then president of Carnegie Mellon University from 1965 to 1972. Following up with Stever about an appropriations committee report that said social science funding should remain at its current level for another year, Kennedy acknowledged that this recommendation could seem "potentially damaging." But he also wrote that it was "certainly understandable." Congress had continued its efforts to induce a reorientation and redirection in NSF social science toward national problems, but "to no avail," claimed Kennedy, although he didn't elaborate with specific examples.[28]

Assessments of RANN from Congress and the nonpartisan Government Accountability Office provided additional criticisms. These included the following: some projects were unnecessary because they duplicated the work of other agencies, or at least other agencies already had sufficient responsibility for carrying out such work; some research was not important with respect to national needs; and the results of some studies could not be generalized, limiting their value.[29]

Making matters worse, RANN's efforts also received poor marks from influential social scientists who challenged the wisdom of trying to make the

NSF a significant player in applied research. As the sociologist Otto Larsen put it, political pressures that made RANN a major presence "startled traditionalists in the academic science community."[30] This concern, already present during the legislative discussions about the Daddario amendment, received ample expression in the mid-1970s when the NSF commissioned a study on its social science programs from the National Academy of Sciences. This study, produced by a committee chaired by Herbert Simon and thus known as the Simon Report, conveyed "substantial reservations about the effectiveness" of NSF support for applied social and behavioral sciences (SBS) research. More specifically, the report found that the NSF directorate for research applications—the organizational unit responsible for RANN—did not include adequate participation by staff with SBS training. Echoing a worry expressed by the NSF social science advisory committee, the Simon Report also judged the agency's applied research efforts to be second-rate, "highly variable in quality and, on the average, not impressive." On top of that, it found that procedures for developing projects supported by RANN did not allow enough participation by the community of applied scientists. Consequently, pressures from RANN staff had resulted in research proposals tailored to the aims of "narrowly specified programs, planned from the top down," while "genuinely unsolicited proposals" received too little attention.[31]

Looking forward, the Simon Report recommended that the NSF return control over the development of research proposals to the scientific community: "More participation of the performer communities is essential for judging the scientific quality of proposed projects."[32] This line of reasoning spoke to pervasive fears among scientists who believed that increased political control of federally funded research favored applied studies at the expense of basic research and led to excessive regulation and oversight. As an article in *Science* warned, the movement to direct science toward problem-related areas threatened to turn the endless frontier of science into a "meticulously planned and managed frontier."[33]

The final blow came in 1977 when, after a thrashing from many quarters, the NSF brought RANN to a close. During a period of general cutbacks in federal science support that began in the late 1960s, RANN had been handsomely funded, for a total of nearly half a billion dollars, $468 million—equivalent to $2,227 million in 2018 dollars.[34] But high expectations and financial support had given way to widespread criticism and abolishment.

It must be noted that RANN's demise did not put an end to the agency's efforts to promote applied science and relevant research. The very next year, the agency created a new applied science and research application program to support work in the areas of "problem analysis, integrated basic research, applied research, and intergovernmental science and public technology."[35] The agency also sought effective ways to promote translation of new ideas into applications with commercial potential, for example, through the new Industry–Cooperative Research Program.[36] In light of the continued political demand for applied research, other NSF research units devoted some funds as well. Thus, as was noted in a retrospective account three decades later, many initiatives supported by RANN lived on in other forms at the agency.[37]

But for the social sciences in particular, RANN's collapse reflected a growing disenchantment with applied social research. We have already seen the pushback against NSF involvement with this research in the Simon Report and in a number of government assessments. In addition, the hopes associated with such research at RANN and elsewhere had rested heavily on a social engineering viewpoint, which assumed—naively, according to its many critics—that value-neutral, apolitical knowledge produced by rigorous basic, problem-oriented, or applied studies would enable policy makers to devise more effective solutions to various problems. By the mid-to-late 1970s, vigorous challenges to this viewpoint appeared in countless publications.[38]

Criticism of social engineering also became a central theme in an emerging subfield of scholarly investigation known as the sociology of knowledge application. Arguably, the single most influential figure in this new field was the Columbia University–trained sociologist and Harvard professor of educational policy Carol H. Weiss. As Weiss pointed out, in earlier decades, the sociology of knowledge, associated with landmark contributions from the German sociologist Karl Mannheim and the American sociologist Robert Merton, had focused on how social conditions shaped intellectual activity and the production of knowledge. But the new field of study went farther by considering how social factors shaped the uses of knowledge, which required investigation of the "conditions under which knowledge is produced, diffused, and applied."[39]

Contrary to the standard social engineering outlook, Weiss argued that the main benefits of social research typically did not come from direct application or straightforward incorporation of new knowledge. Instead, such benefits arose when ideas, information, and viewpoints shaped how people thought, conversed, and acted. Although this sort of diffuse influence might be hard to trace, Weiss claimed it was valuable at a number of levels. In her 1977 edited book *Using Social Science in Public Policy Making*, she argued that social science could provide "a common language of discourse in our fragmented society"; that social science "terms, data, models, and orientations" could strengthen the "coherence" of discussions, debate, and policy formulation; and that by penetrating governmental units, social research could open up "new vistas."[40] Social research, Weiss noted elsewhere, could also provide a foundation for informed social criticism. Here, in what she said was perhaps "its most valuable, and valued contribution," social science offered "officials a conceptual language with which to rethink accustomed practice."[41] Still, none of those benefits conformed to the linear instrumentalist vision of social engineering.

Moreover, Weiss and other scholars identified a host of "congenital difficulties in the application of social science research to policy." For one thing, government policy making favored a strong pragmatic orientation that emphasized "the resolution of aching controversies with minimal pain," rather than "scientifically elegant solutions." In addition, governmental units generally had "an enduring respect for order." When proposals based on research findings threatened "to bring about rapid change," the units involved tended to defend the status quo, with all of its known "ills." It was widely understood as well that knowledge produced by social researchers had a "frail character." Realizing that social science was "beset with fads of attention, with competing theoretical frameworks, and with contradictory empirical evidence," governmental officials naturally hesitated to give such work much weight.[42]

Thus, this new line of study about the uptake of social science brought worrisome news. Apparently, earlier enthusiasm about its practical value and especially its public policy-making uses was naïve.

Amid this barrage of scholarly criticisms concerning how social science was used, misused, or simply not used at all by policy makers, the NSF as well as the NAS jumped on board by blessing this chastened outlook just as RANN's reputation was plummeting. As the opening quote from the 1975 NSF annual report put it, "the results of social research are often disappointing when measured against the immediate practical demands of current and recurring social problems." A few years later, this deflationary observation received further elaboration in a six-volume study commissioned by the Executive Branch's recently established Science and Technology Policy Office and carried out by the NAS. The study's main finding underscored "the limitations of social research as a tool for making social policy or operating social programs." Examination of these limitations revealed specific factors that hampered the "effective application of knowledge," including "doubts as to the quality or relevance of the results of research and other knowledge production activities," "the lack of clear policies on the dissemination and use of results," and "a weak sense of the appropriate audience for many results."[43]

These efforts to rethink and reassess the complicated relationship between the social sciences and policy making had serious implications for their standing in American society. Confidence that social science guidance would make the policy-making process more rational and effective had sunk by the late 1970s. Scholarly literature and national science reports now argued that social science–informed programs and policies fell far short of their goals. Moreover, social science research itself helped to undermine "naïve and simplistic faiths upon which the legislative initiatives of the 1960s implicitly or explicitly rested," noted the Brookings Institution scholar Henry J. Aaron in his respected book *Politics and the Professors: The Great Society in Perspective* (1978).[44] Concurrently, "the belief that the experts in the research community are at least partially to blame" spread like wildfire. So did "rising dissatisfaction" with the social sciences more generally, as the 1978 NAS report mentioned above pointed out. By the decade's end, the social science community faced "increasing pressure" to demonstrate its "usefulness to policy makers, programs, officials, and legislators."[45]

How far the mighty hopes for powerful applications in the social sciences had fallen since RANN's early days!

By the time of RANN's closure, any expectations about this being an exciting new chapter in NSF social science had been thoroughly dashed. To be sure, this worrisome development reflected the fact that the social sciences were caught up in larger institutional and policy challenges concerning NSF support for applied science regardless of the particular field of study. At the same time, however, specific criticisms of RANN-supported social research were part of a broader depreciation of the social sciences' practical value in the political and scholarly communities during the 1970s.

While the saga of RANN was unfolding, a second clamor compounded the difficulties for NSF social science. In this case, harsh scrutiny from the

Democratic Senator William Proxmire raised questions about the wisdom of spending tax dollars on research projects whose practical contributions and social utility he and other critics, including some conservative commentators, found hard to fathom.

WASTING TAXPAYERS' DOLLARS/WINNING GOLDEN FLEECE AWARDS

From time to time, doubts about the wisdom of funding social and psychological research have suffused American political culture. During the postwar NSF debate, Republican Representative Clarence Brown warned natural science leaders that they would not get a new agency if this meant making room for short-haired women and long-haired men studying sensitive topics such as marital relations. During the McCarthy era, the Reece Committee proposed that social science funded by large tax-exempt foundations had advanced a plethora of dangerous isms, including atheism, socialism, and moral relativism. In the early 1960s, skeptical legislators raised questions about irrelevant-sounding research, citing the case of a three-year NSF psychobiology grant for "Ethological Investigation of Bird Sounds."[46] As concerns about wasteful federal expenditures became increasingly widespread during the 1970s, legislators shined an unflattering spotlight once again on certain science projects supported by federal agencies. Nobody drew greater attention to this issue than Senator Proxmire.

First elected to Congress in a special election held in Wisconsin in 1957 to fill the position left vacant due to Senator McCarthy's death, Proxmire became well known for his boundless energy, expertise in public administration and economic affairs, and fiscal frugality. During his 1976 and 1982 Senate election campaigns, Proxmire refused to accept any campaign contributions and spent less than $200 from his own pocket.[47] Still, he won both elections handily. In 1988, after serving five six-year terms in a row, he retired from politics. The author of many books, including two on excessive government spending, *Uncle Sam—The Last of the Bigtime Spenders* (1972) and *The Fleecing of America* (1980), Proxmire's charges of wasteful expenditures on social science projects would pose special problems at the NSF for many reasons.

To begin with, Proxmire chaired the Senate subcommittee with jurisdiction over NSF appropriations during the mid-to-late 1970s. At the time, Senator Kennedy—who had become sharply critical of RANN's social science efforts—chaired the parent Senate appropriations committee. In addition, although the strongest political opposition to federal funding most often came from conservative figures, Proxmire was a maverick Democrat, as seen in his backing of consumer protection legislation, his criticisms of the Vietnam War, and his persistent and ultimately successful effort to get Senate approval for the Convention on the Prevention and Punishment of the Crime of Genocide.

Furthermore, Proxmire could not be considered anti–social science in a general sense. His education included advanced studies in economics and government at Harvard University, where he completed two master's degrees, one at the Business School in 1940 and the other at the Graduate School of Public Administration in 1948. He also served in the Army's intelligence branch during World War II. As a senator, he supported the implementation of the Programming-Planning-Budgeting-System (PPBS) throughout the federal government during the 1960s. First developed within the defense department under the leadership of Robert McNamara and his Whiz Kids from the RAND Corporation, the PPBS promised to place military budgeting on a rational basis by using a combination of systems theory and quantitative cost-benefit analysis. Proxmire declared that this new budgeting system represented the "most basic and logical planning tool which exists: it provides for the quantitative evaluation of the economic benefits and the economic costs of program alternatives, both now and in the future, in relation to analyses of similar programs." This position was consistent with Proxmire's well-known interest in exposing, as one journalist wrote, "costly boondoggles in the military budget."[48]

With regard to the NSF, the drama began with an announcement that Proxmire had created a new award of an unflattering sort. In 1975, a *Chronicle of Higher Education* article observed that the Wisconsin senator was "on the prowl again," searching for ways to eliminate wasteful expenditures.[49] He had already singled out more than a couple dozen NSF projects for criticism, including some involving RANN that he said should be funded by other agencies.[50] While reviewing the NSF's budget for his appropriations subcommittee, Proxmire identified a number of "academic con games." "The American taxpayer," he quipped, "would get a better return on his money if he put it in White Russian bonds."[51] Starting at mid-decade and continuing each month until his retirement in 1988, he drew public attention to "the biggest, most ridiculous or most ironic example of Government

Figure 5.1
Senator William Proxmire talks with President Carter in the White House, January 18, 1978. Photo by White House Photographer, courtesy of Wisconsin Historical Society, WHS-30077.

spending or waste" by bestowing a Golden Fleece Award.[52] In this context, the title brought to mind the verb *to fleece*, that is, to charge an unreasonably high price for something.

Although Proxmire did not focus exclusively or even mainly on the social sciences, he did find a number of research grants in this area problematic. One time he bestowed a Golden Fleece Award on a National Endowment for the Humanities grant of $25,000 for a study about why people in Virginia behaved rudely on tennis courts.[53] Another time he singled out a $260,000 NSF grant to the University of Wisconsin for a project on passionate love. Although Wisconsin was Proxmire's home state, he showed no favoritism to its flagship university or its scholars, as he declared this grant to be "an outrageous waste of the taxpayers' money."[54] Another NSF grant to the University of Pennsylvania for $199,000 to support a quantitative study of linguistic change and variation would "leave most Philadelphians speechless," declared

Proxmire on another occasion. Such a study hardly seemed needed, he added, because many decades before, the great British author George Bernard Shaw had already examined the acquisition, change, and spread of speech patterns in a book that led to the smash Broadway musical *My Fair Lady*.[55]

The dispute over whose judgment should matter—the scientist's or the politician's—came to a head in a case involving the University of Michigan cultural anthropologist Sherry Ortner. An up-and-coming scholar at the time, Ortner had studied with Clifford Geertz at the University of Chicago and published a recent book called *Sherpas through Their Rituals*. She went on to a stellar career marked by a 1980 Guggenheim fellowship, a 1992 MacArthur "genius" award, and the 2004 best-book prize in anthropology for another one of her monographs, *Life and Death on Mt. Everest: Sherpas and Himalayan Mountaineering*. What caught Proxmire's eye was an NSF-funded project by Ortner called "Himalayan Mountaineering, Social Change, and the Evolution of Religion among the Sherpas of Nepal," which involved household surveys, interviews with village people and mountaineers, library and archival study of Buddhist monasteries, and examination of religious rituals. In September 1979, Proxmire lavished unwelcome attention on her grant with a Golden Fleece Award.[56]

Proxmire explained that he did not intend to criticize Ortner's research per se, although he did not speak highly of it. Nor did he believe in censorship. But at a time of "rampaging inflation," he asked whether the government should spend taxpayer money "to send researchers half way around the world to study what is at best an esoteric question." Support for such research would, in his view, be more appropriate if it came from either a "religious order … private foundation … or university."[57]

In Ortner's case and others like it, NSF leaders typically responded to Proxmire's charges by defending the particular grants in question, but they refrained from challenging him more generally. This is not surprising because Proxmire chaired the Senate appropriations subcommittee responsible for NSF funding. According to one journalist, his unflattering awards and wry commentary, together with extensive press coverage, elicited from agency leaders no more than a "turn-the-other-cheek forbearance."[58] But Ortner herself mounted a vigorous defense.

By questioning public support for her research, Proxmire betrayed a dangerous ignorance about the forces shaping international development and U.S. foreign policy, charged Ortner. Here, Ortner placed her research in the

Figure 5.2
Sherry Ortner in Nepal for her dissertation research, ca. 1967. Courtesy of Sherry Ortner.

context of mounting critiques of modernization theory and its uptake in efforts to promote American-friendly development around the world. Proxmire, she said, shared "an attitude prevalent among most economists … most development planners, and even many anthropologists, to the effect that religion is a relatively trivial social force." Due to such blinkered thinking, American leaders had been caught with their "collective pants down" in the recent and distressing case of Iran. After receiving "untold millions of dollars in [U.S.] aid for economic development," Iran had erupted in a "religious revolution" that replaced the shah, a long-time American ally, with the radical Islamic leader Ayatollah Khomeini. In light of Proxmire's "anti-intellectual" and "ethnocentric" views, the future academic star questioned whether he had the capacity to "participate intelligently in foreign policy decisions."[59]

Ortner also appealed to this larger context of cultural misunderstandings and foreign policy missteps to underscore the value of her NSF-funded research. Although the Sherpa society had experienced "extremely rapid modernization," she noted that these people did not suffer from various "social 'pathologies'" associated with rapid social change, including alcoholism and juvenile delinquency. Understanding the surprising absence of such pathologies could have far-reaching implications for the conceptualization and implementation of modernization projects. Moreover, Ortner pointed out that whereas development experts often saw religion as an obstacle to modernization, her work showed that religion played a crucial role in maintaining "cultural stability" in Sherpa society. Hence, such research could help construct a better "developmental model."[60]

Proxmire's proposal for direct political review of individual grants could lead to dangerously naive funding decisions as well, a point suggested by Ortner and further elaborated by Roy A. Rappaport, the chairman of Ortner's home department at the University of Michigan. Regarding Proxmire's proposal, Rappaport highlighted the desperate need for better knowledge about development in policy-making circles. In recent years, it had become increasingly clear that much of the nation's support for so-called Third World regions had been wasted outright or else spent on projects that actually "worsened the lots of those whom they were presumably going to help." Such "lamentable results" were at least partially due to "the planner's ignorance of the cultures, societies and ecosystems." Rappaport thus asked,

> If billions of dollars are to be spent on foreign aid, is it not merely wasteful but downright criminal not to spend a few thousand dollars to support the research of those who would enlarge our understanding of societies in which our interventions have all too often been ignorant, wasteful, brutal, destructive, and stupid?[61]

Seen in this light, Proxmire's cost-cutting crusade threatened to undermine the nation's interests and global responsibilities in a complex and precarious world.

Proxmire's stance also seemed to threaten scientific progress itself. The senator's "crusade" could "intimidate researchers" and pressure agencies to cut their budgets "to a point where basic research" would not "survive," Ortner noted.[62] Rappaport concurred, saying that one would be tempted to "laugh off" the unflattering awards he bestowed except that such "media events" had a "chilling effect" on "serious—and relatively inexpensive—research" as well as on "those who fund it."[63]

Similar points appeared in news outlets around the country, with an article in the *Arkansas Gazette* suggesting that to comprehend and address the frightening damage caused by religious cults and their charismatic leaders, the nation needed a "large-scale scientific effort" worthy of federal support. Unfortunately, however, such an effort would probably inspire "congressional grand-standing about 'wasteful' social science research."[64]

Despite such criticisms, Proxmire remained firm. Budging not even an inch, he called the replies by Ortner and Rappaport surprising because of their "harshness." He deemed "elitist" their claim that the evaluation of research proposals should remain entrusted to scientific peer groups selected by executive agencies such as the NSF that themselves were run by scientific interests. Here, exclaimed the Wisconsin senator, was the all-too-familiar argument: "Leave it to the experts! Papa knows best!" The time had come for people to recognize that the academic community functioned as a "self-interest group," fighting and lobbying for its share of the public pie, just as other interest groups did, aiming to obtain "money without strings, without criticism, and with no public questioning of its priorities."[65]

Criticism of wasteful funding gained additional support from conservative opinion makers, who would soon help to place Ronald Reagan in the White House. Donald Lambro, a journalist who wrote for the *Conservative Digest* and *Free Enterprise*, dedicated his 1980 book *Fat City: How Washington Wastes Your Taxes* to "the American Taxpayer." For this study, Lambro had accepted research assistance from the Heritage Foundation, one of a growing number of conservative think tanks and policy institutes concerned about the relationship of social inquiry to public policy. Stoking worries in American political culture and especially in conservative quarters, Lambro characterized the federal government as a "bloated, extravagant, paternalistic, remote, cluttered, disorganized, inefficient, frivolous, duplicative, archaic wasteland." He then identified the NSF as one of the hundred "most wasteful, inefficient and unnecessary federal agencies, programs and expenditures."[66] Social science programs were among the specific areas he selected for closer examination—and budget cuts.

As had been common among rightwing politicians, intellectuals, and cultural commentators for decades, Lambro found studies on sexuality and gender roles particularly disturbing. Among them was one he dubbed the "pot-porno-penis project," although in this case, the patron in question was not the NSF but the National Institute on Drug Abuse, which had awarded the study $121,000. According to Lambro's description of this study, volunteer male college students would smoke marijuana while watching pornographic movies. Researchers would then examine the effects of these stimuli on sexual excitement, measured by a ring placed over the male sex organ. Thankfully, Lambro noted, Senator Proxmire had already helped to end federal funding for this study.[67] Yet this example was only the tip of the iceberg, as many other government-funded studies also deserved rebuke, including some with NSF funding.

In a chapter on "Love and Passion at the National Science Foundation," Lambro called numerous research grants supported by the agency "laughably ludicrous" and claimed they deserved "almost zero priority." These included a $261,000 grant for a sociological study that aimed to determine if "equity theory" would help predict "human reactions in deeply intimate relations" and another grant of $82,000 for the development of "quantitative measures of masculinity and femininity."[68]

NSF funding for other disciplines also caught Lambo's eye. By the late 1970s, the agency was providing about $3.5 million annually to political science. But the value of certain research studies seemed questionable. For example, one project that received $80,000 involved the examination of decision-making processes in presidential candidates who sought nomination from their parties. Lambro asked why the government needed to spend money for such a study. He did not elaborate, though, perhaps presuming that no further comment was necessary. The agency also spent a much greater amount, roughly $10 million, on research in economics, "one of the most inexact of the sciences," according to Lambro. All in all, many federally funded social science studies seemed "about as scientific as the Wizard of Oz."[69]

From this vantage point, the time for stringent budget cuts was long overdue. In 1980—the year *Fat City* appeared—NSF funding for social research went mainly through its Directorate for Biological, Behavioral and Social Sciences (DBBSS), which was created in 1975 and will be discussed in the next chapter. The agency's FY 1980 budget included $172 million for the DBBSS. Of this amount, Lambro proposed that fully $100 million or nearly 60 percent could be justifiably cut. Funding for "studies like those on lawyers, congressional campaigns, peasant Alpine villages, and the pitfalls of romantic entanglements" simply had to end.[70]

These developments shined an unflattering light on social science supported by public funds and the NSF in particular. Neither Proxmire nor Lambro focused primarily on whether this or that project might contribute to the advance of scientific knowledge per se, although Lambro's comment about the Wizard of Oz resurrected a longstanding concern about scientific legitimacy. But the Democratic legislator and the conservative journalist both asked why the government should pay for studies whose practical relevance seemed minimal or whose content seemed problematic—and sometimes perverse. Furthermore, Lambro himself applauded the wisdom of "the Senate's penny-pinching spending critic."[71] In the mid-to-late 1970s, their criticisms, together with the truckful of charges directed at RANN, fueled growing concern that NSF social science programs, whether for applied or basic studies, were "prone to unproductive ventures."[72]

CORRUPTING THE NATION'S YOUTH

As serious as the charge of wastefulness was, the problems hardly ended there. In the controversy over Ortner's study, Proxmire had also raised the sticky issue of determining whether the NSF peer review system protected the narrow interests of scientists at the expense of the American taxpayers' well-being and more pressing national priorities. Lambro's discussion of the NSF-funded pot-porno-penis project provided additional bite, for it drew on a robust reservoir of conservative opinion that said a perverted form of liberalism ran rampant in the social sciences, which, in turn, compromised the country's religious, moral, social, economic, and political foundations. In the mid-to-late 1970s, these two lines of criticism came together in a third controversy over MACOS, the NSF-funded social science-based grade school curriculum that had emerged amid landmark reforms in science education following Sputnik's launch.

During the late 1960s and continuing into the 1970s, the NSF Division of Pre-Collegiate Education in Science provided MACOS with extensive funding. Along with RANN's applied social research component, this educational project stood out as a major new development in NSF social science, complementing its more established efforts to support research projects, training fellowships, and institutional resources. The 1970 NSF annual report drew attention to MACOS's special appeal. Situating this initiative in the context of cutting-edge, progressive trends in American education, the report reiterated the notion that the development of "suitable" curriculum materials in the social sciences still "lagged rather far behind" those for the natural sciences and mathematics. The problem deserved immediate national attention, as "large gaps" could be found from kindergarten to grade 12.[73]

After reaffirming the importance of questions about human nature and society posed by Jerome Bruner, the report explained that MACOS treated them in two parts. The first part examined the "life cycles and behaviors of salmon, herring gulls, and baboons." Here, students would encounter a number of basic topics, including the "significance of generational overlap and parental care, innate and learned behavior, group structure and communication, and their relevance to the varying life styles of animal species, including the human species."[74]

At first glance, the decision to focus heavily on nonhuman animal species might seem odd. Why not concentrate directly on human beings and human society? The answer was based on evolutionary thinking: if the evolution of our species, *Homo sapiens*, depends on the same sort of naturalistic processes that shape the development of all life forms, then one could presumably learn a lot by studying our similarities and differences with other species. For more than a century, this line of reasoning had been advanced by prominent biologists, most famously by Charles Darwin, but also by leading psychologists, among them the behaviorists John Watson and B. F. Skinner. The move to study human psychology and social behavior within an evolutionary framework also gained enormous academic and popular appeal with the rise of ethology, a field of scientific study that flourished in the post–World War II era. Nothing marked ethology's prominence, with its call to study human behaviors alongside the behaviors of nonhuman species, more clearly than the 1973 Nobel Prize in Physiology or Medicine, awarded to three of the field's European founders, Konrad Lorenz, Nikolaas Tinbergen, and Karl von Frisch.[75]

MACOS's second part focused directly on people, through the "intensive study of man in society—as culture-building, ethical creatures, toolmakers and dreamers." Here, case studies would be central, including a unit on the Netsilik Eskimos of the Canadian Arctic.[76] (Although the term *Eskimo* is often considered derogatory today, it was used regularly in discussions about MACOS.)

This choice to focus on Netsilik culture might seem odd as well. Why not focus on materials more familiar to American schoolchildren? Given that MACOS aimed to teach children to think about themselves and society in a scientific manner, wouldn't the scientific study of American culture and society seem especially valuable? In this case, the answer concerned the assumed benefits of examining a society that seemed comparatively simple.

Not only did the study of a complex society such as the U.S. promise to be more difficult, but MACOS's framework assumed that the features of human society of greatest importance are the features shared by different human societies around the world. In the words of the NSF report, the Netsilik society was "small and technologically simple, yet universal in the problems it faces."[77] The underlying reasoning, then, was that by examining the (allegedly) simple Netsilik society, students would learn to appreciate the fundamental importance of such things as culture building and tool making in any society they might encounter, including their own.

MACOS incorporated other strategies for teaching scientific reasoning and independent thinking as well. Based on the notion promoted by Bruner and other cognitive scientists that human learning and scientific learning had much in common because both depend on creativity and interpretation rather than passive acquisition and regurgitation, MACOS had no place for rote memorization. Rather, it encouraged students to become active learners. Thus, the course required students to explore new ideas and work through problems that involved investigation tailored to their young age. If all went well, they would learn how to discover things on their own, albeit with guidance from teachers who served as facilitators—not transmitters—in the process of discovery. Peter Dow, an expert in educational planning and social policy who was deeply involved with MACOS's development, underscored the importance of these pedagogical ideas: "For us, the goal of the course was to help teachers and children … to question and explore their own preconceptions about what it means to be human in a spirit of inquiry that would permit a diversity of views and invite children to think for themselves."[78]

Beyond educational theory and teaching strategy, MACOS produced course materials for classroom use. For this, the NSF turned to the Education Development Center (EDC)—as noted before, a nonprofit organization founded by MIT physicist Jerrold Zacharias and based in Cambridge, Massachusetts. The NSF had already used the EDC to develop educational projects with a multimedia character for the natural sciences, including its flagship high school physics curriculum. Similarly, for MACOS's development, the EDC produced "a variety of media, including films, filmstrips, records, posters, and booklets."[79]

But after that job was complete, selling and distributing MACOS course materials proved challenging. Not even one of the fifty-eight commercial publishers that considered the new curriculum would sign a contract.[80] According to the NSF, a trial phase in MACOS's development revealed that teachers and children themselves were "enthusiastic." However, "commercial book publishers and film distributors were unwilling to contract for publication." The main reason was economic. Course materials would be expensive to publish and thus costly for schools to purchase. Effective use of these materials would require additional funds to pay for teacher training programs. Furthermore, MACOS's "unconventional" subject matter might limit its appeal at the local level, where school curriculum and budgetary decisions were made across the nation.[81] Thus, the financial risk was considerable.

Faced with the possibility that MACOS would never reach the classroom, the NSF decided to subsidize large-scale production of course materials through a contract awarded to Curriculum Development Associates (CDA).[82] A private company located in the nation's capital, CDA received nearly $5 million. Thanks to this subsidy, schools could afford the course materials sold by CDA.[83]

By 1975, total NSF support for this project amounted to more than $7.3 million, equivalent to about $44 million in 2018 dollars. By that time, some 1,700 grade schools in forty-seven of the nation's fifty states had purchased course materials.[84] MACOS, one of fifty-three NSF precollege curriculum projects since Sputnik, was on a roll. This ambitious venture to bring the social sciences into American schools seemed destined for rousing success.

But in the mid-to-late 1970s, conservative figures took action to rescue impressionable young schoolchildren from liberalism, cultural relativism, atheistic science, and amoral social engineering. In this context, MACOS came under blistering attack from coast to coast. One of the earliest sites of local outrage was Phoenix, Arizona, where the bitter politics of educational reform catapulted the Republican politician John Conlan into the national spotlight—we will consider his specific criticisms of MACOS shortly.[85]

By mid-decade, conservative organizations and think tanks in the nation's capital had set their sights on MACOS as well, with the Council for Basic Education and the Heritage Foundation forecasting considerable damage to young minds and harmful impacts on society. If Jerome Bruner, in his role as one of MACOS's main architects, could "effectively change an individual's understanding of the world he lives in, he can also change society as a whole," warned a 1975 report prepared for the Heritage Foundation by Susan Marshner. Sounding the alarm bells, Marshner further noted that a teacher's guidebook for MACOS included quotations from the Harvard psychologist B. F. Skinner, whose view of "culture and environment is fundamentally deterministic, behavioristic, and relativistic." More generally, her report found it troubling that "in place of God," MACOS's creators were "erecting" the god of "Humanism."[86]

To show how damage to young minds could occur, Marshner singled out a story in MACOS course materials that involved an elderly Eskimo lady named Kigtak, who is left on the ice by her son-in-law Arfek to die alone. For Marshner, this story raised important "moral questions." But she found that the "writers of the curriculum" had avoided them "as though they did not exist." Noting that moral questions involve making a choice between alternatives, she pointed out that either it is possible to determine the morally correct choice or it is not. The answer depends on whether there exist "moral absolutes which determine the choice." However, MACOS course materials taught that there are "no moral absolutes," thereby producing "values obfuscation." For corroboration, Marshner quoted the educational psychologist Rhoda Lorand's view that not only did MACOS force children "to be preoccupied with infanticide and senilicide," but it also encouraged them to be "accepting of these practices." In short, Marshner concluded that the new curriculum promoted the "truth of situation ethics and relativism."[87]

The fact that the government backed MACOS raised further concerns. Federal sponsorship threatened the unfettered activity of the marketplace for educational materials, observed Marshner. To underscore this point, she acknowledged that even though MACOS's aims and content were deeply problematic, it would be wrong for anyone to question "the right of a private company, if it so desires," to put this type of course "on the market." In such a case, the marketplace would determine its value: it would either "sink or swim based on its intrinsic merits and the demand it generates." But in the present case, massive federal funding channeled through the NSF gave a curriculum packed with "material of questionable educational merit" an unfair advantage over "privately developed courses."[88]

Around the time of Marshner's report for Heritage, conservative legislators joined the battle. First among them was the aforementioned John Conlan, a lawyer who served in the Arizona state legislature from 1965 to 1972 and represented his home state in the House of Representatives from 1972 to 1977. It should be noted that Conlan had an extensive education, including degrees from Northwestern University and Harvard Law School, a year as a Fulbright Scholar in Germany, and further advanced studies at The Hague Academy of International Law. He had a brief academic career as well, teaching political science at the University of Maryland and Arizona State University. So, Conlan's disapproving comments on MACOS could not simply be dismissed as the uniformed thoughts of a reactionary or anti-intellectual. In 1975, as a member of the House Science and Technology Committee, Conlan reviewed the NSF budget request, which included funds for MACOS.

Not one to mince words, the Arizona representative portrayed this project as part of a "dangerous plan for a federally backed takeover of American education." Similar to Marshner, Conlan explained that federal support for MACOS undermined the beneficial workings of the free market and hurt the legitimate financial interests of private, commercial textbook publishers. Furthermore, MACOS course materials, with their plentiful references to "adultery, cannibalism, killing female babies and old people, trial marriage and wife-swapping, violent murder and other abhorrent behavior," were offensive and damaging to children.[89]

Such incendiary charges were gulped up by the mass media. Horror flicks: Is your ten-year-old watching "X-rated" films at school? This question appeared in a large headline for a *Washington Post* advertisement about a special NBC news report. The advertisement went on to describe MACOS as an NSF-funded program that involved "experimental classroom activities in fifth grades throughout the nation—including about 80 Washington area public schools." Some activities had "ten-year-olds" watching "films featuring such fare as the torture killing of a giraffe by tribesmen and a small child eating the raw eye of a deer."[90]

Seeking to mitigate such harms, Conlan proposed an appropriations amendment to prevent the NSF from providing further funds for MACOS's development and implementation, unless the agency first obtained support from the relevant House and Senate committees. When this proposal came up for vote, the House narrowly rejected it, 215 to 196—with 182 Democrats and 33 Republicans opposed, 89 Democrats and 107 Republicans in favor.[91] However, the House Science and Technology Committee then pressured the NSF to freeze funding until Congress could review the program. Soon, not one, not two, but three governmental reviews were under way: one by the Ad Hoc Science Curriculum Review Group under the auspices of the House Science and Technology Committee, another by the Government Accountability Office, and a third by the NSF itself.

Just as bad for the NSF, this clamor inspired additional worries about its peer-review system. After voting down Conlan's proposed amendment, the House narrowly approved another one from Maryland Republican Representative Robert Bauman. After suggesting that the NSF had gone astray by venturing into the social sciences and education, Bauman said it should return to a narrower focus on funding basic research in the natural sciences. His proposal also aimed to give Congress greater political oversight of NSF grant making. Specifically, it would have required the NSF to provide Congress with a list of all of its grants—nearly 14,000 at the time—together with the justification for each one. And it would have given Congress veto power over individual awards.[92]

Bauman's proposal thus threatened the longstanding commitment to scientific self-governance and the principle that judgments about the merit of scientific projects should be free from political meddling, a position articulated forcefully by Vannevar Bush in *Science—The Endless Frontier* and repeatedly reasserted over the years by the agency and its supporters. "The autonomy of science" was often "defended by the scientific community in ideological terms, but viewed by others as a form of scientific arrogance which asserts that the scientist's own frame of reference is the only one appropriate for evaluating the output of science," noted a 1973 essay by Harvey Brooks, a physicist and high-level government adviser.[93] So, by suggesting that NSF leaders, staff, and peer reviewers were not always the best judges of whether project proposals deserved federal support, Bauman had reawakened an issue of fundamental importance not only for the agency but also for the science community more generally.

When the House voted 212 to 199 in favor of the Bauman Amendment, NSF leaders were left "dumbfounded," reported an article in *Science*. The prospect of establishing congressional micromanagement of NSF grant making even went too far for Democratic Senators Edward Kennedy and William Proxmire.[94]

Although the Bauman Amendment failed to receive approval from a joint House-Senate conference committee, critical inquiries regarding NSF policies and practices seemed endless. Reasoning that the social sciences posed the main trouble rather than the NSF as a whole, William V. Roth Jr., a Republican senator from Delaware and strong fiscal conservative, introduced an alternative to Bauman's amendment that would have required OMB approval on all federal social science grants over $25,000.[95] Although Roth's proposal also failed to get enough votes, Arizona's John Conlan and North Carolina's Jesse Helms, an uncompromising conservative Republican, put forth other proposals that would have required the NSF to provide national legislators and principal investigators with full peer-review reports of grant proposals, including the names of the reviewers.[96] According to standard practice at the time, the agency only made anonymous and redacted reviews available. Accordingly, when Conlan requested full peer-review reports for grant proposals concerning MACOS, the NSF refused. Outraged, Conlan argued that the proposed changes were needed to prod the agency to "operate in an environment of total openness," in contrast to its "completely arbitrary system that is closed and unaccountable to the scientific community and to the Congress." The agency, in his unflattering words, depended on "an 'Old Boy's System', where program managers rely on trusted friends in the academic community to review their proposals."[97] Following up on these concerns, the House Science and Technology Committee undertook an investigation of the peer-review system.

This episode had a wrenching impact on science education programs as well. Responding to the charges against MACOS, the NSF issued two policy statements designed to clarify its role and, hopefully, ward off further criticisms. One statement said that "prior to undertaking full-scale dissemination and assistance activities for NSF-developed materials," the agency would "undertake a careful review to ensure that the proposed subject matter fits within reasonable limits or norms with respect to educational value." This review would include "opportunities for input … by representatives of the scientific, educational, child development, commercial publishing, and informed public communities." On a complementary note, the second statement said that the U.S. was "deeply committed to pluralism in education." Thus, the NSF would seek to "disseminate as many alternatives as are feasible and necessary, given the diversity of views and needs." The agency would also "ensure that federal funds do not directly or inadvertently lead to the development of a monolithic curriculum structure."[98] In short, the agency agreed that henceforth, it would not promote curriculum reforms that involved substantial departures from educational norms or were designed by elite scientists and science education experts.

Notwithstanding these new policy principles, the agency could not protect its social science–based curriculum project from damaging attacks. Between 1974 and 1975, sales of MACOS course materials fell precipitously, fully 70 percent.[99] And the NSF never resumed funding. By the decade's end, the accusations of moral relativism, mechanistic views of human nature and society, and federal control had succeeded. The generously funded NSF project of turning the social studies curriculum into a crucible for educating grade school children so they would think like social scientists had been derailed.[100]

As if that were not enough punishment, the agency's science education programs took a brutal hit more generally. From the early 1970s to the early 1980s, NSF support for science education in all areas shrunk drastically, from 10 percent of its total budget in 1973 to 2 percent in 1983.[101] Much as Susan Marshner's report had done in 1975, the Heritage Foundation continued to attack federal educational reform efforts and the NSF in particular, as seen in a 1981 document that claimed "during the past 15 years, there has been a concerted nationwide effort by professional educationists to turn elementary and secondary school classrooms into vehicles for liberal-left social and political change in the United States," and warned that public dollars had been used to support "situation ethics" and "secular humanism."[102] NSF funding for educational activities would rise again a few years later, and the agency would create a new Directorate for Science and Engineering Education in 1985. But by that time, the goals had shifted significantly. Rather than concentrating on the cultivation of creative minds through an open-ended process of inquiry and scientific discovery, the agency now supported science education to strengthen economic competitiveness, national defense, and moral character.[103]

In short, the barrage of attacks, legislative proposals, and congressional investigations of MACOS produced awful headaches at the NSF. Harvey Averch, who served as assistant director for RANN and then acting assistant director for science education, claimed that the controversy had produced "the worst political crisis in NSF history."[104] Perhaps this assertion is a bit hyperbolic. Nevertheless, the drawn-out effort to bring the social sciences into the movement to reform American science education and the central role played by the NSF in supporting that effort attracted extensive public scrutiny and voluminous criticism from right-wing opponents.

According to detractors, the construction, content, implementation, and effects of MACOS made it far from a value-neutral and nonideological project, as had been implied by its architects and promoters. While the NSF's Henry Riecken, Harvard's Jerome Bruner, and other figures highlighted the need to teach schoolchildren to think about human nature and society as the modern social and behavioral sciences sciences did, Heritage Foundation author Susan Marshner and Republican Representative John Conlan depicted MACOS as part of a dangerous project to manipulate young minds that was spearheaded by liberal intellectuals, promulgated by the federal government, and designed to alter the nation's character. Thus, this NSF-funded effort to make these sciences relevant in American education had provoked critics who found such relevance pernicious.

As the attack on MACOS gathered force, the agency, its educational programs, and the social sciences found themselves in hot water. Not only did NSF funding for MACOS abruptly end, but the agency's educational programs more generally also suffered severe cutbacks. In addition, the clamor reinforced growing worries about the social relevance of the social sciences, about NSF social science in particular, and about the agency's peer-review system.

CONCLUSION

The events of the 1970s took a worrisome turn. During the previous decade, growing federal science budgets together with a groundswell of liberal interest had supported modest although still significant expansion in NSF social science, including a new mandate to promote applied social research. Although growth in federal science budgets ended by the late 1960s, the new decade began with high hopes and a couple of promising initiatives, including MACOS and RANN. But by the mid-to-late 1970s, mounting discontent in the political arena, national science policy circles, and the broader society placed NSF social science on the defensive.

Three episodes in particular raised trouble. In the case of applied social research, an initial wave of optimism and significant funding gave way to critical assessments and chastened expectations in the nation's political and scientific communities. As became clear through the Simon Report, disillusionment concerning the scientific quality and practical uses of RANN's social research efforts spread, as did skepticism about the NSF's ability to promote applied social science more generally. An NSF-commissioned study together with scholarship in the sociology of knowledge by Carol Weiss and others further undermined enthusiasm for applied social research and social engineering (although Weiss also argued that social science had practical value by providing a common discourse and conceptual language that enhanced the coherence of public policy discussions).

Meanwhile, from a somewhat different angle, Senator Proxmire's Golden Fleece Awards focused critical scrutiny on the wisdom of spending hard-earned taxpayer dollars on particular research projects of questionable practical value. Not only did Proxmire bestow his unflattering award on a number of NSF-funded social science projects, such as anthropologist Sherry Ortner's study of the Sherpas, but his criticisms also added to simmering political discontent that inspired plans to tighten congressional oversight of NSF grants and its peer-review system.

More trouble came from damning conservative criticism—at the local, state, and national levels—that portrayed MACOS as part of a dangerous plan to mold the young. Buttressed by the journalist Donald Lambro's biting commentary about certain social science projects and NSF social science in general, discontent over MACOS contributed to conservative suspicions that a large segment of American social science, along with its advocates in federal agencies such as the NSF, promoted a host of bad things (i.e., moral relativism, secularism, and liberal social engineering).

Under those conditions, the NSF took some measures to rein in its troublesome programs. The agency shut down RANN. Although this did not end its mandate to support applied research, RANN's closure did mean that the major source of NSF funding for social research relevant to national needs was gone, with no comparable replacement in sight. The agency also terminated funding for MACOS, thus bringing an end to the ambitious effort, begun in the early 1960s, to reform American grade school education based on advances in the modern social and behavioral sciences.

Although the worrisome events discussed so far are crucial for understanding the conflicted evolution of NSF social science during this period, we have only examined half of the story. This is because when the 1970s began, the major organizational unit responsible for basic research programs for the disciplines and for some interdisciplinary fields of study was still the social science division—RANN and MACOS were the responsibility of other organizational units. Thus, to really understand the course of NSF social science, we need to consider what happened to that division, why it was closed in 1975, and what happened to the social science programs after that. We also need to consider what difference it made when, in an unexpected twist, a psychologist rose to occupy the NSF director's office.

6

MOMENTUM LOST: REORGANIZATION AND RETREAT, BUT NO RESPITE

> Criticism follows two contradictory lines of argument. In the first, social science research is regarded as irrelevant to societal needs and, therefore, a waste of taxpayers' dollars. The contrary argument is that the social sciences are all too relevant—leading to social engineering and manipulation of moral values—and should not be encouraged, let alone supported.
>
> —Richard C. Atkinson, psychologist and NSF director, 1980[1]

Throughout the 1970s, the NSF remained a major public patron for the social sciences. The agency also retained its special importance in supporting academically oriented scholarship that had little chance of attracting funding from most other private and public patrons because of their practical aims. As of 1979, NSF financial commitments represented 48 percent of the federal government's total obligations for basic social research carried out in the nation's universities.[2] This put the agency first among all federal agencies, followed by the Department of Health, Education, and Welfare at 32 percent and the Department of Agriculture at 20 percent. As for total federal obligations for basic social research both inside and outside the universities, the NSF's contribution was somewhat lower but still substantial at 30 percent. That put the agency in second place, behind the Department of Health, Education, and Welfare at 34 percent but ahead of the Department of Agriculture at 14 percent.[3]

Yet as the conflict-ridden stories of RANN, Proxmire's Golden Fleece Awards, and MACOS have made clear, NSF social science faced a variety of pressures and threats by the mid-to-late 1970s. To better appreciate the depth of the difficulties and their impact, this chapter examines developments inside the agency during these years, including changes in NSF leadership as well as measures to reorganize, repackage, and reorient its social science endeavors. The first section considers changes at the top that put neither a physicist nor a biologist at the helm but a psychologist, Richard C. Atkinson. The second section reviews steps taken by Atkinson to defend and reshape NSF social science programs, including an agency-wide reorganization in 1975 that led to the closing of its social science division. The third section takes a closer look at the 1976 Simon Report and its significance for NSF social science during Atkinson's leadership. The development of substantial funding for big social science is examined in the fourth section. Next, the fifth section considers lingering worries and growing pessimism, both inside and outside the agency, regarding its social science efforts. Rising discontent toward the end of the decade also prompted an intriguing proposal to reconsider the wisdom of the 1975 reorganization, as will be discussed in the sixth and final section.

This chapter reveals more fully how a combination of mounting political pressures, explosive controversies, institutional changes, and growing disillusionment resulted in a loss of momentum. One can find a few bright spots, as certain areas of social research enjoyed modestly strong support. Still, if one had to pick a single word to sum up social science during these years, *retreat* would be a good choice.

THE "HARDEST SOCIAL SCIENTIST" THAT THE NSF COULD FIND

Changes at leadership in the mid-1970s included a major surprise. In 1972, the physicist Guy Stever replaced the biologist William McElroy as NSF director. When Stever took over, he also became President Nixon's top science adviser, giving him a privileged position to oversee the agency's applied research efforts in ways desired by the White House. But in 1976, Stever left the agency—although he remained prominent in national science policy circles and served as President Ford's top science adviser.[4] In August of that year, Richard Atkinson, who had been serving as NSF deputy director, was appointed to the temporary position of acting director. Previously, Atkinson had served as a consultant to the NSF Office of Computing Services (1968–1975) before joining the agency in July 1975 as deputy director under Stever.[5]

When the governing board submitted its initial list of candidates for the director's position in 1976, Atkinson was not on it. However, Senator Ted Kennedy, who was responsible for holding confirmation hearings, knew and liked Atkinson. They first became acquainted with each other in the late 1960s, when Atkinson worked on the ill-fated presidential campaign of

Ted's brother Robert Kennedy, who was gunned down in June 1968. Eight years later, Ted Kennedy said he wanted Atkinson as the permanent NSF director. In response, the board included him in a revised list of candidates. With Kennedy's strong support, the Senate approved his candidacy. Finally, in May 1977, President Carter appointed Atkinson to this post.[6]

Atkinson thus became the first individual from the social or psychological sciences to serve as director. This was an unusual turn of events for the natural science–oriented agency. Without Kennedy's intervention, it seems highly unlikely that the agency would have had a director outside of the physical and biological sciences. At the same time, Atkinson was about the "hardest" social scientist that the agency could find, as the science journalist Constance Holden commented.[7] As explained below, his scientific outlook had been at least three decades in the making.

While pursuing undergraduate studies at the University of Chicago during the early post–World War II years, Atkinson developed a keen interest in questions concerning the nature of the sciences and especially the value of mathematical analysis. A biographical account by Patricia Pelfrey tells us about his participation in a discussion during a class taught by the philosophically oriented, humanistic scholars Robert Hutchins and Mortimer Adler. Regarding the question of what one should study to become an educated person, Atkinson proposed that knowing the calculus was essential—an unorthodox position in that setting. In addition, his undergraduate experiences included a research assistantship for the émigré Russian mathematician Nicolas Rashevsky. Best remembered as the founder of mathematical biophysics, Rashevsky proposed that mathematics could provide the foundation for understanding complex phenomena at different levels: physical, biological, psychological, and social. Atkinson, who graduated from Chicago in 1948, found this viewpoint exciting.[8]

In the next decade, Atkinson became enmeshed in academic psychology's cognitive revolution. During graduate studies at Indiana University, he came under the influence of William K. Estes, a former student of the behaviorist B. F. Skinner and a specialist in the mathematical modeling of human learning. Under Estes's guidance, Atkinson received his PhD in 1955, with a dissertation called "An Analysis of Rote Serial Position Effects in Terms of a Statistical Model." But, following a path similar to prominent leaders of the cognitive revolution such as George A. Miller, Atkinson became enthusiastic about new ideas, methods, and instruments for studying the mind, including the notion that the mind is much like a computer. During two years in the U.S. Army (1954–1956), Atkinson worked at the Human Resources Research Organization in California, where he participated in investigations involving quantitative analysis of large data sets and digital computers on topics of military interest, such as game theory applied to combat scenarios. That year, he also met Patrick Suppes, an eminent philosopher of science and expert in decision making at Stanford University. The two began a long-time collaboration. The connection to Suppes also helped Atkinson obtain a position at Stanford, as a lecturer in the Department of Applied Mathematics and Statistics (1956–1957) and as a research affiliate with the Psychology Department.[9]

Thus commenced a stellar academic career. Following the lectureship at Stanford, Atkinson moved to UCLA as an assistant professor of psychology, from 1957 to 1961. Subsequently, as part of a vigorous effort to strengthen Stanford's excellence in the social and psychological sciences, Atkinson accepted an offer to return, this time as associate professor in the Psychology Department. The conditions at Stanford, which was climbing rapidly into the top ranks of American research universities, provided Atkinson with ample opportunities. Moving steadily up the academic ladder, he became a full professor in 1965 and served as departmental chairman from 1968 to 1973. Atkinson established productive relationships with scholars in other departments as well, marked by his appointments as a professor of education and an affiliate faculty member in the Institute of Engineering–Economic Systems. Atkinson also held administrative positions beyond the Psychology Department: as associate director of Stanford's Institute for Mathematical Studies in the Social Sciences from 1961 to 1975 and dean of Stanford's School of Humanities and Sciences from 1968 to 1973.[10]

In addition, Atkinson's scholarship aligned perfectly with the NSF's strategy for promoting progress. He had two principal scholarly passions, one being the quantitative modeling of cognitive processes, especially learning and memory. His research illuminated the structures and control processes in human memory, including the relationship between short-term and long-term memory. His best-known work, coauthored with his former student R. M. Shiffrin and published as a long book chapter (more than 100 pages) in 1968, was "Human Memory: A Proposed System and Its Control Processes." This piece became the basis for the Atkinson-Shiffrin model, credited with placing the "theory of memory on a mathematical basis for the first time."[11] This model also identified features of attention and control processes that govern memory and are fundamental to cognitive functioning more generally. The model's analysis of strategies for retrieval and decision rules became standard in the field.[12]

Atkinson's other main scholarly passion was computer-aided teaching, a field of scientific inquiry and practical application that led, in 1968, to Atkinson's first position with the NSF, as consultant to its Office of Computing Services. His work in this area could not have been timed better. During the push to reform American science education in the post-Sputnik years, the notion that grade school education could be improved with the help of computer technologies attracted widespread national attention and generous funding. Atkinson's projects received support from many sources and garnered acclaim, including a 1967 article in *Life* magazine that featured his contributions. The following year, Atkinson presented his work in an article published in *Science*.[13] Attracted to the new field's commercial possibilities, Atkinson and his collaborator Suppes went into business together, as founders of the Computer Curriculum Corporation.[14]

Atkinson's publications reveal a deep drive for mathematic expression and precision in psychology, along with a more general commitment to the unity of the sciences based on the notion—reiterated often at the NSF—that scientific progress depends on the application of a rigorous and universal method. Atkinson's understanding of this method came through clearly in *Human Memory and the Learning Process* (1980), an edited collection of his papers that was translated into Russian. In the book's preface, Atkinson highlighted "the close interplay between theory and experimentation" as follows:

> Whenever possible, the theory is stated in formal terms either as a mathematical model or as a computer program; predictions are then derived from the theory; the predictions are used to design an appropriate experiment; the experiment is conducted and data collected; discrepancies are identified between theoretical predictions and experimental outcomes; the theory is revised to take account of the discrepancies; and the cycle of events is repeated.

Atkinson further claimed that this cycle of events "characterizes the scientific method whether in psychology or any other field of sciences."[15]

Atkinson's views placed him squarely within the camp of cognitive psychologists who saw themselves as the vanguard of a new and improved scientific psychology following behaviorism's precipitous decline. Also important for establishing his place in the history of psychology, Atkinson's scientific commitments and scholarship reveal little engagement with other developments at the time, including the rise of social constructivist and humanistic viewpoints as well as certain approaches within feminist psychology that were less committed to, or even downright critical of, a disciplinary vision grounded in quantitative rigor, controlled experimentation, and an underlying unity with the natural sciences. As Atkinson himself acknowledged, he had a "mathematical background," wasn't "very qualitative," and in general "fell very much on the hard end" of the social sciences.[16]

During the 1960s and 1970s, Atkinson became one of the most prominent psychologists in the country. He received the Distinguished Research Award from the Social Science Research Council (SSRC) in 1962. Together with other scholars, including Estes and Suppes, he helped found and became the first editor of the *Journal of Mathematical Psychology*, which published its first issue in 1964. He received fellowships from the Center for Advanced Study in the Behavioral Sciences (1963–1964) and the Guggenheim Foundation (1967–1968). He was a member of the center's Mathematical Social Science Board (1968–1973) and chairman of its governing board (1971–1973). He chaired the psychology section of the American Association for the Advancement of Science (1975), served as president of the experimental psychology division for the American Psychological Association (APA; 1974–1975), and was also president of the Western Psychological Association (1976–1977). And in 1977—the year when Atkinson became NSF director—he received the APA's distinguished scientific contribution award.

Moreover, Atkinson's reputation and activities reached well beyond psychology. In 1974—one year before his move to the NSF as deputy director—he was elected to membership in the prestigious National Academy of Sciences. In the late 1970s, Atkinson was a member of the U.S.–People's Republic of China joint commission on scientific and technical cooperation, which in 1979 produced a groundbreaking memorandum of understanding on this subject. In fact, this was the first such memorandum concerning any area of joint activity between the two countries. In his post-NSF years, he served as president of the American Association for the Advancement of Science (1990) and president of the University of California system (1995–2003).

During the mid-to-late 1970s, Atkinson supported some initiatives at the NSF that deserve brief mention here because, even though they did not affect the social sciences directly, they shaped the agency's character in

Figure 6.1
Richard Atkinson speaking with President Carter in the White House, March 15, 1977. Photo by White House photographer. Courtesy of the NSF.

ways that reflected his priorities as director. In response to attacks on its peer-review system, Atkinson commissioned a major study by the National Academy of Sciences, published under the title *Peer Review in the National Science Foundation* (1978).[17] He also encouraged the agency to forge stronger ties between the nation's research universities and industry. This included funding for studies about how to strengthen technology transfer, about incentives to promote industry investment in science, and about how investments in R&D contributed to economic development. NSF-funded work in this latter area had important policy implications and influenced the 1980 Bayh-Dole Act, landmark legislation that allowed universities, businesses, and nonprofit organizations to pursue patents on inventions that grew out of federally funded research, whereas previously, ownership of such inventions was assigned to the federal government.[18]

In addition, Atkinson focused on strengthening NSF support for the engineering sciences, complementing his efforts to forge closer ties between universities and industry. In 1979, he advocated creating a Directorate for Engineering and Applied Sciences. After some reconsideration, he concluded that it would be better to distribute support for applied studies throughout all of the science directorates. Atkinson thus proposed giving engineering its own directorate—although that didn't happen before he left the agency in 1980.[19] When it came to the social sciences, however, Atkinson's time at the agency was thick with difficulties.

CRISIS MANAGEMENT AND REORGANIZATION

Having someone from the "soft" sciences in the director's office was a novelty and hardly what many natural scientists expected. Shortly after Atkinson moved into this position, a meeting with the Nobel-Laureate physicist I. I. Rabi underscored this point. Recall that during the postwar NSF debate, Rabi had said that in contrast to natural scientists who arrived at quite objective results based on controlled experimentation, social scientists had a hard time demonstrating that their results were sound. Now, in 1976, Rabi told Atkinson that he had heard only good things and was glad that he had been named the new director. Not knowing much more about him, Rabi asked, What field of physics do you work in?[20]

The series of damaging episodes concerning the social sciences discussed in the previous chapter—concerning RANN, the Golden Fleece Awards, and MACOS—also marked Atkinson's terms as acting and permanent director. Upon his arrival at the agency, Atkinson thus became deeply involved in crisis management including a major reorganization of NSF social science.

In August 1975, after having spent just one month at the NSF, Atkinson told his disciplinary peers that he was reasonably confident about the future of the agency's social science efforts. Speaking at the APA's annual meeting, the new deputy director acknowledged some difficulties, mentioning that the social and behavioral sciences had "received a few nicks and bruises perhaps." But he claimed these sciences had not been subject to a "battering." Thus, despite "some important qualifications," he presented an overall "prognosis of good health and well-being."[21]

"In most instances," negative attitudes were "based on a poor—and faulty—impression of research, rather than on actual facts," Atkinson added.

As an example, he pointed to criticisms of "zany" NSF grants. Although he believed that such grants were "generally not 'zany' at all," they could appear so to the uninformed "layman" because of something in the grant titles.[22] One case involved a project called "A Theory of Necking Behavior." After a critical story in the *Chicago Tribune* mentioned it, Atkinson and his staff "tried in vain" to locate it in the agency's social science projects list. Eventually someone did find it, but among the engineering projects. Necking, it turned out, referred to the behavior of a metal, not to human romantic behavior.[23] So, perhaps negative attitudes stemmed not from any real understanding of social research but from superficial impressions and even just an odd-sounding project title.

Nevertheless, attacks in the mass media and political sphere regarding wasteful expenditures on science projects, including Senator Proxmire's Golden Fleece Awards, had to be addressed. By the mid-1970s, this matter, as Atkinson put it, had become "serious business" at the NSF, playing "havoc with the Foundation's public image and relations with Congress."[24] He thus sought to change Proxmire's mind, by engaging in a "running exchange" with him.[25] According to Atkinson, this effort succeded, as Proxmire "became a solid friend of the Foundation." Atkinson also said that during his years as director—from May 1977 to the summer of 1980—the agency did not receive a single Golden Fleece Award.[26]

Atkinson's recollection, however, was not entirely accurate. For one, in June 1977—Atkinson's second month as director—Proxmire gave the agency a Golden Fleece Award for three of its research grants, including one on how the brain recovers after damage.[27] More important for the social sciences, in September 1979, Proxmire bestowed another award on the NSF for a $39,600 grant in support of Sherry Ortner's anthropological research on religion and social change in Sherpa society. As discussed last chapter, this episode included a heated dispute in which Ortner raised doubts about Proxmire's ability to contribute effectively to foreign policy matters and also questioned the broader mind-set that had led U.S. leaders to underestimate the importance of religion in social and political affairs in Iran and elsewhere. In response, Proxmire held his ground, claiming that the agency was run by scientific elites who insisted on retaining control over public funding for research without proper democratic oversight. Nevertheless, the main point here is that soon after Atkinson's arrival at the NSF, he was directly involved in trying to contain the damage done by Proxmire's criticisms of wasteful spending.

***

Atkinson also became involved in protecting the agency from harm due to the MACOS controversy, which first erupted on the national scene in 1975—the year that Atkinson began working at the NSF, which was also the year that Proxmire began issuing his Golden Fleece Awards. In response to mounting congressional pressure and public criticisms of MACOS, the NSF decided, in the spring of 1975, to conduct its own evaluation of the program. According to Atkinson, the resulting report was the first thing he read after arriving at the agency as deputy director in July. Upon finishing it, he was convinced that the agency had "done its business in an orderly and thoroughly appropriate way," and anticipated that "the cloud of criticism hovering over NSF would soon be dispersed."[28]

But when he testified before Congress on this matter, Republican Senators Robert Bauman from Maryland and John Conlan from Arizona said the agency's internal report amounted to a bunch of lies. The harsh allegations "stunned" Atkinson. Their charges also prompted another investigation of MACOS, this time by the Government Accountability Office (GAO) and which turned out to be sharply critical.[29]

To respond to the deepening crisis, Director Stever placed Atkinson in charge of a small group charged with reviewing the GAO report along with relevant agency files. After doing some research, Atkinson's group became alarmed, for some problems they discovered seemed more troubling than the GAO report itself, which, in Atkinson's words, had "merely scratched the surface." None of these problems concerned the way MACOS handled questions of morality that had so outraged conservative critics. Instead, the Atkinson's group found that the NSF had not followed good business practices in supporting MACOS. For one, some NSF-funded, multiyear MACOS projects had not been subject to proper assessment. A more troubling issue concerned the process of getting approval for large MACOS grants from the governing board. As Atkinson explained, the peer review reports sent by program officers to the board had been redacted, making them "highly selective, emphasizing positive assessments and deleting negative ones."[30]

In light of these findings, agency leaders, including Atkinson himself—in his roles as deputy director and then acting director—took steps to put their house in order. Rather than sweeping the problems under the rug, the agency admitted its mistakes to Congress. It placed a couple individuals on administrative leave. It made a number of changes to its science education programs and policies. It formally ended its support for MACOS. And it established an audit office to conduct random checks of peer-review materials, to ensure that the agency used them properly.[31]

Meanwhile, in the words of former MACOS project director Peter Dow, Atkinson had become a "veteran of the Conlan wars."[32] Atkinson did not believe that the mounting charges about damaging young minds were credible. Quite the contrary: after having examined MACOS materials closely, he saw "no basis for such complaints."[33] Nevertheless, he noted that attacks on MACOS caused serious damage, including a "staggering loss of over $9 million for all pre-college curricular implementation." This controversy also revealed how "the critical attitude toward the social sciences" could "snowball."[34]

***

In addition to his running conversation with Senator Proxmire and his efforts to deal with the awful mess surrounding MACOS, Atkinson became involved in a mid-decade restructuring of the entire agency that had important implications for the social sciences, including a decline in their organizational position and status.

At the start of 1975, the agency had five high-level organizational units, for: research, education, national and international programs, research applications, and administrative operations. And the research unit had divisions for: mathematical and physical sciences, biological and medical sciences, engineering sciences, computer science, materials research, environmental sciences, and social sciences. Remember that this last division's establishment in 1961 marked a key development in the agency's deepening involvement with these sciences, as it meant a major upgrade in their organizational standing at the natural science–dominated agency as well as a welcome vote of confidence in their scientific status. Subsequently, this division served as the central organizational unit for them in the expansionary years that followed. As other programs for applied social research and social science curriculum building came under intense scrutiny, the division remained the home for the agency's basic social research programs.

But in July 1975, the social science division was closed, part of a major organizational overhaul that Atkinson, shortly after his arrival at the NSF,

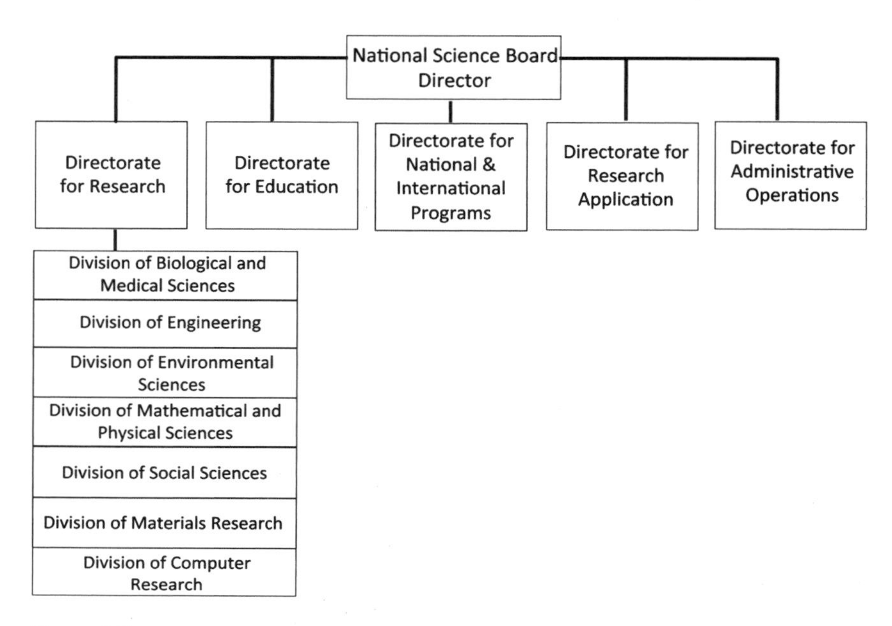

Figure 6.2
National Science Foundation organizational chart as of June 1975. Nondivisional administrative offices have been omitted; only the divisions under Directorate for Research are shown.

became "deeply immersed in."[35] In the new organizational structure, the agency had seven directorates for: mathematical, physical, and engineering sciences; astronomical, earth, and ocean sciences; biological, behavioral, and social sciences; science education; scientific, technological, and international affairs; research applications; and administration. Thus, not only had the established division for the social sciences been closed, but these sciences were placed together with the biological sciences in the new Directorate for Biological, Behavioral and Social Sciences (DBBSS).[36]

Atkinson served briefly as acting director for the new DBBSS, which had four divisions.[37] Two were dedicated to the biological sciences: one for physiology, cellular, and molecular biology, and the other for environmental biology. A third division for the behavioral and neural sciences included programs in neurobiology, sensory physiology and perception, psychobiology, social psychology, anthropology, and linguistics. The fourth and last division handled all of the other social sciences, with separate programs for economics, human geography and regional science, law and social sciences,

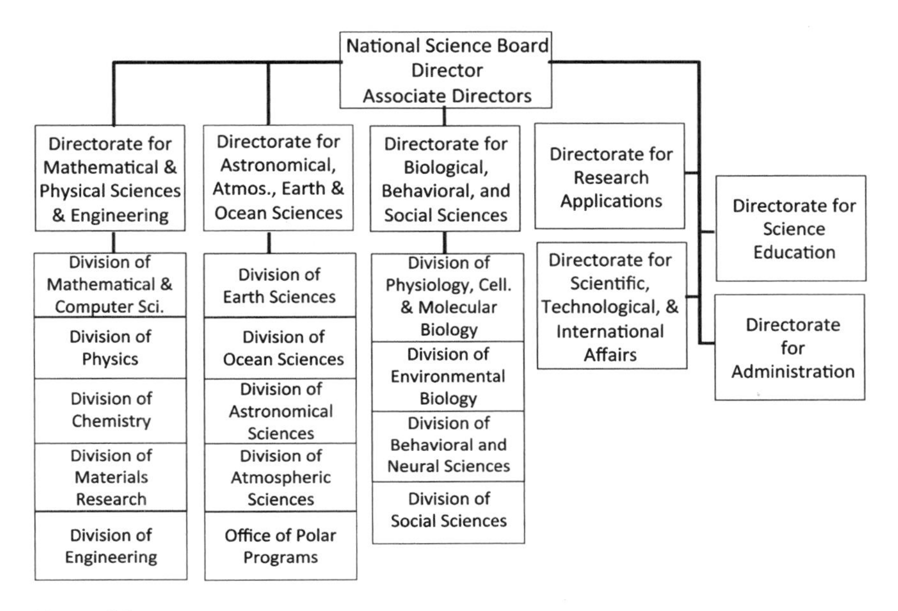

Figure 6.3
National Science Foundation organizational chart as of July 1975. Nondivisional administrative offices have been omitted; only divisions under the disciplinary Directorates are shown.

political science, sociology, science policy research, history and philosophy of science, and special projects in social indicators.

In his 1975 APA talk, Atkinson anticipated that this reorganization would have positive implications for the social sciences. He claimed that the new arrangement "elevated" their position. More specifically, he pointed out that these sciences now shared "an assistant directorship with biology." By supporting this reorganization, the agency's governing board had "indicated its full approval of strengthening the social sciences," he added. Looking forward, Atkinson said he had "every reason to believe" that the new arrangement would "meet the new demands of the time."[38]

Despite Atkinson's optimism, the reorganization really did not benefit the social sciences in a broad sense. In fact, it is fair to say that, on balance, it reduced their organizational standing along with their visibility and status inside the agency. Most obviously, they no longer had a high-level organizational research unit of their own.

In addition, within the DBBSS, they were not on par with the biological sciences in terms of resources allocation. This can be seen in figure 6.4. In

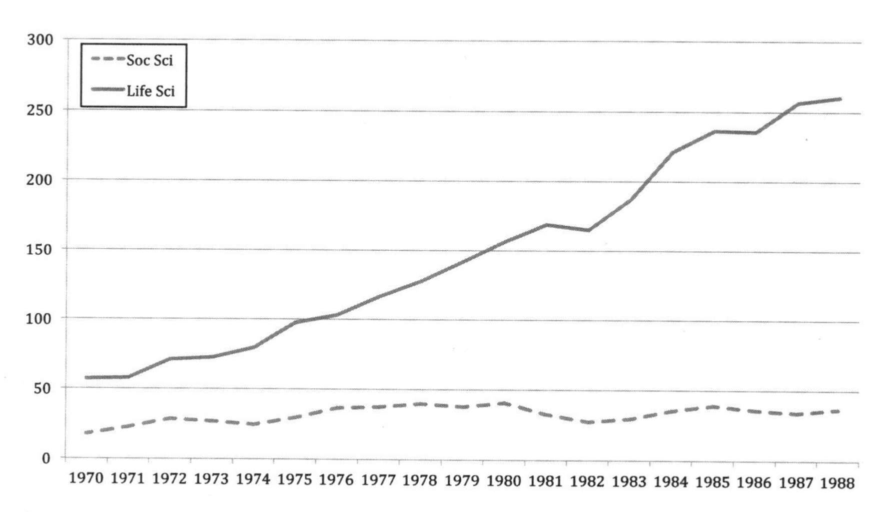

Figure 6.4
Federal obligations for basic research, social and behavioral sciences vs. life sciences, National Science Foundation, 1976–1988 (dollars in millions). Data from table 2L, federal obligations for basic research, by detailed fields of science and engineering: National Science Foundation, fiscal years 1970–2003. In National Science Foundation, Division of Science Resources Statistics, *Federal Funds for Research and Development: Fiscal Years 1970–2003*, NSF 04-335, Project Officer Ronald L. Meeks (Arlington, VA, 2004). Amounts for Life Sciences computed by adding separate amounts for Biological Aspects of Psychology in table 2L; amounts for Social and Behavioral Sciences computed by adding separate amounts for Social Aspects of Psychology in table 2L.

the dozen or so years following the mid-decade reorganization, funding for life sciences research increased from about $100 million to more than $250 million, while funding for the social and behavioral sciences remained at less than $50 million.

Furthermore, although Atkinson briefly served as DBBSS acting director, its first full-time director, Eloise E. "Betsy" Clark, did not come from the social or psychological sciences but from biology. A specialist in developmental biology, Clark completed her PhD at the University of North Carolina in 1959 and held a professorship at Columbia University during the 1960s. At the end of the decade, she moved to the NSF, where she served in various posts within the biological sciences division: as program director for developmental biology (1969), program director for biophysics (1970), head of the molecular biology section (1971–1973), and then director of the biological sciences division (1973–1975). Following the 1975 restructuring, she served briefly as DBBSS deputy assistant director under Atkinson, followed by her appointment in 1976 as DBBSS leader, a position she held until 1983.[39] Clark was also the first female to lead a major division at the NSF, a full quarter century after its founding. Later, in her post-NSF career, she served as provost of Bowling Green State University and president of the American Association for the Advancement of Science.

In contrast to the former leaders of the original social science division, Clark had no training or expertise in this area. Nor did she have any standing within the relevant professional scholarly communities for sociologists, political scientists, and the like. Neither did her deputy assistant director Robert Rabin, a biologist specializing in microbiology and biochemistry.[40] However, at least scholars with relevant expertise were placed in charge of the two main DBBSS divisions for the social and behavioral sciences (SBS). The division for the behavioral and neural sciences was led by the psychologist Richard T. Louittit. A specialist in physiological psychology, Louittit had been chairman of the psychology department at the University of Massachusetts–Amherst and then head of the behavioral sciences research branch at the National Institute of Mental Health. Meanwhile, the sociologist Herbert Costner was placed in charge of the new social sciences division. A faculty member from the University of Washington, Costner specialized in quantitative methods and the measurement of social phenomena.

The timing of the decision to move the social sciences into the new directorate with biology suggests that it was done, at least partially, to protect them and the agency from criticism. After all, the agency could have chosen to give the social sciences their own directorate, rather than placing them alongside biology in a directorate led by a biologist. According to Otto Larsen, Atkinson himself acknowledged that fallout from the MACOS controversy influenced the agency's decision to reorganize and reposition its social science programs. In effect, the strategy of "protective coloration" from the Alpert era had been fully revived.[41]

In terms of status and resources, the 1975 reorganization gave some SBS fields a natural advantage over others. In general, fields that overlapped with the biological sciences, including certain lines of research in the behavioral sciences and cognitive sciences, enjoyed greater opportunities for funding and scientific validation. Presumably, this advantage strengthened the standing of their work within the scientific community more generally.[42]

Given Atkinson's scholarly expertise and scientific outlook, it's no surprise that he took a special interest in those areas. He explained that with the aim of advancing research on human cognition, the agency sought to bring "research on cognitive process, human biology, developmental biology, and linguistics together." Similarly, he supported the move "to push the behavioral sciences more in the direction of neurobiology and the cognitive sciences."[43]

Assessing the reorganization's impact on anthropology—located in the behavioral and neural sciences division—is complicated because of the discipline's hybrid character. Certain fields such as archaeology, physical anthropology, and biological anthropology had strong ties to the biological sciences. But other fields such as cultural anthropology and anthropological research in linguistics arguably had more in common with sociology. In current dollars, NSF funding for the anthropology program climbed from $3.5 million in 1970 to $4 million in 1975 and then to $6.1 million in 1979. But when measured in constant 1972 dollars, the trend in funding just missed keeping up with inflation, going from $3.8 million to $3.2 million to $3.7 million.[44]

The budgetary trajectories of the three other main social science disciplines, which had few ties to the biological sciences, suggest a mixed bag. In the years mentioned above—1970, 1975, 1979—funding for the political science program rose in current dollars from $1.1 million to $1.6 million and then to $3.4 million, representing an increase in constant 1972 dollars from $1.2 million to $1.3 million and then $2.0 million.[45] Considered by itself, this trajectory is impressive. But of all the disciplinary programs, Political Science had been created last and its origins had been especially contentious. It also started the decade at the bottom of the pack.

By the end of the decade, Political Science had at least pulled even with Sociology. However, pulling even also reflected the fact that Sociology, which began the 1970s with three times more funding than Political Science, experienced a dismal decline. In current dollars Sociology's budget went from $3.1 million to $2.3 million and then to $3.4 million. But its budget in constant 1972 dollars fell sharply from $3.4 million to $1.8 million before rising, but only a bit to $2.0 million.[46]

The economics program fared better than Sociology and Political Science and also better than Anthropology. During those years, the budget for Economics rose in current dollars from $4.1 million to $6.8 million and then to $9.7 million. The increase in constant 1972 dollars was still notable but much less impressive, going from $4.5 million to $5.4 million and then $5.8 million. More striking, for the entire decade, Economics accounted for

26 percent of the total NSF funding for social science research. The other main disciplinary-based programs were Anthropology at 17 percent, Sociology at 10 percent, and Political Science at 7 percent. The remaining funds went to the programs for Social Psychology at 9 percent, History and Philosophy of Science at 5 percent, Special Projects at 11 percent, Geography at 4 percent, Social Indicators at 6 percent, Law and Social Science at 4 percent, and Science Policy at 1 percent.[47]

In 1979, the agency also singled out economics for special attention by inserting it into the title of the DBBSS's Social Science Division, making it the Social and Economic Sciences Division.[48] The elevated status and funding for economics would continue through the 1980s, as will be discussed in chapter 8.

In sum, Atkinson's efforts in crisis management had helped the NSF to address serious troubles associated with the MACOS and Proxmire controversies. This included a repositioning of its social science programs as part of the larger 1975 agency-wide reorganization, which Atkinson himself supported. However, the end result did not actually improve funding for these sciences in any consistent manner. Moreover, in terms of organizational standing and visibility, they had lost ground. As explained in the next section, Atkinson also played a key role in the story of the Simon Report and its role in the repackaging of NSF social science.

THE SIMON REPORT

Just like the other cases discussed above, Atkinson's involvement in this matter began shortly after his arrival. In fact, he was only in his first week as deputy director when, in July 1975, he began the process of commissioning a report on the agency's social sciences activities from the NAS.[49]

The following month, Atkinson told his fellow psychologists that he welcomed the upcoming study as a means of putting the NSF's house in order. As he saw it, the goal was to "ensure" that "at least one Federal agency can provide strong support for basic research in the behavioral and social sciences." This seemed all the more urgent because other government agencies had been pulling back. At the National Institute of Mental Health, funding for basic research had peaked at $48 million in 1972 and then fell to $20 million. Financial support for basic research from the Office of Education and from the National Institute of Education took a nosedive as well, from a combined high of $11.5 million in 1973 to just $2.5 million in 1974. Although Atkinson did not give any figures for the Defense Department, he mentioned that it had recently "curtailed activities in artificial intelligence and other areas relating to psychology."[50]

Just as importantly, Atkinson had good reason to anticipate that the commissioned report from the NAS would advance a hard-core scientific orientation. Such an orientation had long characterized the Academy's involvement with the social sciences, a point that was common knowledge in federal science policy circles and at the NSF. Having been elected to NAS membership in 1974, Atkinson also had firsthand knowledge of that.

Ever since the early twentieth century, natural scientists had dominated NAS leadership. From all of the social and behavioral sciences, only psychology and anthropology had permanent albeit relatively small roles at the Academy up through the 1950s. Starting in the 1960s, and following much discussion colored by fervent objections from some natural science members, the NAS decided to extend membership to a larger number of scholars in the social sciences and welcomed a broader range of disciplinary backgrounds. However, these sciences remained subordinate in status and influence.[51] Academy membership figures are a good indicator of substantial disparities that existed between the natural and social sciences. During the 1970s, social scientists accounted for only about 11 percent of the total membership. Moreover, this figure was well below the percentage of social scientists in the wider scientific and engineering communities, which, based upon the number of doctorates awarded between 1936 and 1978, was roughly 32 percent.[52]

Furthermore, the Academy had a special interest in social scientists who had a great deal in common with their natural science counterparts. By and large, members of the Behavioral Sciences Division (est. 1963) and its successor, the Assembly of Behavioral and Social Sciences (est. 1973), embraced this stance. Besides Atkinson and Simon, this group included Leon Festinger and Neal Miller from psychology, Heinz Eulau and William Riker from political science, Kingsley Davis and James Coleman from sociology, and Zvi Griliches and Duncan Luce from economics.[53]

Moreover, NAS practices ensured that its SBS projects were reviewed and approved by physicists and other natural scientists. This peculiar type of oversight followed from a requirement that NAS reports had to be supervised by a board of eleven members. In practice, these members came overwhelmingly from the hard sciences, although this was not a written requirement.[54] Meanwhile, NAS leaders apparently never entertained the idea of having its natural science reports reviewed by social scientists.

In addition, the fact that the NAS committee for this study chose Herbert Simon as its leader suited Atkinson. Simon later recalled that ever since his undergraduate studies at the University of Chicago during the 1930s, he had believed that the social sciences "needed the same kind of rigor and the same mathematical underpinnings that had made the 'hard' sciences so brilliantly successful." Thus, he had decided to become a "mathematical social scientist."[55] The parallel with the development of Atkinson's own scientific interests at Chicago a decade and a half later is uncanny.

In the post–World War II era, Simon became one of the most remarkable and versatile scholars from the social sciences' hard-core wing, again making him a kindred spirit to Atkinson even if the latter's contributions were concentrated more narrowly in his home discipline of psychology. During the 1950s, Simon joined the young Turks leading the behavioralist revolution in political science. He served as a consultant for the Ford Foundation and its behavioral sciences program. He established a long-time consulting relationship with RAND, which, among other things, pursued mathematically based social and behavioral inquiry such as systems analysis and game theory in an effort to place strategic decision making on a rational basis. At the SSRC, Simon joined the board of directors (1958–1971), where he focused on improving interdisciplinary work and mathematical training for social scientists and also served as council chairman in 1967. Simon became a member of the physical science–dominated President's Science Advisory Committee as well, although not because of his varied social science contributions but because of his expertise in artificial intelligence. And in 1978, Simon received the Nobel Memorial Prize in Economics for his pioneering work in decision theory.[56]

At the National Academy of Sciences, Simon was known as a powerful advocate of what he himself called the social sciences' "hard wing."[57] First elected as a member of its Behavioral Sciences Division in the early 1960s, Simon joined the division's governing board, served as its chairman, and was then elected to membership in the full Academy. Commenting in 1967 on Senator Harris's NSSF proposal, which as we saw included support for humanistic studies and critical forms of social inquiry, Simon said he favored sticking with the NSF's more restrictive approach and declared its social science program to be "basically sound"—although he also advocated dramatic increases in its budget.[58] Simon also opposed a separate idea to create a social science academy, which would have been parallel to the existing NAS, citing his concern that such an entity would "inevitably end up 'softer' in the balance of its composition than would a behavioral science class of appropriate size within NAS." He also worried that the proposed academy would "add to the confusion of voices addressing the congress and the public that purport to be the voice of 'science.'" The cultivation of "a united front with their hard-science colleagues" thus appealed to Simon as "the politically wise approach." He also saw it as conducive to healthy scientific progress: "Collaboration with physical and biological sciences ... provides important support to the 'hard' and 'modern' component of the behavioral science disciplines, and gives them additional leverage against their 'soft' and 'traditional' colleagues." "This leverage," Simon claimed, would encourage these sciences "to continue to grow in the healthy directions they have been taking for the past generation."[59]

Figure 6.5
Herbert Simon giving a lecture, ca. 1972. Courtesy of Carnegie Mellon University.

So, when it came to federal policies to stimulate social science progress, Simon and Atkinson were on the same page. Moreover, they were on good terms personally. I do not mean to imply that in the case of the NSF-commissioned study, Atkinson—who had initiated the study during his stretch as deputy director—worked with Simon behind the scenes to ensure that he, Atkinson, would be happy with the final report. However, the striking similarities in their views about the social sciences together with their scholarly friendship made it highly probably that Atkinson would find much to like in that report.

The 1976 Simon Report began by underscoring the NSF's key role in supporting top-notch scholars doing high-quality, *basic* social science research. That year, while the United States celebrated its 200th anniversary and the NSF celebrated its 25th anniversary, the agency provided nearly one-fourth of the federal government's total support for basic SBS research. As noted in chapter 5, the Simon Report presented sharp criticisms of RANN. However, it found the quality of NSF basic research projects to be "generally excellent."[60]

As evidence, the Simon Report noted that of the five most recent American scholars to win the Nobel Prize in economics, four had benefited from NSF support. So had all of the following from other disciplines: six of the last eleven presidents of the American Economic Association, six of the eleven winners of the American Sociological Association (ASA) Sorokin Award for distinguished research in the discipline, every winner of the ASA Stouffer Award for methodological contributions, and forty-three of fifty-seven winners of the American Psychological Association Distinguished Scientific Contribution Award. The Simon Report emphasized as well that many outstanding scholars had first received NSF support before winning their awards, implying that the agency's review process had identified their strong scientific promise at an early date and its grants had helped them to realize their potential.[61]

Based upon that admirable track record, the Simon Report proposed a major expansion of the NSF's basic social science program to provide adequate resources for important work that otherwise would probably suffer from underfunding. From this perspective, it seemed that the real problem facing the agency did not concern those studies whose scholarly contributions and practical value might be questioned—or ridiculed—by journalists, partisan organizations, or politicians. Rather, the report asserted that inadequate funding had forced the agency to reject a "large number of excellent proposals." In addition, "projects that require large-scale support (several hundred thousand dollars per year or more) or that need to be supported over a long period to produce results" had a particularly difficult time. The agency therefore needed much more money to fund a greater number of worthy social science projects, including big-budget projects.[62]

Furthermore, comparisons between the numbers of funded proposals in the different sciences indicated a worrisome pattern of preferential treatment. In the social and psychological sciences, only about one-third of all proposals received funding. But in the physical and life sciences, nearly half succeeded, and in the atmospheric sciences, about three-quarters did. One might wonder if the unfunded social science proposals were simply of inferior quality. But the Simon Report argued against this view, claiming that a smaller proportion of social research projects received NSF support due to the "limitations of funds and not to a deficiency in the numbers of good proposals submitted."[63]

Last but not least, the Simon Report reaffirmed the view that the social sciences belonged together with the natural sciences in a unified scientific enterprise. As presented here, this view presumed that scientific inquiry carried out across the many fields of study had much in common, including a fundamental commitment to value neutrality and an associated disinterested scholarly stance with respect to the social uses of scientific knowledge. During congressional testimony before an appropriations subcommittee, Simon underscored the importance of these issues: "Scientific inquiry has progressed much farther in some fields than others, but the basic characteristics of the activity are the same in all. The social sciences are an integral part of the total scientific endeavor." The goal of "social science, like all science, is to explain the phenomena, not to advocate actions to change them." Thus, scientific knowledge is "neutral; it only loses its neutrality when it combines with values and goals."[64]

Simon's insistence on this last point is worth highlighting, not only because value neutrality had been a longstanding concern for the social sciences and their patrons, including the NSF, but also because the criticism that social science projects reflected the values of private or federal funding sources was once again prominent in national discussions. In the MACOS controversy, conservative detractors from Phoenix to Washington, D.C. charged that the NSF-funded social science-based curriculum was marinated in a constellation of secular humanist and liberal values. In another case, a U.S. Senate investigation in the mid-1970s exposed previously secret CIA programs that supported mind control experiments. This investigation revealed, among other things, that the CIA had provided funding for researchers who used hypnosis, sensory deprivation, verbal and sexual abuse, and other powerful forms of mental and physiological manipulation. Revelations about the origins of such research, the conditions under which it was supported and carried out, and the anticipated uses of research results suggested, at least to some critics, that the CIA's commitment to certain goals, such as the ability to control thoughts and behaviors, at the expense of other goals, such as strengthening individual autonomy and dignity, had shaped the research process from beginning to end.[65]

Nevertheless, Simon conveyed unqualified confidence in the value-neutral standpoint, telling Congress that "well-trained social scientists achieve a high level of it [value neutrality], even when they are dealing with value-laden or emotionally explosive topics." He also noted that the challenges involved in remaining neutral were not unique to the social sciences. They also arose when natural science research had a bearing on controversial policy issues, as in the case of environmental research and protection. Nevertheless, just as natural scientists strove for neutrality in their analysis and often achieved it, Simon said that "working social scientists have by and large made their peace" with this matter. They had "learned how to separate facts from values in the research they do and the conclusions they reach." Simon even told his legislative audience that the alleged threats to value neutrality were so well known and so effectively managed that they were really not worth worrying about any longer. The NAS committee that produced the Simon Report did not even give the matter much attention. It simply did not arise as a "salient" issue.[66]

In the fall of 1976, shortly after Atkinson had become NSF permanent director, the Simon Report received a warm reception inside the agency. A social science subcommittee of the governing board was responsible for providing a summary and assessment of the study. In an overview of the subcommittee's work, Roger Heyns, a board member and a psychologist, praised the scholars who produced the Simon Report as "wise, sensible, and experienced." Heyns added that the subcommittee agreed "in general" with the report's recommendations regarding basic research. It also "took seriously" the report's criticisms, including the shortcomings of applied social research programs. According to an article in *Science*, Atkinson himself called the Simon Report "tremendously useful" as well.[67]

Following the report's analysis and recommendations, Atkinson "took steps to phase out the RANN program," which he said had been "reasonably productive" but "its approach to the support of research was not appropriate for NSF and did not live up to our standards."[68] In her history of NSF engineering programs, Dian Belanger points out that whereas the two previous directors, William McElroy and Guy Stever, strongly supported RANN, Atkinson did not.[69]

Upon Atkinson's urging, the NSF also redoubled its emphasis on basic research. Just one year before RANN's closure, an NSF committee explained to Congress that barriers to the practical applications of social science advances were, as recent studies had shown, often "essentially political and turn on sharp conflicts over social values or goals." Furthermore, especially in the case of basic research, which the NSF was best equipped to support, practical payoffs were hard to foresee. To convey this point vividly, Herbert Simon explained that supporting basic social research projects was like planting trees. In both cases, a great number wither away and fail to produce anything impressive. Although one might wish it were otherwise, identifying ahead of time which plants or which studies would flourish was impossible. Nevertheless, a small number do thrive, reach full maturity, and prove to be valuable. From this perspective, the agency's basic research efforts remained vitally important to the social sciences and the nation.[70]

By the decade's end, the renewed emphasis on basic social research had solidified. According to the 1979 NSF annual report, its determination to focus mainly on scholarly advances meant that concerns about applied payoffs had receded: "Practical benefits will come later, if at all." As had always been the case, the agency indicated no interest in scholarship that embraced a humanistic, normative, or critical perspective. But it now downplayed the importance of practically oriented work as well, in order to concentrate more fully on supporting "better methods and more comprehensive data resources so that research findings are more valid, reliable, and generalizable."[71] At the time, this position, although hardly a new one, fit well with the Carter administration's interest in boosting federal support for basic research across the sciences.[72]

In short, the story of the Simon Report needs to be seen in the context of the mid-to-late 1970s crisis management efforts that Director Atkinson so actively participated in. Not only was this NAS study commissioned by the NSF at Atkinson's request, but Atkinson anticipated that the resulting report would be valuable in reaffirming the special importance of the agency's basic social science efforts. Indeed, with the like-minded Simon in charge of the study team, the final report emphasized a hard-core strategy for advancing the social sciences and especially basic research. Atkinson and the agency then used the Simon Report to argue that support for such research should once again be the top priority, while practical applications and the high hopes previously associated with it would receive diminished attention and support.

BIG SOCIAL SCIENCE—A NATIONAL RESOURCE

At this point, the Simon Report's charge that the NSF failed to provide adequate support for large-scale basic social research projects deserves closer consideration. The report presented this inadequacy as one of many missed opportunities, a claim consistent with its more general recommendation that the agency needed to beef up funding for basic social research. However, in a brief discussion, the sociologist Otto Larsen pointed out that "big social science" was actually one area that did reasonably well at the NSF during the 1970s.[73]

What, then, did big social science involve? And why did it attract support within the agency at a time when the social sciences, on the whole, were losing ground?

Here, I am using the term *big social science* to refer to projects that were much larger than the typical NSF-funded project in terms of the number of scholars and staff members involved; the requirements for material resources such as computer equipment; the nature of the work involved at the levels of collecting, analyzing, and preserving data; and the duration of the project. By the late 1970s, a big social science project might receive hundreds of thousands of dollars, in contrast to the average social science project, which received roughly $55,000.[74]

In fact, support for big social science had been growing since the mid-1960s. Much of the funding in this area was channeled through two units within the original social sciences division: a special projects program, created in 1965, and a social indicators program, created in 1970. When that division was closed in 1975, the two programs were moved into the new social sciences division within the DBBSS. And in 1978, these programs were combined and renamed the measurement, methods and data resources program.[75] Besides nurturing big social science projects, these programs reflected a growing recognition that social science progress sometimes depended on research not confined to a single discipline.

Those programs were all led by one individual, Murray Aborn. After completing his PhD in psychology at Columbia University in 1950, Aborn held a series of positions: instructor at Michigan State University until 1953, research psychologist with the U.S. Air Force until 1957, and science administrator at the National Institutes of Health until 1963. At that point, he moved to the NSF as a senior scientist, a position he held for nearly three decades, until he left the agency in 1990. An effective science administrator and program leader, Aborn promoted support for many big social science projects that involved a number of disciplinary and interdisciplinary fields of study.[76]

One group of projects involved the development, preservation, and use of national databases for the social sciences. Among the most important was the National Election Studies, which had roots in an earlier research endeavor known as the Michigan Election Studies. First carried out in 1952 with funds from the Carnegie Corporation and Rockefeller Foundation, the Michigan Election Studies were conducted by the University of Michigan Institute for Social Research's Survey Research Center and later by its Center for Political Studies. That project generated massive amounts of data on election results, voting behavior, voting preferences, and contextual factors. It also contributed significant advances in the methodology of social survey research. Starting in the early 1960s, data from what was by then called the National Election Studies (NES) became widely available to researchers through Michigan's Inter-University Consortium for Political and Social Research (ICPSR).[77]

In 1970, when the NSF began to provide regular funding, the NES was already a well-established and highly regarded research project with considerable support from private sources, including the Russell Sage Foundation, the John and Mary Markle Foundation, and the Carnegie Corporation. By the end of the 1970s, however, private organizations had withdrawn their support. Henceforth, starting with the 1978 national elections, the NSF was the only major patron for such research. The agency's political science program benefited as well because, just one year before, it had taken over responsibility for handling NES funding requests and awards.[78]

In terms of scholarly importance, no project of this type anywhere in the world could compete with the NSF-funded NES. By 1982, more than 260 institutions of higher education in the U.S. and abroad had joined the ICPSR as members. Thus, they all received NES data. In addition, the NES served as the primary data source for over 850 scholarly studies, while many more articles used NES data as a subsidiary resource. As had long been the case, political scientists took a special interest, using NES resources to study such things as "opinion and attitude changes, public support for congress and the presidency, the dynamics of political representation, the causes of political participation and the decline in voter turnout, the evolution of political values, the role of the media in American politics, the structure of policy thinking among the public, and the dynamics of social choice." But scholars from many other areas, including economics, cognitive science, social psychology, sociology, and history, also used NES data.[79]

Inside the NSF, database projects acquired significant support during the Atkinson years and remained a high priority after he left the agency in 1980. In 1977, the governing board referred to the NES as a "national resource for the social sciences analogous to high energy accelerators, telescopes, or oceanographic laboratories."[80] A few years later, a report on NSF social science asserted that databases deserve "special attention and continuing support." Furthermore, the agency had a unique role in cultivating this work:

> The social science programs of NSF have taken the lead in fostering the establishment of computerized social science data archives which are accessible to research workers in all the disciplines nation-wide and they are used extensively ... to develop and test social and behavioral science measurement techniques and theories of societal functioning.[81]

By this point, NSF supported eleven such facilities, including the NES as well as the Center for Demography and Ecology at the University of Wisconsin, the Computer Research Center for Economics and Management Science at MIT, and the Center for Coordination of Research on Social Indicators in Washington, D.C.[82]

As suggested by the presence of this last center in the nation's capital, the NSF also became involved with big social science through its support for social indicators. Back in the 1960s, interest in social indicators had flourished within scientific circles, private foundations including the SSRC, national science organizations including the NAS, and the federal government including the

Department of Health, Education and Welfare (HEW), NASA, and the President's Commission on Federal Statistics. An influential edited volume from 1966 called for the development of social indicators as "a means by which our society can assess where we are now and where we have been," thus providing "a basis for anticipation—rather than prediction—of where we are going in a number of areas critical to our national welfare."[83] That same year, President Johnson directed HEW to find better ways of charting the nation's progress toward social goals, moving beyond standard economic measures to better illuminate other aspects of national life: health and illness; social mobility; the physical environment; income and poverty; public order and safety; learning, science, and art; and participation and alienation. The result was a 1969 study called *Toward a Social Report*.[84] In the late 1960s and early 1970s, the Minnesota Democratic Senator Walter Mondale proposed the establishment of a national council of social advisers that would promote the development of social indicators and use them to produce an annual social report on the state of the nation.[85]

Against this backdrop, the NSF became a major sponsor of social indicators research. In 1971, the agency held a planning conference, began making grants, and created a social indicators program within the social sciences division. Three years later, the governing board noted that "inadequate information on the current state of society and lack of detailed data about particular individual and social problems" impeded the development of effective public policy. Overcoming these limitations would require "the expansion of effort in the social indicators area, as well is in large survey research."[86]

The major recipient of NSF funding was the Center for Coordination of Research on Social Indicators (CCRSI) in Washington, D.C. Founded in 1972 under SSRC auspices, the CCRSI was staffed by professional social scientists and had three central activities. First, it helped to build "networks among individuals and institutions working or interested in the field through its library, newsletter, and participation in meetings and conferences." Second, it brought "suitable analytic strategies to bear on the development of social indicators." Third, the CCRSI worked to improve "the accessibility and availability of the data base for measurement of social change."[87] By 1977, NSF grantees had contributed to two national reports issued by the Office of Management and Budget: *Social Indicators 1972* and *Social Indicators 1976*.[88] Moreover, by this point, the NSF had, as one congressional report observed, "probably supported more fundamental work in social indicators than any other Federal agency."[89]

Social indicators had broader ramifications within the NSF as well. For one, the agency's involvement contributed to the development of national science indicators, which had strong support from the agency's natural science leaders and included an important biennial publication, *Science and Engineering Indicators*. In addition, positive assessments of social indicators helped, as Murray Aborn put it, to persuade "both NSF's top management and the National Science Board" that the social sciences had legitimate needs requiring "large-scale, long-term funding." In this way, social indicators helped to move the social sciences into "the realm of 'big science' at the NSF—albeit on a very modest scale compared with most of the natural sciences."[90]

As with so many fields and approaches, big social science had its share of disagreements and conflict. Consider the case of social indicators. One contentious point concerned whether the field had arrived at a stage of intellectual maturity such that it could contribute effectively to policy making. Although some believed the field had reached this stage, Eleanor Sheldon, the first CCRSI director, spoke for many skeptics in a 1975 *Science article*:

> The concepts which focused much of the early enthusiasm gave exaggerated promise of policy applications and provided an unproductive basis for research. The essential theoretical prerequisites for developing a system of social accounts—defining the variables and the interrelationships among them—are missing.[91]

Over the years, this NSF-funded center placed greater emphasis on scholarly research than studies with a practical, policy-oriented focus. This trend helped reaffirm the agency's commitment to basic social science research during the Atkinson years. In the words of Richard C. Rockwell, a sociologist and SSRC program officer, the NSF-funded CCRSI emphasized "the statistical and conceptual craftsmanship required for sound social indicators" while resisting "the pressure for premature efforts to carry out the agenda of those who sought measures of policy and political use."[92]

Regardless of the specific field of research, funding for big social science also encountered challenges from those who feared for the health of other types of scholarship. Inside the NSF, differences of opinion arose over how to strike the right balance between big social science and smaller-scale studies that required fewer resources. In addition, funding for big projects, Otto Larsen notes, "sometimes irked" those who were more interested in disciplinary-based programs.[93]

Nevertheless, such complaints did little to halt the momentum of big social science. Any explanation for its continuing appeal must include the crucial part played by social science program managers, especially Murray Aborn, who worked hard to convey the nature and value of large databases and social indicators to agency leaders. In addition, the agency's natural science–oriented leadership viewed such projects favorably because they seemed, in a sense, familiar. The tremendous growth of big science in the physical, biological, atmospheric, geological, and engineering sciences piqued their curiosity about the possibilities of comparable work in the social sciences. In the case of social indicators, Aborn himself observed that such work enjoyed a "happy life" because it fit well with "the NSF ethos of modeling the social sciences after the natural sciences."[94]

So, while social science activities came under sharp criticisms and encountered a number of major setbacks during the mid-to-late 1970s, big social science remained, for the most part, above the fray. Agency leaders took a special interest, as they found big social science, such as the National Election Studies, analogous to work in the natural sciences. None of the crisis management measures taken under Atkinson's leadership were designed to dampen progress here. To the contrary, such work continued to gather strength. By the early 1980s, "no other federal agency" had a comparable program.[95]

SIMMERING DISCONTENT

In his posts as deputy director, acting director, and permanent director, Atkinson sought to address the troubles associated with RANN, the Golden Fleece Awards, and MACOS head on. In addition, under his leadership, the new behavioral sciences division did reasonably well, giving a boost to research closely related to the biological and cognitive sciences. Within the new social sciences division, big social science also enjoyed considerable support, and economics fared reasonably well. The agency also drew on the Simon Report to reassert its commitment to basic social science research. However, in the overall picture, the reputation of NSF social science had fallen considerably since the expansionary 1960s and even since the early 1970s. Moreover, the second-class standing of the social sciences remained striking with no signs of improvement at the levels of funding, organizational standing, and representation. Against this background, a gloomy outlook became common among those responsible for the social science programs—apart from the few favored areas.

In 1978, Atkinson told a science policy journalist that the social and behavioral sciences were "slated for healthy increases" in basic research funding. However, when program staff members heard such hopeful news, their reactions were muted. As Atkinson himself explained, the NSF budget for these sciences was "so low compared to NSF support for other scientific disciplines" that the staff found it "hard to get too excited."[96]

Indeed, entrenched disparities between the natural and social sciences persisted. During the 1970s, the percentage of total NSF research funding for the latter remained low, hovering around 5.5 percent, with minor fluctuations from year to year. So, out of every twenty research dollars, the natural sciences still received roughly nineteen.[97] In addition, from 1976 to 1980 only 33 percent of social science proposals judged to be competitive actually received support, compared with 46 percent of proposals from all other scientific fields.[98] These percentages were roughly the same as those already flagged as worrisome by the 1976 Simon Report.

The broader implications for basic social research seemed worrisome as well. According to a 1980 study from the NSF planning office, only 9.7 percent of all applicants to this agency also applied to another agency, presumably because they believed their chances elsewhere would be poorer or because they were confident about receiving NSF funding. These data strongly suggested that the overall pool of social science researchers pursuing basic studies had come to depend heavily on the NSF. "Most" of them, however, would not succeed in getting the support needed to conduct basic research.[99] Indeed, at the NSF, the overall success rate of social science applicants was, as noted above, just 33 percent.

In addition, although Atkinson remained director throughout the late 1970s, the number of social science representatives on the governing board fell. In 1972, the twenty-four-member board included four representatives: the political scientist James G. March (1968–1974), the psychologist Roger W. Heyns (1967–1976), the anthropologist F. P. Thieme (1964–1976), and the economist W. Glenn Campbell (1972–1978). But their memberships had all ended by the end of 1978. The following year, the cultural anthropologist Ernestine Friedl, from Duke University, began a six-year appointment, making her the board's only social scientist as the decade ended—as the director, Atkinson was an ex officio member.[100]

Under these conditions, social science program leaders complained that natural science leaders still did not support them strongly. In 1976, political science program director David C. Leege published an article under this disconcerting title: "Is Political Science Alive and Well and Living at NSF?" According to Leege, a number of issues compromised support for political science and the social sciences more broadly. Regarding his program in particular, agency leaders found certain aspects of its origins in the 1960s "distasteful." That unpleasantness lingered "in their memories" and had been "passed on in institutional memory." Some NSF officials understood that "not scientific merit but political pressure" had forced the program's establishment. As a result, pressure from Congress to fund political science had produced "resentment." The notion that calls for additional funding were rooted in pork barrel politics rather than confidence in scientific progress persisted. As Leege saw it, "despite sometimes heroic efforts by past program directors and panelists, many Foundation managers" really did "not understand what scientific progress political scientists are making."[101]

Leege also had more general concerns about NSF leadership. Just one year before, Guy Stever, still the director at the time, acknowledged that the social sciences presented special difficulties: while the "physical sciences, mathematics, engineering, and computer sciences" were "generally accepted and understood," "government" involvement with the social sciences raised many questions. According to Leege, "the feeling lingers that, under severe congressional pressure, the Foundation would abolish social science programs to salvage support for physical and biological sciences and engineering." The fact that some NSF managers still referred to the social sciences as "social studies" did not help. Unfortunately, physical scientists, who had often "dominated" the agency's "planning and budgetary processes," were still "not at home" with these sciences and "frequently revert to presumed analogies with intellectual problems in their own disciplines." With the agency facing "tighter budgets," Leege anticipated that "increased support to study subject matters foreign to budgetary planners" was unlikely.[102]

Because Leege's article appeared when Atkinson was still deputy director, one might wonder if such worries subsided after he became director. Apparently, they did not, as seen in disheartened comments from Herbert Costner, another social science staff member. In 1976, Costner, a sociologist, was placed in charge of the new social science division. By early 1978, he had concluded that the situation there was bad. In private correspondence with the sociology program leader, Otto Larsen, Costner identified many sources of difficulty, including sharp congressional criticisms and a failure by social scientists to defend their work well. Costner also found fault with the "NSF hierarchy": the agency "gives poor support (and sometimes active opposition behind the scenes) to the social sciences." Under demoralizing conditions, Costner no longer wanted to continue in his position. And in 1979 he left the agency.[103]

The psychologist Kelly Shaver also bemoaned the weak position of the social sciences. Shaver had been a member of the social psychology advisory panel from 1973 to 1975, then served as director for the social and developmental psychology program from 1977 to 1979. In August 1979, Shaver told Atkinson that the NSF, unfortunately, had "not done the sort of job that it could do in bringing the case of social science to the public." Some of his fellow program directors put the "blame" on the NSF's "hard science" orientation, while Shaver himself had "the distinct impression that many people in the Foundation secretly (in some cases not so secretly) share the public view of social science as a trivial endeavor."[104] Similar to Costner, Shaver left the agency in 1979.

Meanwhile, criticism from conservative quarters in American political culture remained strong. Attacks by Ohio Republican Representative John M. Ashbrook, who was no fan of the social sciences, were especially worrisome. "Let us strike a blow for common sense by sending a message to NSF that it is time to stop awarding Federal research funds for 'intellectual welfare,'" Ashbrook declared in 1979. Accordingly, he put forth an amendment to an appropriations bill to cut $14 million from the directorate for biological, behavioral and social sciences. Ashbrook had previously put forth a similar proposal but without success. This time, however, the House voted 219 to 174 in favor.[105]

The amendment really did not threaten "basic, valuable research" carried out by scientists in fields outside of the social sciences, emphasized the conservative Republican. Instead, he intended his legislation to deal with "the foolish, fringe folly of researchers who use our tax money like the dilettante squanders his inheritance—recklessly and with little meaning or value except to pander to their own snobbish tastes." In the end, a Senate-House conference committee decided to reduce the actual budgetary damage from $14 million to $2 million.[106]

The following year, Ashbrook renewed his attack. Noting that "there is a fine line between what is 'nice' to know and what is vital to know when you begin assessing basic knowledge," he argued that NSF policy had erred on "the side of 'nice,'" with undesirable effects on the economy. A few grants seemed especially worrisome for other reasons as well, including one $88,000 grant for research on "homosexual couple formation." Ashbrook understood that most Americans still considered "homosexuality a perversion or a disease and would object to any research that might aid in legitimizing such a thing." Seeing "deep problems" with government social science funding more generally, the Ohio Republican proposed that perhaps it was time to end it "altogether."[107] Ashbrook's criticisms provide further evidence that Atkinson's efforts in crisis management had not succeeded in protecting NSF social science from conservative hostility.

By the decade's end, Atkinson himself recognized that something had gone wrong. There is no reason to think he reconsidered the premise that support for basic research and hard-core studies deserved pride of place. Nevertheless, he told one reporter that the agency had "fallen behind in behavioral studies." He even said the agency had been "incredibly short-sighted," although what he thought it should have done differently remained unsaid.[108]

In any case, as the opening quote in this chapter from him suggests, Atkinson was worried that the agency's ability to support SBS research was undermined by external pressures grounded in strong and sometimes conflicting criticisms. One line of criticism said that social science research was "irrelevant to societal needs and, therefore, a waste of taxpayers' dollars." Another and rather different line of criticism held that "the social sciences are all too relevant" in ways that led to "social engineering and manipulation of moral values." The upshot was that these sciences "should not be encouraged, let alone supported." These criticisms, Atkinson added, led to constraints on federal agencies, including the NSF, which would probably support social research where "relevance" could be "easily justified" and seemed to pose "no threat to society's values."[109]

As discontent inside and outside the agency simmered in the final years of the decade, the following questions arose: Would the social sciences be better off if they had greater organizational status and independence, as had been the case before the 1975 reorganization and the closure of the original social science division? Should the social sciences be separated from the biological sciences and given their own directorate?

REORGANIZATION RECONSIDERED

In September 1980, social science leaders took up those questions at a meeting in New York City organized by the SSRC. Participants included representatives from the major professional social science associations, the NAS, the NSF, and the Center for Advanced Study in the Behavioral Sciences.

In a report on the meeting to the NSF governing board, the cultural anthropologist Ernestine Friedl—the only board member from the social sciences by this time—emphasized that many of her peers felt that "their interests were not always viewed with sympathy by the natural scientists." Some of them therefore found the idea of establishing a separate social science directorate appealing. However, others feared that having their own directorate in an institutional context where they lacked strong support from their natural science colleagues would leave them "very exposed," meaning they would be "more vulnerable to public criticism and to congressional budget reductions." Still, on the whole, the discussion in New York favored a new social science directorate.[110]

This proposal attracted some high-level interest within the NSF as well. In fact, two or three months before the New York meeting, Atkinson told Otto Larsen that he wanted to correct "the disruption" that he himself had "brought about by separating the social and behavioral sciences." Placing them together in a stronger and separate organizational unit now seemed appealing. However, Atkinson did not want to upset the head of the existing directorate, Eloise Clark, whom Atkinson suspected would not favor this idea. Atkinson thus told Larsen that he would consider this possibility again at a later date. But he never did.[111] And soon after his conversation with Larsen, Atkinson left the agency. In July 1980, he returned to the West Coast to take up the chancellorship at the University of California–San Diego, a position he held until 1995, at which point he became president of the entire University of California system.

Nevertheless, the possibility of establishing a unified SBS directorate remained a live option in the mind of the new acting director, Donald Langenberg. A physicist and expert on superconductivity, Langenberg considered a broad reorganization to strengthen the agency's capacity to support applied research, an issue that had remained problematic ever since RANN's closure.[112] In this context, Langenberg noted that a reorganization along the lines suggested at the New York meeting would be risky, because increased visibility of the social sciences could provoke increased criticism of them, as others had also noted. Yet he said he saw "no evidence" that the existing structure and practices at the NSF had done much to insulate and protect these sciences: they were not "so small a target" that the Congress had any "difficulty identifying them." Hence, taking the risk might be worth it. Langenberg also anticipated that a new directorate might empower social scientists, so that they could better "assist the foundation in defending" their work.[113]

Any serious consideration of this option subsided quickly, however. By November, the agency had a new permanent director, John Slaughter, an electrical engineer who was also the first African American in this top position. Previously, Slaughter had been a researcher, a high-level science administrator for the U.S. Navy, the director of the Applied Physics Laboratory at the University of Washington, and the provost for this university. He first joined the NSF as assistant director in charge of the astronomics, atmospherics, earth and ocean sciences directorate (1977–1979). Now, as the agency's director, Slaughter disagreed with Langenberg's recent assessment. At a board meeting in late November, Slaughter argued against the proposed reorganization, saying that a separate SBS directorate would not be in the agency's "best interest."[114]

According to Slaughter, the proposed reorganization would separate the behavioral from the biological sciences, a move that he believed had "no clear intellectual basis." This point also concerned the governing board, which had recently heard "convincing presentations from the concerned scientific and educational communities that the neural sciences are much more strongly coupled to the biological than to the social sciences." Before stepping down as acting director, Langenberg had also concluded that the behavioral and neural sciences division should remain in the directorate with the biological sciences. In this context, Slaughter agreed that the social sciences needed "increased visibility." But he anticipated that likely "negative fallout" from a separate directorate would outweigh any benefits.[115]

Exactly what type of negative fallout Slaughter had in mind he did not say. But just a couple weeks before, in early November, national elections resulted in a landslide victory for the conservative Republican presidential candidate Ronald Reagan, providing NSF leaders with good reason to fear a continuation and perhaps intensification of political attacks.

CONCLUSION

In 2003, the NSF governing board presented Richard Atkinson with its Vannevar Bush Award. This award, created upon Atkinson's initiative in 1980 shortly before he left the agency, honored "outstanding contributions toward the welfare of mankind and the nation through public service activities in science and technology."[116] The 2003 award citation noted that at a time when Congress was "questioning the appropriateness of many of the Foundation's research grants and the peer review system," Atkinson had "skillfully defended the agency, won congressional support, and maintained the Foundation's commitment to basic science and the peer review system." Of special relevance here, the citation also claimed that under his guidance, NSF work in the social sciences "grew."[117] On a similar note, in a 1997 article Atkinson himself stated that during his tenure as director, the agency "expanded" its efforts in the "behavioral and social sciences."[118] In a 2007 interview, Atkinson added that before the 1975 reorganization, the NSF had "a very limited role for the social sciences," including some work in economics, social psychology, and anthropology. He thus saw the reorganization as an important step because, among other benefits, it enabled the agency to cover "a broader range of activities, including the behavioral sciences like perception, memory, cognition, and so forth."[119]

In those documents, neither the board members nor Atkinson elaborated much on how he had strengthened the agency's work, but they could have pointed to three fields as evidence: the strong emphasis on biologically oriented work and cognitive science in the behavioral and neural sciences division, the relatively strong standing of economics within the social and economic sciences division, and support for big social science, nurtured under psychologist Murray Aborn's leadership, first through the programs for special projects and social indicators and later through the measurement methods and data resources program. In the case of the National Elections Studies, big social science also became the responsibility of the political science program starting in 1977. In addition, one must recall that the NSF retained a special place in the national funding system throughout the 1970s, especially when it came to supporting academically oriented social research. After three physicists and one biologist at the helm, Atkinson was the first NSF leader from the social or behavioral sciences as well.

However, a string of developments examined in the previous chapter and this one suggests a more complex and less upbeat view of NSF social science and Atkinson's impact specifically. Through his efforts in crisis management, Atkinson supported a number of measures that, arguably, diminished the overall standing of social science inside the agency (although there is no reason to believe that this was his intention):

1. In the case of MACOS, Atkinson did not side with conservative critics in their charges about this program's abhorrent content, perverted aims, and harmful impact on American students. But he did lead the internal review that resulted in the admission to Congress that the agency had made serious mistakes in its support for MACOS, thereby helping to solidify the agency's withdrawal from social science curriculum building.
2. He supported the 1975 agency-wide restructuring that resulted in the closure of the original social sciences division and the repositioning of NSF work in this area alongside biology within the newly created DBBSS, which was led not by a scholar from sociology, political science, or a neighboring discipline but by a developmental biologist, Eloise Clark, and her assistant Robert Rabin, a microbiologist. This arrangement promised to provide the controversial social sciences with a measure of protective coloration—a strategy reminiscent of the Harry Alpert era.
3. With Atkinson's encouragement, the NSF commissioned the NAS study that led to the 1976 Simon Report, whose findings Atkinson then used as support for the decision to close RANN. Subsequently, Atkinson backed the agency's effort to reemphasize its unity-of-science outlook, its commitment to supporting basic research from the social sciences' hard wing, and a corresponding pivot away from relevant social research and applied studies.
4. Meanwhile, Atkinson's efforts to win over Senator Proxmire may have succeeded to some extent, but Proxmire did bestow at least one highly publicized and extensively discussed Golden Fleece Award on the NSF during his years as director—for the support of Sherry Ortner's research on the Sherpas. Furthermore, the more general suspicion in American political culture that NSF social science funding was making a monkey out of the American taxpayer had hardly been contained.

On top of those developments, glaring funding disparities between the social and natural sciences persisted, even though the former's share of the overall NSF budget remained roughly the same, around 5.5 percent, throughout the decade. Moreover, the level of representation on the agency's governing board fell to a bare minimum by 1979, with only one social scientist, Ernestine Friedl, included among the twenty-four members.

Not surprisingly, frustration with the NSF intensified within the social science community. Program leaders from political science, sociology, and social psychology all expressed their discontent, in the process raising concerns about the impoverished position of the social sciences inside the agency and the questionable level of support provided by its natural science leadership during an increasingly difficult stretch of time. Seeking to reverse the downward spiral, some prominent figures suggested that the social sciences should be given a directorate of their own. No doubt, this was a rather modest suggestion compared to the much more ambitious proposal, put forth by Senator Harris a dozen years before, to give the social sciences an agency of their own. Still, in 1980, the suggestion of creating a separate directorate entailed a significant upgrade in their standing. This suggestion also attracted some interest among agency leaders, from Richard Atkinson and his successor, Donald Langenberg, in his role as acting director. But further consideration by the agency's governing board and its new permanent director, John Slaughter, resulted in a decision to maintain the status quo.

At this historical juncture, the recent past and current state of NSF social science seemed troubling for many reasons. As for the future, that was impossible to predict with any certainty. But with Ronald Reagan riding a surging conservative wave into the White House, the days ahead did not seem inviting.

7

DARK DAYS: SOCIAL SCIENCE IN CRISIS DURING THE EARLY REAGAN YEARS

> [The Reagan administration] is about to destroy some of the most important social research work in the nation for "economic" reasons and the pursuit of ideological purity.
>
> —Journalist Roger Witherspoon, 1981[1]

In June 1981, *Washington Post* science journalist Philip Hilts reported that the Reagan administration had "begun a social science hour at the White House." Hilts continued,

> Twice a month, the president, vice president, Cabinet and senior White House staff view charts and graphs. They listen to statisticians sketch profiles of our changing society. The object is to understand the background social facts against which policy will be mapped.

At the same time, however, "budgeters at the other end of the White House are ordering huge cuts in the programs that produce the very data the administration wants to use." "Where the hell do they think these numbers come from?" wondered one perturbed social scientist working for the new administration. Presumably, "they don't fall out of the sky."[2]

Pointing to a worry that became pervasive among social scientists and their advocates during the early 1980s, Hilts reasoned that the huge cuts would probably "kill" the "hardest, most neutral and most useful basic work in the social sciences" but not the looser, partisan work carried out by some mission agencies. Hence, it seemed particularly worrisome that the proposed cuts included substantial funding decreases for NSF social science programs. Although these programs "cost little more than half the price of maintaining the Pentagon's military bands," the new administration seemed bent on decimating them, while leaving the bands in place.[3] As Hilts's article suggests, the onset of the Reagan years posed significant threats to the social sciences, to their position within the federal government, and to the federal science system's efforts to promote the health of basic social science research, especially through the NSF.

The largest source of trouble was the Reagan Revolution, a realignment that made conservatism dominant in American politics and public policy discourse. The previous chapter documented the mounting political pressures and especially the resurgence of right-wing criticisms that plagued NSF social science during the latter half of the 1970s. As the political landscape lurched farther rightward during the 1980s, an alliance between religious conservatives and business interests controlled the Republican Party, which, in turn, controlled the White House. In a curious twist of fate, the new president had majored in sociology and economics during his undergraduate years at Eureka College in Illinois. Nevertheless, during Reagan's two presidential terms, conservative attacks on liberal social policies and ideas together with damning assessments of their intellectual supporters in the social sciences and elsewhere—in education, the humanities, and the mass media—reached a heightened pitch. As historians of the social sciences Roger E. Backhouse and Philippe Fontaine have recently observed, during the Reagan era, many "decision makers were hostile to social science, seen as tainted by association with the left."[4]

Unrelenting partisan attacks came from an increasingly powerful conservative "counterintellectual" establishment. During the preceding years, conservative intellectual leadership had established nodes of influence in the mass media, universities, policy institutes, think tanks, business, and politics. Reagan's ascendancy to the White House enabled a burgeoning conservative intellectual network to more fully realize its power. Again and again, figures from this network charged that American social science promoted run-away growth in the size and powers of the federal government, including crushingly burdensome economic regulations, cripplingly high tax rates, and terribly costly social welfare programs that did more harm than good to those they were designed to help. Conservative critics further charged that federally funded social research frequently turned out to be useless when it came to designing effective social programs, thus accelerating the demise of the social engineering ideal. Conservative attacks on universities and their professors for promoting leftist viewpoints that undermined good old American values, the family, Christian religion, and other sources of established authority gained renewed momentum as well. A clutch of influential works in this vein were penned by professors, including the best-sellers *Losing Ground: American Social Policy, 1950–1980* (1984) by Charles A. Murray, a political scientist at the conservative Manhattan Institute when he wrote this book, and *The Closing of the American Mind* (1987) by Allan Bloom, a University of Chicago philosophy professor.[5]

The counterrevolution also drew sustenance from changes within the social sciences themselves. As late as 1968, a note from the editor of *Social Research*, a scholarly journal published by the New School for Social Research, could make the following observation without being accused of exaggeration: "it may safely be said that conservative approaches are not currently in vogue in the human sciences and that the general climate of opinion in these disciplines ranges from liberal to left."[6] But in the following years, research supporting right-of-center viewpoints about human nature, the social order, and public policy matters gained ground. Of particular interest to the Reagan administration was a macroeconomic framework called supply-side economics, whose advocates championed the benefits of the free market, deregulation to minimize government interference with the supposedly natural flow of economic resources, and lower tax rates as the keys to healthy capitalist accumulation and economic productivity. These ideas reached a broad audience through the mass media and popular writings, including yet another best-seller, *Wealth and Poverty* (1981), by George Gilder, a successful investor, economic pundit, and Republican political speechwriter.[7]

For American science and science policy, the conservative ascendancy had far-reaching implications. Here, the dominant framework during the Reagan years privileged matters of economics and national security. In a 1984 book called *The New Politics of Science*, the journalist David Dickson observed that amid the "reemergence in the U.S. of an almost religious belief—dormant for much of the 1970s—in the powers of science-based technology," industrial leaders claimed that only worldwide supremacy in science and technology would allow the nation to thrive economically. At the same time, U.S. military leaders together with the White House argued for greater spending on military research, which was needed to develop increasingly sophisticated weapons. Of special interest to the NSF story, the Reagan administration also placed a "renewed emphasis" on supporting basic research, although this emphasis was now coupled with a belief that input from the private sector, especially from business, would be valuable in shaping the directions of scientific investigation.[8]

The new science policy framework had implications for the social sciences, too, as it gave short shrift to work oriented toward "social objectives," such as the protection of the natural environment, public health, and social welfare. Consequently, federal support for "various aspects of science, in particular science education and social science research" suffered. Not long before, the federal science system had given those matters greater emphasis. But by the early 1980s, one could hear only a "few echoes of the calls for relevance to social—as opposed to military and industrial—needs in federal research programs."[9]

The broad changes sketched above are crucial for understanding how NSF social and behavioral sciences fared during the early 1980s. The first section of this chapter analyzes the White House's 1981 proposal for a massive reduction in funding for those sciences and an ensuing widespread sense of crisis in American social science. The rest of the chapter considers the harried responses by social scientists and their supporters, as they tried to comprehend what was happening and developed strategies to curtail the impending damage. The second section turns to the Consortium of Social Science Associations (COSSA), which, in a surprising manner, emerged as a formidable force in the struggle against the proposed budget cuts. The last section considers how social scientists and NSF leaders defended the agency's work in the face of sharp challenges from the White House and conservative circles more generally.

The 1970s had been a difficult transitional decade. But the early Reagan years promised to be even more problematic. So much so, in fact, that the very survival of the social sciences at the NSF emerged as a serious question for the first time in the agency's history.[10]

EYE OF THE STORM

In the 1980 presidential election, the team of Ronald Reagan and George H. W. Bush received 489 electoral votes, crushing the Democratic contenders Walter Mondale and Geraldine Ferraro, who received only 49. Four years later, Reagan and Bush won again, in an even bigger landslide with an astounding 525 electoral votes. In the 1980 national elections, Republicans also gained 12 seats in the Senate to obtain a majority, 53 to 46. They gained 33 seats in the House as well, although Democrats retained a strong 243 to 192 advantage.

In February 1981, just a few weeks after Reagan was sworn in as president, the new administration presented its budget plans in *America's New Beginning: A Program for Economic Recovery*. Concentrating on "the most serious set of economic problems since the 1930s," including run-away inflation, high unemployment, and anemic economic growth, this report identified "the most important cause" of these problems as none other than "the government itself." Promising to revive the nation's ailing economy, the White House proposed a four-part plan, including "a substantial reduction in the growth of Federal expenditures ... a significant reduction in Federal tax rates ... prudent relief of Federal regulatory burdens ... and ... a monetary policy on the part of the independent Federal Reserve System which is consistent with those policies."[11]

The White House also called for deep cuts in NSF social science. At that time, its Directorate of Biological, Behavioral, and Social Sciences (DBBSS) had five divisions. Compared to the agency's FY 1981 budget, the administration's plans for FY 1982 included increases of 6.4 percent for the physiological, cellular, and molecular biology division; 6.7 percent for the environmental biology division; and 1.1 percent for the informational sciences division. In contrast, the social and economic sciences division —SES, which at this point had programs for economics, geography, history and philosophy of science, law, measurement, methods and data resources, political science, and sociology—would be shrunk by a stunning 70 percent, from $33.6 million to $10.1 million. In addition, the behavioral and neural sciences division—BSN, with programs for anthropology, cognitive and behavioral science, and neuroscience—would be shrunk by 26.3 percent, from $39.6 million to $29.2 million. On top of this, the administration singled out specific BSN programs for larger cuts, including a 33 percent reduction for anthropology and a 61 percent reduction for cognitive and behavioral science. Yet neuroscience, the division's only other program, was targeted for an 8.2 percent increase, making it crystal clear that the administration had singled out the social and behavioral sciences for aggressive cuts.[12]

With MACOS still a target of conservative wrath, the new administration also took aim at NSF programs for science education, by proposing a gigantic reduction in funding from $112 million to $10 million. At a conservative gathering in March 1981, Reagan himself called for an end to "the manipulation of schoolchildren by utopian planners."[13]

Shortly before the announcement of the proposed budget cuts, a *New York Times* article reported that the Reagan administration was taking aim at the "soft" sciences.[14] Now that the precise magnitude of the cuts was known, an article in *Science* reckoned that "dark days for social research" had begun.[15]

To appreciate the importance of the proposed cuts, we need to remember that the NSF still provided a substantial share of federal funding for basic research in the social and behavioral sciences, much of which continued to be carried out in the nation's universities and colleges. According to a 1981 NSF document, the percentage of total federal research support provided by the agency for the established disciplines remained significant: 25 percent for psychology, 36 percent for sociology, 37 percent for economics, 91 percent for political science, and 95 percent for anthropology. NSF support also accounted for 49 percent of federal funding for "other" social research, a nondescript category for studies outside of the established disciplines.[16]

Not surprisingly, many experienced social scientists and national science administrators found the proposed cuts startling. "Reductions of comparable magnitude in federal funding for specific fields of science were unprecedented in the history of the relationship between the federal government and the academic community," claimed the psychologist and former high-level NSF administrator John T. Wilson.[17] The University of Chicago political scientist and SSRC president Kenneth Prewitt opined that the Reagan administration's actions threatened to "make the social sciences an even smaller corner at the NSF," perhaps "just a few desks."[18] Although the president of the American Association for the Advancement of Science D. Allan Bromley was not a social scientist but a nuclear physicist, he too observed that the anticipated "brutal reductions suffered by the NSF social science programs" had produced a "shock wave of uncertainty and confusion" in the relevant scholarly communities.[19]

The administration's stated rationale for the cuts concerned the elimination of nonessential federal expenditures. On the surface, this seemed straightforward enough and fell in line with the president's overall economic message: in his inaugural address, Reagan promised to confront "an economic affliction of great proportions."[20] With this in mind, the Office of Management and Budget (OMB) explained that social science funding had "relatively lesser importance to the economy than the support of the natural sciences."[21]

In the early 1980s, newspapers of various types kept the matter before the American people, often accompanied by sarcasm and sensationalism that readers would have been familiar with from Senator Proxmire's Golden Fleece Awards and the jibes of conservative journalists such as Donald Lambro. A 1981 article in the respectable, business-oriented *Wall Street Journal* noted that it was unclear whether funding for social science produced anything of much value. Inspired by blistering attacks on "eggheads who get lucrative contracts for fuzzy theoretical studies and then write jargon-laden reports that few people read and fewer comprehend," the nation's new president was thus "clipping the wings" of the social sciences.[22] The less-respectable but still widely popular *National Enquirer* jumped on the bandwagon with a 1982 article "They're Really Monkeying Around with Your Taxes," which stated that "ape-brained federal bureaucrats have gone bananas with your tax dollars again—flushing more than $70,000 down the drain on a study of a monkey's sex life!" That year, another article, "Playing Possum with Your Taxes," proposed that instead of using federal money to support scholars "studying opossum brains," this money "could be channeled in some other, useful direction—like trying to find out if the bureaucrats who authorized this inane study have any brains at all!" In a letter to the NSF, a self-identified "irate octogenarian taxpayer" from Shreveport, Louisiana, asked how he could apply for a $50,000 grant to study the reasons why the agency "dissipated taxpayers' money in such lackadaisical ways," adding that the results would "benefit President Reagan."[23]

At one point, the electrical engineer and NSF director John Slaughter indicated that he too accepted the administration's cost-cutting rationale. Speaking before a congressional appropriations committee in 1981, Slaughter observed with dismay that the OMB had not asked NSF leaders for their views before making its budgetary recommendations. Nor had the OMB bothered to ask anyone at the agency if the cuts should be distributed in another manner, rather than targeting a particular area such as the social sciences. Despite this irritating failure to consult, Slaughter stated that in light of the "severe financial crisis," NSF leaders understood that the proposed cuts were "not motivated by either a lack of understanding of the importance of these areas or by any plan for anything more than a temporary reduction in these areas."[24]

Yet many others deemed the cost-cutting rationale fishy. After all, given the enormous size of the federal budget and the administration's promise to cut all excess expenses, the anticipated savings in this particular case were, as the journalist Hilts noted, miniscule. In the months and years to come, this point, together with recent criticisms of social science from conservative quarters, would lead many to conclude that the administration's real concerns emanated from partisan ideology, not neutral cost-cutting imperatives.

Efforts to identify the individuals most responsible for the proposed cuts reinforced that conclusion. Presumably, Reagan himself was not involved in determining the specific level of social science funding. So, who was?

According to the sociologist Otto Larsen, "after much detective work," national social science leaders had identified the libertarian economist and prominent Republican policy adviser Martin Anderson "as the person ... most responsible for pinpointing the severe cuts."[25] An industrial engineer and management expert by training, Anderson taught finance at Columbia University's business school during the 1960s, worked on Nixon's 1968 presidential campaign, and obtained an appointment, starting in 1971, as a senior fellow at the Hoover Institution. Subsequently, as an assistant to President Reagan from 1981 to 1982 and as a member of the President's Economic Policy Advisory Board from 1982 to 1989, Anderson helped shape economic and national security policies. Anderson was also a prolific author of works attacking big government. His first book, *The Federal Bulldozer*, published in 1964, provided a blistering critique of government-supported urban renewal projects whose "admirable goals," Anderson wrote, could really only be accomplished by "free enterprise."[26] In addition, Anderson approached social research from a perspective with substantial appeal in conservative circles. For example, he advocated the use of evaluation research to measure the costs and benefits of government programs, with the expectation that the results would be used not to expand programs judged to be working well but to scale back or eliminate programs judged wasteful.[27]

Although circumstantial evidence made Anderson a likely suspect, the weight of evidence indicates that it was David Stockman who played the key role. Raised in a midwestern Republican family, Stockman developed a taste for leftist thinking and protest activities during his student days at Michigan State University in the late 1960s. He soon embraced a more conservative outlook, however, as he came under the influence of scholars known for their critiques of liberal social engineering, including Nathan Glazer and Daniel Moynihan; at one point, Stockman worked as a live-in babysitter for the Moynihan family. From 1977 to 1981, Stockman served in the House of Representatives. Toward the end of that period, he prepared a model counterbudget to the Carter White House's plans that would have eliminated all NSF social science funding, an idea suggested by the Ohio Republican Representative John Ashbrook as well. Stockman also became a major advocate of supply-side economic policy and an adviser to Ronald Reagan.[28] President Reagan appointed Stockman to the powerful position of OMB director. According to William Wells, a science policy analyst at the American Association for the Advancement of Science, "Stockman issued detailed, specific 'guidance' on reducing or eliminating social and behavioral research in the NSF and other agencies."[29]

Years later, John Slaughter recalled how he learned of Stockman's role in this episode. After becoming NSF director in early 1981, Slaughter and his deputy director, Don Langenberg, walked over to the OMB. There they met with Hugh Loweth, the person who handled the NSF budget and someone with whom Slaughter had worked well before and even considered a good friend. Loweth explained to them that Stockman said the NSF would have to close its programs for science education and for the social and behavioral sciences. Hoping to control the damage, Slaughter asked if it would be possible, instead, to reduce budgets elsewhere within the agency, thereby leaving some money to keep these programs alive. Looking Slaughter straight in the eye, Loweth replied, "John, any effort on your part to do that will be unquestionably denied."[30]

Known in some quarters as the grim reaper, Stockman, who remained Reagan's OMB director until 1985, defended the proposed funding cuts by appealing to standard conservative positions. "Overreliance on the pet theories of econometricians, educators, and social science 'fixers'" had "created the vast gulf between federal spending and resultant social benefit," claimed Stockman. The Reagan administration aimed to close that gap. "Given present fiscal realities," Stockman continued, "such research is a very low priority, and funding should be cut back drastically in the short term."[31] As can be seen in figure 7.1, a cartoon published in the *Washington Star* presented Stockman's plans with dark humor.

Given the prevalence of such concerns about social science fixers, other people deduced that the proposed cuts represented a veiled ideological attack on the social sciences. The biochemist and National Academy of Sciences (NAS) president Philip Handler suspected that the large reductions were "dictated not so much by financial constraints as by social philosophy."[32] His statement was supported by a NAS resolution expressing "deep concern

'Do you have a study there showing that nine out of ten social scientists are dispensable?'

Figure 7.1
David Stockman, looking like the Grim Reaper, scrutinizes a social scientist who is hopelessly tangled up in his studies, while a chopping block nearby awaits its next victim. The cartoon caption reads, "Do you have a study there showing that nine out of ten social scientists are dispensable?" Cartoon appeared in the *Washington Star*, June 30, 1981.

over the proposed severe reductions in federal support for basic research in the behavioral and social sciences."[33] The anthropologist Robert McCormick Adams pointed out that conservatives harbored a "political vendetta" against these sciences because of their contributions to Lyndon Johnson's Great Society.[34] Roberta Balstad Miller, a historian with extensive knowledge of federal funding policies, added that the proposed cuts were inspired by the identification of the social sciences with liberal social programs, lifestyles, and ideas.[35]

Concerned scholars also noted that the proposed cuts represented a potent challenge to the social sciences' scientific identity. Over the decades, the NSF had been "symbolically important" in this regard, Miller emphasized. Continuous NSF support had "provided institutional acknowledgment" of the social sciences' "scientific nature." The proposed cuts thus challenged the notion that they belonged together with the natural sciences in a unified scientific enterprise.[36] The British sociologist Martin Bulmer, whose writings documented the enormous importance of Rockefeller philanthropy in advancing the social sciences during the early twentieth century, understood that the Reagan administration's effort to "emasculate" the NSF's budget amounted to a "direct attack upon the [scientific] standing, credentials and funding of the social sciences."[37] Skepticism about their scientific bona fides and their place within the federal science establishment threatened to haunt these sciences for the foreseeable future.

Social science leaders, their supporters, and NSF officials also questioned whether the proposed cuts made much sense if the real goal was to revive the economy. After noting that social science was "scorned by natural scientists as 'soft'" and "lampooned by yahoos in Congress," a *New York Times* letter to the editor challenged the notion that it was primarily natural science research that led to greater economic advances. Better "understanding of how an economy works" also improved economic performance, while behavioral scientists contributed "to efficiency and safety in many businesses," asserted the anonymous author.[38] But even when commentators challenged the administration's cost-cutting rationale, their remarks, as seen in this case, often reinforced the salience of economic considerations in research funding policy. These terms of discourse thus helped to legitimate the new politics of science and its relevance to NSF specifically.

Some wondered as well how cutting the agency's social science budget, which provided 37 percent of the government's total support for basic research in economics at the time, could seem reasonable. Similar to the anonymous author above, U.S. Representative Robert Shamansky, a middle-of-the-road Democrat from Ohio, identified a troubling contradiction within the new administration. President Reagan regularly claimed that professional economists supported "his economic message." And policies based on that message would surely have an enormous impact on "the whole country, if not the whole world." Yet the president also planned to cut funding for these "very fine economists because they represent," as the Ohio Democrat put it, "a soft science."[39]

Though he accepted the need to eliminate unnecessary federal expenses, Director Slaughter underscored the value of maintaining adequate support for economics as well, a point of the utmost importance in light of the White House's questionable embrace of the new-fangled "supply-side" economics. The heart of the supply-side approach was the Laffer curve, named after the Chicago economics professor Arthur Laffer. According to the basic idea, often illustrated visually with a curve on a graph, "high marginal tax rates, at some [critical] point, destroy incentives to work, save and invest, and thereby cost the Government more revenue than they generate." Although in the abstract, this idea made intuitive sense, its actual relevance remained uncertain. In congressional testimony, Slaughter explained that "the curve, as it applies to the U.S. economy," had "not yet undergone extensive empirical tests." Without such a "sound scientific basis," it was impossible to say whether the nation's economy fell within "the prohibitive range of the curve (as Laffer contends) where tax rates stifle work activity so that total tax revenues fall" or whether the economy fell below "that negative segment of the curve." So, without additional research, debate over proposals for substantial changes in taxation and spending would continue "without an adequate factual or analytic base."[40]

Further elaboration came from leading economists, including Zvi Griliches, Lawrence Klein, and Robert Lucas, all of whom had received extensive NSF funding. Hoping to educate legislators and the public about the agency's great importance, each of them invoked longstanding claims about its special role in the national funding system and its support for cutting-edge work in their discipline.

Griliches had been a member of the NAS since 1975, the chairman of Harvard's economics department from 1980 to 1983, and a world leader in mathematical and statistical analysis in economics. Through his work on committees of the American Economic Association and the NAS as well as his collaboration with the group that produced the 1976 Simon Report, Griliches had ample knowledge of federal funding programs, including the NSF, which as of 1979 had given him fifteen grants. When called to testify in 1981, Griliches explained that the agency had become "the major source of financial support of fundamental—both basic and applied—disinterested research in economics." The administration's proposed cuts thus threatened to "cripple economic research for years to come." More generally, the Harvard scholar attacked "the proposed differential deep cuts" in NSF funding for economics and the social sciences more generally as "both unjust and unwise."[41]

That threat, Griliches emphasized, seemed particularly grave in light of broader changes in the extra-university funding landscape. During an earlier era, large philanthropic patrons, including the Rockefeller and Ford foundations, had played leading roles in supporting academically oriented, basic studies. But more recently, these large foundations had abandoned any general commitment to support scholarly resources per se. Meanwhile, the interests of nearly all federal agencies that provided social science funding were "very topic specific, directed to today's fashion, and with many strings attached." As an example, Griliches pointed to the Social Security Administration, located in the Executive Branch. Could one really believe that this agency would fund research that seemed likely to reveal that the Reagan administration's policies had a "negative effect on the U.S. saving rate"? Other funding sources had limitations as well, added Griliches, especially "corporate sources or specific foundations" that tended to support research favorable to certain viewpoints.[42]

Griliches then came to the punchline: when seen as part of the overall national funding landscape, "the NSF is one of the very few sources of research money whose allocation is made almost entirely on the basis of scientific merit." Such support had already been crucial in advancing many important fields, including one of his specialties, econometrics. The "detailed statistical and economic analysis of large data sets" simply could not "be done on a shoe-string, after hours, on a pocket calculator." A sharp decrease in NSF funding for such work could be—and he believed would be—devastating.[43]

Lawrence Klein, another accomplished economist, extended the argument by discussing NSF contributions to the nation's economic health and global leadership. Awarded the 1980 Nobel Prize in Economics for his work on econometric models, which had received extensive NSF funding since the early 1960s, Klein was a University of Pennsylvania professor and past president of the American Economic Association. Like fellow econometrician Griliches, he was also a NAS member. Testifying before Congress, Klein claimed that in economic science, the U.S. was "preeminent," playing a "leading role" in "economic thinking" around the world. Furthermore, the NSF had contributed to this exalted position by providing the discipline with valuable support. Unfortunately, however, the level of federal funding, including NSF funding, had deteriorated in recent years.[44]

In addition, regarding the worrisome state of the economy, Klein placed the blame partly on the federal government's recent neglect of economic science: "some significant part of our productivity slowdown and general loss of competitiveness" could be traced "to the disappointing program of federal research support of the 1970's." If the Reagan administration insisted on additional cuts, then "the productivity of economists" would deteriorate further during a time of many “challenging problems.” Klein asked, “How can we finance social security?” “How can we deal with stagflation?” “What are the causes of the productivity slowdown?” “How can we stabilize the exchange value of the dollar?” “What should be the shape of the future world monetary system?” “How can we best preserve our exhaustible resources?” Although the NSF had funded valuable research on these very problems, the new administration’s indiscriminate, reckless attack on wasteful federal spending placed this crucial source of support in jeopardy.[45]

Robert Lucas, a third star in economics, inveighed against the proposed cuts as well by deploying the catchy rhetoric of capital accumulation. As a graduate student at Chicago in the early 1960s, Lucas had been profoundly influenced by Milton Friedman’s approach to economics and his libertarian social philosophy. Later, while Lucas was a professor at the Carnegie Institute of Technology and then the University of Chicago, he worked on the rational expectations hypothesis—that is, the notion that people make economic decisions based on a rational utilitarian calculus that includes their expectations about future events, such as the course of inflation, interest rates, and government policies. The extension of this hypothesis in macroeconomics, which included work by many other NSF-funded scholars, provided a severe challenge to classical Keynesianism and the notion that government could manipulate and control the economy in desirable ways. As with Griliches and Klein, Lucas, who would receive the Nobel Prize in Economics in 1995, benefited from NSF funding over many years, starting with his first grant back in 1964. Of special interest here, his NSF-funded work provided support for the Reagan administration’s supply-side economic policy. Nevertheless, the administration’s proposed cuts now threatened continued funding of such work.[46]

Writing in the *New York Times*, Lucas challenged the cuts at a more general level as well, by advancing an imperative to provide “incentives for ideas” and adequate resources to promote the accumulation of knowledge. Making clever use of concepts and rhetoric common within the Reagan administration, he declared that the government was about to commit a “serious policy blunder on an issue involving the incentive to accumulate a centrally important kind of capital,” namely, capital in social science knowledge. Growth in this area, Lucas explained, depended on the pursuit of basic studies and the availability of ample funding, needed to entice first-class researchers to engage in this pursuit. The results of such work could have tremendously valuable practical applications.[47]

In short, the new Republican administration's proposed cuts had set off alarm bells. While the White House presented funding reductions as motivated by the goal of reigning in a massively bloated federal budget, other pieces of evidence suggested that the more fundamental goal had a partisan undercurrent and involved slashing support for the social sciences in particular. With the NSF budget lying at the eye of storm, a number of supporters rose to its defense, including the economists Griliches, Klein, and Lucas, who argued that NSF funding for high-quality and nonpartisan scholarly studies had tremendous value. Moreover, it seemed very unlikely that any other public or private patron could take over the agency's special role.

With so many individuals from the political, academic, and science policy arenas expressing their opinions forcefully, the case against the proposed funding cuts at the NSF became clearer. It was also clear that mounting an effective case against the cuts (and addressing associated threats to the scientific identity of the social sciences and their standing in the scientific community) would require an effort that was better organized and sustained. But were the social sciences themselves up to this task? Arguably, the historical record since World War Two did not provide much reason to be hopeful.

COSSA TO THE RESCUE

On June 4, 1981, just four months after the proposed cuts were announced, SSRC president Kenneth Prewitt chaired a symposium on "strategies for the social sciences." Participants at this event, held at Rockefeller University in New York City, included officers from the NSF and the National Institutes of Health, representatives from NAS's Assembly of Behavioral and Social Sciences, and other figures from the social sciences and private funding organizations. According to Prewitt's summary, the discussions focused on "the importance of demonstrating that social science is an integral part of the national science system" and the need to strengthen "cooperation with natural scientists." Social science leaders also agreed that the future of federal funding had become crucial, especially since funding policies at private foundations emphasized programmatic goals, a point mentioned by the economist Griliches and confirmed at the New York meeting.[48]

Figure 7.2
Kenneth Prewitt, former U.S. Census Bureau director (1998–2001) and former SSRC president (1979–1985), receives the SAGE-CASBS Award, in recognition of his important contributions to the understanding and advancement of the social and behavioral sciences as applied to important social issues, November 6, 2015. Left to right: Margaret Levi, Center for Advanced Study in the Behavioral Sciences executive director; Kenneth Prewitt; and Sara Miller McCune, cofounder of SAGE Publishing. Courtesy of Center for Advanced Study in the Behavioral Sciences.

But containing the impending damage seemed difficult, partly because social scientists appeared poorly equipped for the task. Later that year, in his SSRC president's report, Prewitt observed that these scholars, by and large, were not well organized or very persuasive when it came to promoting their work in the political arena. He called them "politically naïve," for they had not realized "the need to be a political presence in Washington," a startling failure in light of the more vigorous public relations efforts undertaken by engineers and natural scientists. Prewitt also accused social scientists of being "indifferent toward their own intellectual and practical accomplishments and correspondingly timid about telling their own story."[49]

Commentaries along these lines proliferated, for example, in an essay on "politics and the uses of social science research" by the political scientist Donna Shalala. In the late 1970s, Shalala served as assistant secretary for policy development and research in the Department of Housing and Urban

Development, then became president of Hunter College in New York. Regarding the beleaguered state of the social sciences in the early 1980s, Shalala said she found it hard to think of "a more politically disorganized—and so, politically impotent—group."[50]

At the same time, such worries provoked social scientists to organize themselves on an unprecedented scale. A number of advocacy organizations formed or became more active around this time, including the Consortium of Professional Associations in Federal Statistics and the Federation of Behavioral, Psychological and Cognitive Sciences. But the largest and most important one for the social sciences was the Consortium of Social Science Associations (COSSA).

COSSA's origins lay in the early 1970s, when executive secretaries from professional social science associations began meeting to discuss common interests and concerns. In early 1981, after learning about the proposed cuts, COSSA convened an emergency meeting. Subsequently, the historian Roberta B. Miller and Joan Buchanan, a classicist, came aboard as staff members to run the organization on a continuing basis and develop a strategy to oppose the cuts. Miller also became COSSA's first executive director, from 1981 to 1984—and subsequently held positions at the NSF, as will be discussed. As a scholar, she published on urban and regional development, science indicators, and the history of social science policy. She taught at Catholic University, the University of Minnesota, Oberlin College, and Hiram College. In addition, she worked at SSRC's Washington office on social indicators from 1976 to 1981, giving her a deep understanding of the politics of funding and using social science in the nation's capital.[51]

Several changes elevated the consortium's stature and made it a leading national advocate for the social and behavioral sciences. In 1982, COSSA was incorporated and became a permanent, nonprofit advocacy organization. After incorporation, its staff members registered as lobbyists with Congress. COSSA headquarters moved from temporary housing in Washington, D.C., provided by the American Psychological Association to a more permanent space in an annex of the Brookings Institution provided by the SSRC. COSSA now had a president, a twenty-four-member board of directors, a ten-member executive committee, and a three-tier membership structure: with ten primary member associations, including all of the major social science associations; twenty-two affiliate organizations, including specialty social science associations; and twenty-one contributors, including research

Figure 7.3
Roberta B. Miller participating in a COSSA-sponsored international outreach event held in Paris, 1992. Left to right: SSRC president David Featherman; former COSSA executive director and former director of the NSF Division of Social and Economic Sciences Roberta B. Miller; Center for Advanced Study in the Behavioral Sciences director Philip E. Converse; and COSSA executive director Howard Silver. Courtesy of COSSA.

universities and organizations. Henceforth, COSSA represented a huge number of social scientists, about 140,000 in the early 1980s and 185,000 a decade later, by which point the original ten-member associations had been joined by another thirty-two affiliate organizations and fifty-seven contributors.[52]

COSSA's mission was to establish a "unifying voice for the social sciences in relation to legislation, public education, and science policy." To do that, COSSA worked on various fronts: promoting and protecting federal social research funding programs, increasing the visibility and use of the social sciences in policy making, mobilizing members to take action when needed, and keeping them informed of federal actions that had an impact on researchers and research. Through various activities and annual meetings COSSA facilitated scholarly "communication about such issues across disciplinary lines."[53] Publication of a biweekly newsletter, the *Washington*

Update, provided interested parties with more timely information. Although the SSRC engaged in such activities before, its effectiveness in the science policy arena had been sporadic and limited, a fact made painfully clear during the postwar NSF debate four decades earlier. COSSA thus set out to do what no other broad-based social science organization in the postwar era had managed to do—or even seriously attempted to do on a sustained basis.

Functioning as a political lobby, COSSA made defending the NSF's budget a high priority, for many reasons. Not only did the agency support research in all of the disciplines and fields represented by COSSA, but the proposed cuts here were the deepest of any federal agency. In addition, NSF programs were widely considered to be, as Roberta Miller put it, the "flagship of the social science enterprise." So, if those cuts were implemented, funding at other agencies would probably be more vulnerable as well. In preparation for the challenges ahead, COSSA cultivated relations with several congressional committees that dealt with social science appropriations and arranged for scholars to present testimony at congressional hearings.[54]

During the summer of 1981, COSSA focused mainly on the so-called Winn amendment, put forth by Kansas Republican Representative Edward Lawrence Winn Jr. during NSF authorization discussions. Back in the late 1960s, Winn had been a member of Emilio Daddario's science and research subcommittee—which put forth and promoted the successful 1968 Daddario amendment. Now Winn served on the House Science and Technology Committee. As the early months of 1981 passed, the White House, the NSF, and congressional committees in both legislative branches engaged in the labyrinth process of budget making. At one point, it looked like social science funding cuts might not be as deep as the White House had proposed. But Winn, seeking to restore the full weight of the cuts, called for a $70 million reduction in NSF social science funding, accompanied by the explanation that his amendment embodied "the spirit of a responsible federal budget."[55]

In confronting this threat, COSSA aimed, in Roberta Miller's words, to "depoliticize the issue." Staking out the nonpartisan high ground, the consortium concentrated on persuading recently elected members in the House of Representatives—since new members might not yet have a firm position on this issue—and also all other legislators who seemed undecided. COSSA also targeted legislators who had major research universities in their congressional districts. To reach this last group, COSSA encouraged social scientists to reach out to their political representatives. COSSA also asked legislators who opposed the Winn amendment to write "Dear Colleague" letters signed by both Democrats and Republicans. To further strengthen the bipartisan message, COSSA arranged for supportive Republican and Democratic legislators to stand at the voting room entrance door so that they could encourage colleagues heading inside to vote against the amendment.[56]

It would have been impossible to eliminate all opposition. During the hearings, Republicans Edwin Forsythe from New Jersey and Walter Johnston from North Carolina expressed support for Winn's initiative. Citing the right-wing journalist Donald Lambro's 1980 book *Fat City*, Johnston claimed that Americans were "sick and tired of letting some intellectual dilettante dispense money to some intellectual ineffectual who wants to maintain his sojourn in academia for another year at their expense." In an approving manner, Johnston added that a memorandum circulating in the government called for the total elimination of "all basic social science research" within the next several months—perhaps a reference to the alternative budget prepared by David Stockman in the late 1970s that recommended an end to NSF funding for social research.[57]

But many more legislators spoke out against the cuts, including James Leach, who, moreover, was not a Democrat but a centrist Republican from Iowa. Leach had done graduate studies in international relations at Johns Hopkins University and attended the London School of Economics, educational experiences that one suspects inspired his appreciation for modern social research and its policy uses. Echoing the worries voiced by others, Leach warned that "to eliminate virtually all remaining NSF support for the behavioral and social sciences carries grave consequences." "The health of our citizens, our commercial vitality, and our national security" would be compromised. A "profound imbalance in scientific inquiry in this country" would be created. In addition, Leach reminded his fellow legislators that the "politicization of scientific inquiry" in the Soviet Union had "stunted economic and social progress" and "led to the stultification of free thought and free inquiry." Congress should thus ensure that "scientific merit alone rather than religious and social philosophy remains the basis for research funding decisions." Furthermore, Leach found it unrealistic to think that "profit-making enterprises" could replace federal funding for "most research at the 'basic' end of the spectrum"—precisely where the NSF's vital importance lay.[58]

Joining the centrist Republican Leach was the Michigan Democratic Representative Jerome Traxler, who declared that the Reagan administration was seeking "drastic, ideological, and completely disproportionate cuts" in the social sciences. The administration's proposal to battle run-away federal expenditures by reducing the NSF's already modest social science funding seemed absurd, like a promise to slim down "an overweight giant by removing his eyes, ears, and brain."[59]

In the end, COSSA's campaign paid off nicely. The Winn amendment was defeated decisively, with 264 votes against, 152 for, and 17 abstentions.[60]

This victory also revealed modest success in the effort to establish bipartisan support for social science funding. Legislators who voted against the amendment included an unusually large number of Republicans, sixty-nine to be exact. This included more than one-third of all Republican Representatives in the House, which was the largest Republican opposition to any White House budgetary proposal during Reagan's first year in office.[61] One telling case involved New York Republican Representative John LeBoutillier. Although he had completed his college degree at Harvard in 1976 and his MBA at Harvard in 1979, LeBoutillier was hardly a cheerleader for the country's oldest and most famous university, as seen in his scathing anti-liberal, anti-elitist 1978 book entitled *Harvard Hates America: The Odyssey of a Born Again American*. Yet in 1981, even he claimed that the proposed social science funding cuts "went too far."[62] In another case, Massachusetts Republican Representative Margaret Heckler promised that she would vote against the Winn amendment if COSSA would "tell them [the advocates of social science funding] to stop calling me."[63]

Placed in historical perspective, this episode marked a big advance in the efforts by scholars and politicians to address persistent challenges regarding effective organization and public relations in American social science. COSSA's response to the 1981 budgetary crisis became, in the words of the sociologist James McCartney, a "watershed in the politics of the social sciences." The newly energized and better-organized community of schoars, professional associations, and academic organizations had "moved from a condition of political naivete … to one of surprising effectiveness."[64] Consider the case of psychology. According to one report, mobilization against the proposed cuts had resulted in "thousands of letters, calls, and telegrams." Individual psychologists had made more than fifty visits to their representatives. Communication with congressional staff members and analysis of the voting record revealed that of the seventy-one representatives singled out for attention, forty-six of them (65 percent) ended up opposing Winn's amendment.[65]

In a letter to COSSA's Roberta Miller, the chairman of the House Science, Research and Technology Subcommittee Doug Walgren expressed his "thanks and admiration." In his view, COSSA played "an invaluable role" by helping Congress understand various issues regarding the "proposed deep cuts" and by mobilizing the social science community "to reach directly to their own local representatives," which, Walgren added, had been "enormously successful." Obviously impressed, the subcommittee chairman said he would be "citing" COSSA efforts "as a model for participation in the budget process by other scientific groups."[66]

Yet the Winn amendment's defeat was not the end of this story. When the discussions, debates, and voting finally finished, the amount set aside for NSF social and behavioral sciences in 1982 was only $32.6 million, a notable drop from the 1981 budget of $43.7 million. And this latter figure was already considerably lower than the previous year's $52.5 million.[67] Moreover, social scientists still faced major hurdles in defending themselves and their work within American political culture and the national science policy arena.

To address these problems, COSSA pursued public relations, lobbying, and educational activities, directed at politicians, federal agencies, and the science policy community. For example, the consortium sponsored breakfast and luncheon seminars with legislators and their staffs to educate them about the nature of the social sciences and their contributions to society. These events focused on such timely issues as "Innovation and Productivity: A Human Perspective" and "Black Youth Unemployment: A National Crisis." Some of these events naturally took place in the D.C. area. But many others took place throughout the country, in hundreds of congressional districts.[68] COSSA also invited high-ranking members of the federal science policy community to luncheons, where they learned about the consortium, its activities, and its goals.[69]

In addition, COSSA leaders recognized that social scientists themselves needed knowledge of political realities on a continuing basis. To help scholars who, as Roberta Miller put it, were generally "not adept in politics on their own behalf," COSSA produced instructional materials about how to do basic things such as calling a member of Congress, visiting a legislator, and writing a press release. To keep members informed and in communication with one another, COSSA published the *Washington Update* and held countless seminars at universities and scholarly meetings. And at a time when Democrats controlled the House of Representatives but Republicans controlled both the Senate and the White House, COSSA regularly reminded social scientists to cultivate bipartisan support for their work.[70]

THE UNITY OF SCIENCE, ONCE AGAIN

Although COSSA's transformation could not have been anticipated before the 1981 budget crisis, a second line of response by social scientists was predictable. As noted by SSRC president Kenneth Prewitt, during the New York strategy symposium held in June, the "importance of demonstrating" that "social science is an integral part of the national science system" emerged as a central concern. The difficulties ahead were serious, because national science leaders had often "failed to view the social sciences as part of the scientific resources of the society" and thus "assumed no responsibility for protecting and enhancing them."[71] Hence, the funding crisis prompted social scientists and those who represented them at the NSF to focus, once again, on seeking support from natural scientists and establishing their position within a unified scientific enterprise.

This focus on scientific unity came across clearly during a symposium discussion about the 1960s' effort to create a National Social Science Foundation (NSSF). The general understanding among leaders from the social sciences, federal agencies, and private foundations was that Senator Fred Harris's initiative had been set aside "primarily on the grounds that there is a single standard of competent scientific inquiry and that [having] separate federal foundations ... would contradict this conception of the unity of the sciences." However, as we know from the discussion of his initiative in chapter 4, that general understanding failed to acknowledge the complexity of the NSSF debate, especially the extent to which social scientists disagreed among themselves over several questions. These concerned the similarities and differences between the natural and social sciences, the advantages and disadvantages of trying to fit social research into a natural science mold, and the ways in which giving the social sciences an agency of their own might bolster or undermine their influence within the federal science policy arena and wider society. Nevertheless, such a blinkered perspective of the late 1960s' debate reinforced the dominant sentiment that social scientists needed to redouble their efforts to strengthen "cooperation with natural scientists."[72]

Social scientists reasserted the unity-of-science stance for the benefit of NSF leaders as well. At a board meeting in April 1981—two months before the New York symposium—a group of scholars tried to reassure the natural science–oriented leaders that the agency had historically followed a wise path, by concentrating on the hard-core end of the social research continuum. Philip Converse, director of the University of Michigan's Institute for Social Research, declared that present-day social science theory was "harder, clearer and less ambiguous" due to the advance of "more quantitative efforts." Herbert Simon declared, as he had done many times before, that he favored close ties between the physical sciences and the social sciences, among other reasons because he believed the former had contributed admirably to the research directions of the latter.[73] On a complementary note, a June document prepared for the governing board asserted that "all sciences, including the social and behavioral, share a commitment to norms regarding evidence, verification, inference, rigor, etc."[74] In this view, the criteria for deciding what counted as good science were the same in all of the sciences.

In presenting their case to NSF leaders, social scientists also took care, once again, to separate their work from social action and social reform, an issue that had acquired special urgency due to mounting conservative criticisms and interest in defunding the left.[75] The June document mentioned above explained that the social sciences are "neither therapies, nor social engineering, nor action programs." In a passage reminiscent of numerous previous statements designed to draw a sharp boundary of this sort, the document declared that "the social and behavioral sciences are not social studies, social work, social action, social reform, or socialism. They are sciences. As such, they produce conditional 'if-then' propositions, not normative 'though-shalt' statements."[76]

The NSF itself, in turn, reasserted its commitment to this approach. That same June, the agency's governing board approved a statement that emphasized its special role in providing "the major support for all social and behavioral sciences where the focus is enhancing the objectivity of the sciences and improving the quality of data collection and analysis." The agency thus continued to concentrate on the advance of useful techniques and tools needed for measurement and analysis of social phenomena, though also anticipating that the results would have significant practical applications. The latter included helping federal agencies to improve "the quality and usage of national statistical information" and strengthening activities in the private sector that used "economic forecasting, demographic projections, survey research, cost benefit analysis, marketing analysis, and personnel selection and training." In September 1981, this NSB-approved statement appeared in *Science* for all interested parties to see.[77]

As had happened many times in previous decades, this effort to clarify and purify the social sciences' scientific identity included a valuable assist from the NAS. In response to the latest in a long list of requests from the NSF over the decades, the Academy published a 1982 report that reaffirmed the value of the scientistic approach. Addressing the matter head on, the introduction included the following quotation from Philip Morrison, a physicist and member of the NAS commission that produced the report: "it does not seem possible to draw any clear line between the scientist looking out at the physical or biological world and another kind of scientist concentrating instead on his or her own species." Thus, Morrison continued, "the natural sciences must be extended in the same spirit … to the study of the extraordinary qualities of the species Homo sapiens and its richly diverse works."[78] The report went on to define basic research in the familiar ways, claiming that its "primary aim" was to advance "the understanding and explanation of human behavior and social arrangements," which involved establishing "empirically verified descriptions" and discovering "laws that govern their occurrence." In short, "the behavioral and social sciences" had to be "judged by the same criteria as other sciences."[79]

As if to accentuate the point that this approach followed a well-worn path, the report's title—*Behavioral and Social Science Research: A National Resource*—bore a striking similarity to the title of the late 1940s' report by Talcott Parsons, "Social Science, A Basic National Resource." The latter, as we saw in chapter 1, had been commissioned by the SSRC during the postwar NSF debate, had followed the council's strategy of presenting the social sciences as junior partners to the natural sciences in a unified scientific enterprise, and had emphasized that the modern social sciences were maturing in healthy ways that fortified their nonideological, apolitical, and technical character.

This unity-of-science framework was also central to NSF efforts to defend its embattled social science programs before Congress. In explaining the agency's approach, Director Slaughter emphasized that he and other agency leaders saw a "great unity" in scientific research and scholarship, which rested on "intellectual adherence to the standards of objectivity and verifiability." Slaughter added that this general viewpoint, rather than serving as mere rhetoric, had a powerful impact on funding practices as seen in many examples, including NSF support for quantitative social science databases and archives.[80]

Slaughter also highlighted the agency's strong support for the "behavioral revolution" in political science. "Marked change" in this field had occurred in the past twenty years, resulting in major advances in the "scientific understanding of basic political processes," asserted Slaughter. Due to such progress, the discipline as a whole had changed dramatically: scholars now studied politics "as an outcome of human interaction processes and decisions rather than merely as a product of legal statements and formal rules." According to Slaughter, this new type of work was steeped in "rigorous modes of explanation and analysis," offering a sharp "contrast to conventional descriptive and literary approaches."[81]

Slaughter's remarks are especially interesting because of what we know about the story of NSF funding for political science in relation to the discipline's contentious development. Behavioralist studies did indeed fall in line with the agency's unity-of-science commitments, dating all the way back to the early 1960s when the first research grant in political scientist went to Duncan MacRae for a quantitative analysis of congressional voting behavior. Yet Slaughter's testimony obscured deep disputes about the intellectual foundations, political uses, and moral implications of such work that had wracked the discipline ever since. Indeed, his comments would have seemed curious to anyone knowledgeable about the discipline's fractured character and the sizable body of literature critical of the very behavioralism that he spoke so highly of. In a curious twist of fate, even MacRae no longer championed this approach. His views evolved so much that he had become a strong critic of the value-neutral, disinterested scientistic approach associated with behavioralism, as seen in his assertion that social science research should be "guided by clear notions of social health, i.e., by ethical criteria," and in his call for "valuative discussion among the disciplines."[82]

By the 1980s, a sense that the discipline had entered a postbehavioral era had become widespread, supported by extensive historical investigations carried out by political scientists themselves. In his 1984 book *The Tragedy of*

Political Science: Politics, Scholarship, and Democracy, David M. Ricci, a Harvard PhD and professor at Hebrew University in Jerusalem, documented a startling lack of common ground among his colleagues: with political scientists "increasingly unable to choose among proffered ideas," it had become impossible to "certify some of them so authoritatively as to persuade practitioners to work along common lines." The result was "vocational incoherence, literally a failure to cohere."[83] In 1985, another book by Raymond Seidelman, a political scientist at Sarah Lawrence College, drew attention to the growing multiplicity of research subfields within the discipline. Similar to Ricci, Seidelman pointed to "a widespread sense" among practitioners that their discipline had "lost its identity and a common focus."[84]

CONCLUSION

Two decades earlier, during the Kennedy and Johnson administrations, relations between the social sciences and the federal government had become tighter, federal funding had increased, and optimism that scholarly advances would lead to considerable benefits had reached their post–World War II apogee. But in the following years, growing discord and disenchantment on the left in combination with the resurgence of conservatism in American political culture and national science policy making led to mounting difficulties and eventually a widespread sense of crisis. The Reagan administration's 1981 budgetary proposal, with its draconian cuts in NSF funding, became a focal point of national concern about the place of the social and behavioral sciences within the agency. For social scientists and their advocates, the assault on NSF funding naturally led to deep worries about other issues: about the future of the uneasy partnership between the federal government and the social sciences, about the position of these sciences in American society, and about their standing within the scientific community.

If participants in the mid-1940s' NSF debate had been able to observe the crisis of the early 1980s, they would have been struck by two similarities. First, as the contours of the new politics of science became increasingly legible, social scientists once again found themselves vulnerable to partisan attacks and especially conservative hostility that associated them with a leftist agenda. This time around, the main point of reference was not so much the New Deal but more recent developments associated with the Great Society and War on Poverty, which, in David Stockman's unflattering terms, had privileged the "pet theories of econometricians, educationists, and social science 'fixers.'" No doubt the underlying motivations went well beyond the officially stated concern with budget tightening. As COSSA's Roberta Miller and many other commentators recognized, the plan to reduce federal funding was part of a broad-based conservative attack on liberalism and its influence on everything from economic policy to social programs, education, academic culture, cultural values, and morality.

The second similarity concerns the manner in which social scientists and their supporters responded. Most obviously, in both episodes they argued that federal funding and adequate support from the NSF in particular was crucial to the health of the social sciences. Furthermore, with precious few exceptions, influential figures from national organizations, including the SSRC, emphasized the need to act strategically by building alliances with natural scientists, by presenting social science as part of a unified scientific enterprise led by the natural sciences, and by drawing a firm distinction between scholarly social research, on one hand, and social reform, social engineering, or socialism, on the other.

Comparing the two episodes also reveals a striking difference, namely, the emergence in the early 1980s of well-organized and moderately effective public relations activity on behalf of the social sciences. Amidst a rapidly spreading sense of crisis, social science leaders and organizations banded together to transform COSSA, creating what Roberta Miller aptly described as "a permanent Washington-based infrastructure for political advocacy."[85] For legal reasons, individual professional social science associations could not engage in lobbying, but COSSA could and did with gusto. Tens of thousands of social and behavioral scientists now had an active lobbying organization with a mandate to convey the "virtues" of their work "to other audiences, particularly congressional and administrative ones."[86] With the defeat of the Winn amendment in 1981, COSSA's efforts to strengthen bipartisan political support and protect NSF funding yielded some tangible benefits. Yet budgetary problems persisted, as did other related challenges. In this context, COSSA followed up by developing programs and activities for social scientists, natural scientists, and politicians with the aim of strengthening the standing of social science within the national science policy arena and American political culture.

Another significant difference between the two periods concerns the prominence of economistic lines of reasoning, advanced—not surprisingly—by leading economists themselves. Although economists including Wesley Mitchell and Edwin Nourse had played a pivotal role in developing the case for inclusion during the postwar debate, that case did not give any special place to economic reasoning—though it was common to believe that investments in basic research led to practical benefits. But in response to the cuts proposed by the Reagan administration, prominent economists who had received NSF funding in the past, including Zvi Griliches, Lawrence Klein, and Robert Lucas, emerged as vigorous defenders of the agency, especially its special commitment to support basic social scientific inquiry and its singular role in looking after the overall health of the sciences, including the social sciences. In explaining the NSF's importance, Griliches, Klein, and Lucas identified support for economic science in particular as crucial to the nation's economic health and global leadership in economic thinking. Furthermore, in discussing the NSF's importance more generally, they deployed concepts marinated in economic discourse—accumulation, incentives, and productivity.

At this point, we need to consider what happened after the crisis of the early 1980s had subsided. How did new pressures, opportunities, and constraints associated with the two-term Reagan presidency and the new politics of science influence the agency's evolving engagements with the social sciences?

8

DEEP AND PERSISTENT DIFFICULTIES: COPING WITH THE NEW POLITICS OF SCIENCE THROUGHOUT THE REAGAN ERA

> [The Reagan administration exercised] great courage and wisdom ... in squashing the daylights out of NSF's social science programs.
> —George A. Keyworth II, physicist and President Reagan's chief science advisor, 1984[1]

> We face a grave peril. We face the undoing of American social science.
> —Otto Larsen, sociologist and NSF senior associate of the social and behavioral sciences, 1984[2]

According to a common view, the troubles facing the social sciences during the Reagan years were formidable at first, but they quickly dissipated. In his 1990 book on American science policy since World War II, Bruce L. R. Smith, from the Brookings Institution, stated without qualification that the "administration abandoned its opposition to the social sciences."[3] More recently, in a 2012 essay, historians of economics Tiago Mata and Tom Scheiding have claimed that in response to the administration's proposed cuts, social scientists "coalesced in a lobbying campaign that successfully reversed those plans." Mata and Scheiding further argue that controversies over funding during the 1970s and early 1980s inspired the development of a "network of advocates" who managed to keep "the peering eye of the Congress at a distance" and thus "safeguarded the interests of the disciplines."[4]

The basic idea here is that when social scientists and their supporters refuted the logic of the proposed cuts, organized effectively to defeat the Winn amendment, and strengthened their relations with politicians and natural scientists on a nonpartisan basis, the crisis of the early 1980s passed. However, here I offer a more complicated and less upbeat story. I agree that the social sciences managed to avoid the worst-case cuts in the early 1980s. I also agree that in subsequent years, they did not face such a severe threat again. At the same time, we will see that at the NSF, the social sciences continued to face deep challenges as the Reagan years unfolded.

To begin with, the new politics of science—and the difficulties it posed—remained in place throughout the decade.[5] After the journalist Paul Dickson's book on this topic appeared in 1984, subsequent studies confirmed that the core principles that he identified had staying power. In 1988, historian John A. Remington observed that the "reorientation" in national science policy promulgated by the Reagan administration saw science's value "almost exclusively" in relationship to a small number of specific issues, including "the national economy," "corporate competitiveness," and "military preparedness." Meanwhile, other ideas continued to be marginalized, included the notion that science should help the nation to address its "social and environmental problems in open and pluralistic ways" and the view that science was "a cultural institution with an integrity of its own." Similarly, in his aforementioned book, Bruce Smith observed that a "strong belief in basic research (at least in the so-called hard sciences), in more spending for defense R&D, and in steps to strengthen technology to improve the competitiveness of American industry were important parts of the administration's policy from the start."[6]

Keeping in mind the power of conservative political culture and the new politics of science, this chapter examines what happened to NSF social science throughout the Reagan era—beyond the challenges posed by the 1981 budgetary threats and the short-term responses to them examined in the previous chapter. The first section considers three key issues that made the position of the social sciences perpetually problematic: federal funding trends and NSF funding trends in particular; the status, influence, and position of these sciences in the agency's organizational structure; and views expressed by prominent natural science leaders. The second section takes a closer look at the case of economics, which fared better than the other social sciences and for telling reasons during these years. The third and last section addresses a crucial question about the changing politics of knowledge and its impact at the NSF: should we conclude that there was a tight alignment between the agency's social science efforts and the conservative agenda of the Reagan administration, or should we conclude that the nature of this alignment was only partial and limited?

As we will see, the dark clouds never really dissipated during Reagan's presidency. Under these conditions, the relative success of economics had much to do with the discipline's special relevance to the administration's economic and science policy agendas. Equally significant, however, the impact of conservative politics and ideology was limited, for the agency never became an advocate of social science inquiry with a conservative or partisan slant per se.

PERENNIAL WOES: MONEY, STATUS, RESPECT

Let's start by considering three components of the social sciences' predicament that have reappeared throughout this study. The first concerns financial challenges.

During the Reagan years, federal funding for research and development increased considerably. When measured in current dollars, the amounts rose spectacularly, from $12.2 billion in 1981 to $20.4 billion in 1989. When measured in constant 1982 dollars, the increase is considerably smaller but still impressive, from $13.2 billion to $15.9 billion.[7]

But the social and behavioral sciences did not participate in this upward trend. During Reagan's first term (1981–1985), federal support for basic and applied studies in the social sciences (not including work classified as the behavioral sciences) declined slightly in current dollars, from $497.4 million to $460 million, although the decline in constant 1982 dollars was much greater, from $539.1 million to $411.2 million. As a share of the total federal science budget, social science funding also dropped considerably, from 4.1 to 2.8 percent.[8] The share of federal funding for the social and behavioral sciences (SBS) considered together was only marginally higher and fell in a similar fashion, from 4.5 to 3.1 percent. In terms of the federal budget for basic research only, the social and behavioral sciences' share fell from 4.9 to 3.6 percent.[9]

In 1984, F. Thomas Juster, an economist who directed the University of Michigan's Institute for Social Research, pointed out that the federal social science research budget seemed unhealthy. The level of funding was slightly less in current dollars than it had been in 1980, and it was "substantially less in real [constant dollar] terms." Juster thus found no reason to believe that the Republican administration's outlook had softened. He saw "no evidence" that it had "really backed off from the view that social science research is not very important."[10]

A mid-decade assessment prepared for Congress demonstrated, in addition, that when measured in constant dollars, social science funding had declined considerably in the past twenty years. Not only was the level of federal support in the mid-1980s much lower than it had been in the late 1970s, but it was also lower than it had been back in the mid-1960s. In constant 1972 dollars, the figures were $307.8 million for 1965, $454.2 million for 1978, and $295.5 million for 1986.[11]

The case of the NSF reflected and contributed to this general pattern of decline. From 1980 to 1985, the agency's overall budget increased in current dollars from $873 million to $1,346 million. During these same years, SBS funding went from $52.5 million in 1980 to $52.1 million in 1985. But when measured in constant 1982 dollars, SBS funding fell drastically, from $63.6 million in 1980 to $46.6 million in 1985. Furthermore, the share of total NSF funding for these sciences fell drastically, from 6 to 3.9 percent.[12]

Under these conditions, William Mishler, from the political science program, remarked that a sharp and recent decline in applications to this program reflected a widespread belief that NSF meant Non-Sufficient Funds.[13] Early in the decade, the social psychology program also experienced a "dramatic drop" in applications.[14]

Consider as well the new federal program for Presidential Young Investigator Awards, which each year gave 200 scholars up to $100,000. In 1983, the NSF became responsible for administering this program. The following year, the new NSF director Erich Bloch, an engineer, explained that through this program, the agency sought to identify "the brightest young faculty, chosen solely for their research promise," with the goal of supporting and retaining them on the nation's university faculties.[15] One might wonder: what place did the social and behavioral sciences have in this handsomely financed, high-profile effort to advance the careers of outstanding young scientists?

The simple answer is none. Social research had already been excluded from the Economic Recovery Tax Act of 1981. Based on that legislation, corporations were eligible for a 25 percent tax credit for expenditures on scientific research.[16] Now, the social sciences were also excluded from the presidential young investigators program, reported the *Chronicle of Higher Education*.[17] Moreover, Director Bloch seemed supportive of this arrangement: he explained that "the emphasis" was on "engineering and the physical sciences, where shortages" were considered the "greatest."[18]

During Reagan's second term in the White House, funding challenges persisted. A 1987 congressional report observed that the administration continued to propose cuts in social research programs dealing with social policy and social services considered to be in conflict with conservative goals. Specific targets included numerous government agencies: the Health Care Financing Administration, Federal Trade Commission, Office of Research Statistics and International Policy in the Social Security Administration, Statistics Income Division of the IRS, Centers for Disease Control, National Institutes of Health, Office of Policy Development and Research in the Department of Housing and Urban Development, Employment and Training Administration in the Labor Department, Bureau of Industrial Economics, Civil Aeronautics Board, Interstate Commerce Commission, Immigration and Naturalization Service, Office of Planning and Evaluation in Health and Human Services, and the Justice Department.[19] According to the economist Robert Haveman, the administration's cuts in social spending and its "antagonism to social science research in particular" had led to "large reductions in the budgets, staff, and morale in policy and evaluation research offices throughout government."[20]

The following year, a National Academy of Sciences (NAS) report added that from 1972 to 1987, federal support for SBS research declined 25 percent in constant dollars. Meanwhile, federal support for other scientific areas increased 36 percent.[21]

Once again, we find a similar pattern at the NSF. After rising impressively during Reagan's first term, the agency's overall budget continued on a sharp upward trajectory. Measured in current dollars, it went from $1,346 million in 1985 to $1,827 million in 1989. Meanwhile, SBS funding rose from $52.1 million to $59.7 million. But this is not a story of steady recovery or even moderate improvement. Even after SBS funding reached its low point for the entire decade at $32.6 million in 1982, these sciences experienced a couple budgetary dips in the following years, including one from $52.1 million in 1985 to $49.5 million in 1986.[22] As part of a wider effort to comply with the 1986 Gramm-Rudman-Hollings balanced-budget law, the NSF sustained an overall budget cut from 1985 to 1986 as well. But that amount represented less than 1 percent of the overall budget, whereas the SBS budget fell 5 percent. Thus, from year to year these sciences could still not count on maintaining the previous year's funding level.[23] Furthermore, when measured in constant 1982 dollars, SBS funding only managed to remain nearly steady, going from $46.6 million in 1985 to $46.5 million in 1989.[24]

Against this background, Ronald Overmann, a long-time director of the NSF history and philosophy of science program, noted that the challenges involved in maintaining an adequate social science budget remained severe. The financial facts revealed that the "struggle" to increase that budget "beyond inflation" had become "increasingly difficult." During the two Reagan administrations, the natural sciences once again enjoyed significant funding increases. Nevertheless, some of their leaders were fighting for what they apparently believed was "their 'fair share' of the pie," at the expense of the social sciences.[25]

Budgetary woes also stand out when considering the entire Reagan presidency. Measured in current dollars, federal funding for social science research rose from $497.4 million in 1981 to $561.2 million in 1989. But in constant 1982 dollars, the figures dropped from $539.1 million to $437 million, or almost 19 percent.[26] The percentage of total federal science funding allocated to the social sciences declined dramatically as well, from 4.1 to 2.7 percent.[27] As indicated in figure 8.1, the percentage of the total NSF research budget allocated to the social and behavioral sciences taken together also fell markedly. In 1980, they received 6 percent. But in 1981, during Reagan's first year in the White House, their share fell to 4.6 percent. And by 1989, the end of his presidency, their share had sunk to 3.3 percent.[28]

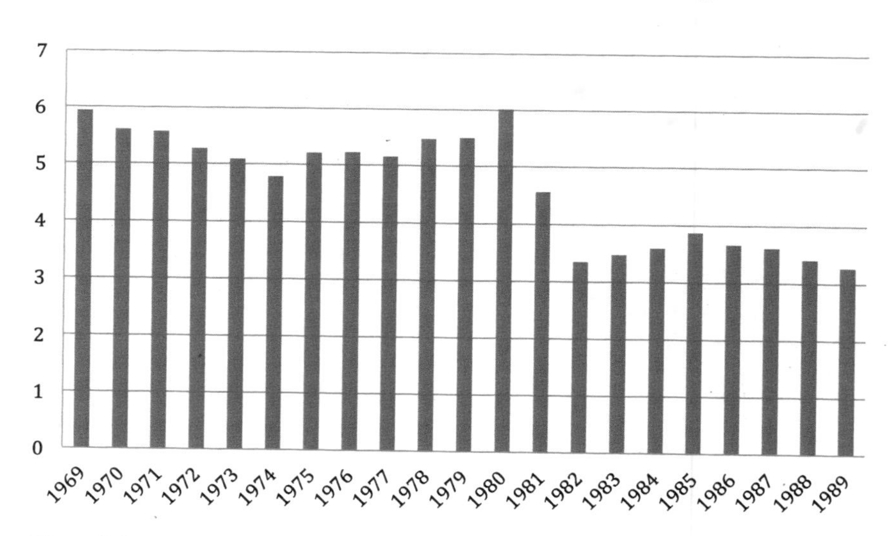

Figure 8.1
NSF social science research obligations as percentage of NSF total research obligations, 1969–1989. Data compiled from tables 5.2 and 6.1 of Otto N. Larsen, *Milestones and Millstones: Social Science at the National Science Foundation, 1945–1991* (New Brunswick, NJ: Transaction Publishers, 1992).

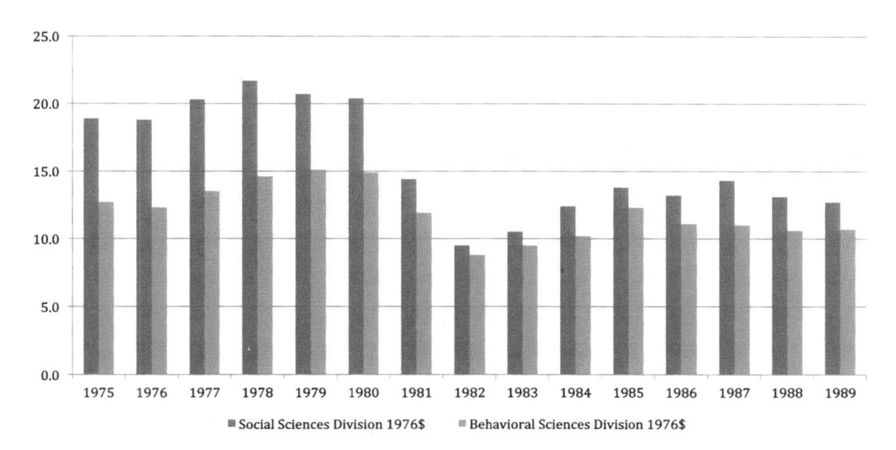

Figure 8.2
NSF research obligations for Social Sciences Division and Behavioral Sciences Division, 1975–1989 (1976 dollars, millions). Data compiled from tables 5.3 and 6.2 of Otto N. Larsen, *Milestones and Millstones: Social Science at the National Science Foundation, 1945–1991* (New Brunswick, NJ: Transaction Publishers, 1992). Current dollars converted to 1976 dollars using cpi calculator from https://data.bls.gov/cgi-bin/cpicalc.pl.

In addition, funding in constant dollars for both the social sciences and the behavioral sciences was lower in 1989 than it had been in 1981, and much lower than it had been in 1980, as can be seen in figure 8.2.

The downward funding spiral had ominous long-term implications, warned the previously mentioned 1988 NAS report. Although early in the century, large private foundations had played a key role in supporting the social sciences, government had assumed a much greater role since World War II. So much so that by the mid-1980s, total private support was merely 8 percent of the federal total. Recently, however, federal commitments themselves had declined considerably, with no replacement in sight. This, despite the fact that the social and behavioral sciences now accounted for 30 percent of all science and engineering doctorates in the U.S., and SBS scholars accounted for 22 percent of all full-time scientists and engineers. These facts made the shrinkage in federal support all the more worrisome.[29]

In a worst-case scenario, dwindling funding threatened to compromise American leadership worldwide. According to the report, a strong federal support system had helped to place American efforts on the "forefront of virtually every behavioral and social sciences field since the 1960s." But if recent trends persisted, other countries would soon assume "leading roles." Compared to the U.S., "Japan and, to a lesser extent, some European countries" were already spending "substantially larger proportions of their national resources" on these sciences.[30] To mitigate this danger, the report recommended robust increases, including additional government support of $240 million per year, $60 million of which would go to NSF SBS programs.[31]

By the decade's end, however, additional support had failed to materialize, leading a large group of scholarly organizations to band together in a harsh letter of complaint. Among the signatories were the new umbrella organizations, COSSA and the Federation of Behavioral, Psychological and Cognitive Sciences, as well as the main disciplinary associations for anthropology, economics, and so on. Their letter blamed the NSF itself for having "made little or no attempt to address the important initiatives highlighted in the NAS report." "Rather than making social and behavioral science a higher priority," NSF plans would contribute "to the deepening void between the resources necessary and the resources provided to make critical advances."[32]

★★★

If funding challenges were one persistent problem, organizational status was another.

While struggling to recover from the funding crisis of the early 1980s, the social sciences acquired a new position at the NSF, created to help represent their interests inside the agency and within the broader science policy arena. In January 1983, the sociologist Otto Larsen became the NSF's first "senior associate of the social and behavioral sciences." In this position, he served as a special assistant to the Directorate of Biological Behavioral and Social Sciences's (DBBSS) assistant director.[33]

Larsen's background provided him with appropriate qualifications. After serving in the Army Air Corps during World War II, he completed a doctorate in sociology at the University of Washington in 1955. During his graduate studies, he came under the influence of George Lundberg, the high priest of scientistic sociology—previously Lundberg had been at Columbia University, where he served on Harry Alpert's PhD committee. Larsen then became a professor at the University of Washington. A specialist in studies of the mass media, he developed an extensive publication record, including a number of books that he wrote, coauthored, or edited: *The Flow of Infor-*

mation (1958), *Violence and the Mass Media* (1968), *Social Policy and Sociology* (1975), and *The Uses of Controversy in Sociology* (1976).[34]

Larsen also had significant advisory, leadership, and administrative experience. He had been a consultant to many federal bodies, including the National Institute of Mental Health and the Commission on Obscenity and Pornography under President Johnson. He served as the American Sociological Association's executive director in the early 1970s. He also became a member of the Social Science Research Council (SSRC) and chaired its governing board from 1978 to 1980. Of special relevance to his NSF work, in mid-1980 Larsen became the director of the DBBSS's social and economic sciences division, a position that he held for two years, until mid-1982 and thus during the dark days of the early Reagan years. Larsen then returned to the University of Washington for a short time, before taking up the new NSF position, which began in January 1983.

At another moment in time, the creation of this position may have signaled the social sciences' increasing importance. However, their standing within the agency had fallen considerably since the brighter days of the 1960s and early 1970s. Nobody understood these difficulties better than Larsen.

The heightened level of positive attention devoted to the social sciences during the Kennedy and Johnson presidencies together with the 1968 Daddario amendment had not really done much to solidify their position at the agency over the long term, Larsen wrote in 1985. Explicit mention in the revised NSF charter had not led to "even minimally appropriate levels of financial support." In addition, inside the agency, there remained widespread concern about the social sciences' "political troublemaking capacity," accompanied by the old fear that their presence might undermine political support for the NSF more broadly. Nor had these sciences managed to obtain adequate "institutionalized representation at the highest policy and managerial levels."[35]

It is telling that the new position held by Larsen did not become permanent. In 1986, after three frustrating years, he would leave the agency, this time for good. The position disappeared with his departure.

Although the position was created to alleviate the problem of institutionalized representation, this problem remained serious where it counted most, namely, at the top of the NSF hierarchy. The psychologist Richard Atkinson had been director in the late 1970s. But during the Reagan years, the top post was occupied by two engineers and one physicist. In addition, whereas the electrical engineer John Slaughter was a Democrat who expressed at least some discontent with the brutal funding cuts proposed during Reagan's first year in the White House, the next two directors were Republicans and strong advocates of the administration's science policy agenda, which included increased support for the natural sciences and engineering alongside funding cuts for the social sciences.

When Slaughter left the agency in 1982—to become chancellor at the University of Maryland—he was replaced by the physicist Edward Knapp. A specialist in particle physics and thermonuclear power, Knapp had worked at the Los Alamos Scientific Laboratory for many years (1958–1982). He then served briefly at the NSF as assistant director for the Mathematical and Physical Sciences Directorate, before becoming director in November 1982. Described by the science writer Wil Lepkowski as "a new, conservative, back-to-basics director," Knapp was handpicked for the agency's top post by Reagan's chief science advisor George Keyworth. Keyworth chose Knapp because he wanted somebody who, as Lepkowski put it, would not pull any "liberally inspired surprises on a conservative White House."[36] Knapp himself explained that under his leadership, the agency would emphasize the "areas of greatest potential promise to our future economic well-being and technological capabilities."[37]

When Knapp left the NSF in 1984, his replacement was Erich Bloch, another electrical engineer who embraced the administration's science policy agenda as well. An accomplished and "seasoned technocrat," Bloch had worked in the private sector at IBM for some thirty years. Based on his involvement with innovative engineering projects, he was credited with revolutionizing the computer industry and received the 1985 National Medal of Technology and Innovation, presented by President Reagan.[38] Bloch's appointment was a glaring departure from the well-established tradition of recruiting NSF directors from academia and government. He had never pursued a PhD. Nor had he held a university position. But neither of these facts undermined the White House's enthusiasm. Bloch's "appointment was welcomed ... because of his accomplishments at IBM and his passion for a science policy that was in complete synch with the Administration," noted Keyworth.[39]

None of these three directors knew much about the social sciences before arriving at the agency. Nor were they ever a main priority—although

Slaughter did not want to see their funding decimated. Knapp, for one, admitted how little he knew: "I really don't understand them. ... But I'm learning. I'm a novice. I've never looked carefully into the social sciences to understand their successes, their failures, the criticisms."[40] Additional evidence comes from NSF annual reports, where each year, the director's statement presented highlights of agency-funded projects. The social and behavioral sciences received precious little coverage. Consider just one such statement from Bloch at the very end of the decade: "Who could fail to be impressed by achievements like the ones described in this [1989] report: ocean-bottom and celestial discoveries, work with super-conducting materials, advances in mathematics, robotics, supercomputers, and many others."[41]

The social sciences' low status was reinforced by spotty representation on the agency's twenty-four-member governing board. In most years, these sciences had but one representative: the anthropologist Ernestine Friedl from 1979 to 1984, the economist John H. Moore from 1982 to 1985, and another economist, Annelise G. Anderson, from 1984 to 1990.[42]

The SBS programs remained subordinate within the agency's organizational structure as well. After the original social science division's closure in 1975, these programs were located in the DBBSS, which, as noted in previous chapters, focused predominantly on the biological sciences and was always led by a biologist. Starting in 1976, that person was Eloise Clark, a developmental biologist. When in 1984 she left to take a job at Bowling Green State University, President Reagan filled the position with David T. Kingsbury, a specialist in medical microbiology, biochemistry, virology, and genetics. Previously, Kingsbury had been a professor at the University of California's Berkeley and Irvine campuses, the scientific director of the Naval Biosciences Laboratory at Berkeley, a biotechnology expert for the Navy, and a visiting scientist at the National Institutes of Health (NIH) Laboratory for Central Nervous System Studies. Kingsbury held his NSF position for four years, until he was replaced by Mary F. Clutter. A cell biologist who had previously been a faculty researcher and lecturer at Yale University, Clutter became well known for her efforts to promote the participation and advancement of women in science.[43]

As the new DBBSS leader, Kingsbury realized that attending to the social and behavioral sciences presented a special challenge. Here's how he explained that to a science policy group in the D.C. area:

> Remember we've got a biochemist (or whatever I am) at one level, and an engineer [Bloch] at another. And juxtaposed, not between us, but peripheral to us, are chemists who are playing significant roles in policy; and ultimately, we talk to physicists. Somewhere in that milieu you've got to introduce some knowledge and challenge about behavioral science.

This was "not always easy," Kingsbury added.[44]

Only at the lower levels did SBS scholars have important positions as division and program directors. For a very brief period of time in the early 1980s, the Social and Economic Sciences (SES) division was led by long-time staffer Bertha Rubinstein, whose experience in this area went all the way back to the Alpert years. SES leadership passed in succession to the sociologist Otto Larsen from mid-1980 to mid-1982, the economist James H. Blackman from mid-1982 to mid-1984, and then the historian and former COSSA executive director Roberta B. Miller from mid-1984 through the rest of the decade. As for the Behavioral and Neural Sciences (BNS) division, Richard T. Louttit, a specialist in physiological psychology, started as director in 1977 and remained there throughout the 1980s. Scholars from the relevant SBS areas of study also led disciplinary and interdisciplinary programs in the two divisions.

Although these individuals had vital managerial jobs, they had a tough time conveying the value of SBS work to agency leaders. During Bloch's tenure as director, that challenge proved especially difficult because of Bloch's preference for a strict chain of command when it came to channels of communication and advice. This meant that top NSF decision makers typically did not have discussions with social and behavioral scientists.[45]

To better appreciate such difficulties, it is instructive to compare the social sciences' low status with the rising status of the engineering sciences. To begin with, for most of the decade, the NSF directorship was held by two engineers: Slaughter from 1980 to 1982 and Bloch starting in 1984. This represented an important alteration from the previous three decades, when the agency had been led by three physicists (Waterman, Haworth, and Stever), one biologist (McElroy), and one psychologist (Atkinson). In addition, after Atkinson established the groundwork for strengthening engineering's organizational standing with a separate engineering directorate, Slaughter secured its establishment. In Slaughter's view, up to that point, the agency had not given engineering adequate attention and resources.[46]

Furthermore, in 1984, the agency—now under Bloch's leadership—created a program to support the development of Engineering Research

Centers on university campuses. Each center would focus on "a major interdisciplinary area of interest to both industrial and academic researchers."[47] This program, in turn, became the model for another generously funded program for the development of Science and Technology Centers. In principle, this program included all scientific fields. But in practice, the social sciences participated only as minor players during the rest of the decade. In 1986, another new directorate for Computer and Information Science and Engineering gave additional visibility and resources to engineering.[48]

Figure 8.3 reveals that from 1970 to 1988, NSF support for basic research in the social sciences rose in current dollars just slightly. Meanwhile, NSF funding for basic research in the engineering sciences rose about sixfold. NSF funding amounts for basic research in the life science, physical sciences, environmental science, and math and computer sciences all increased dramatically as well.

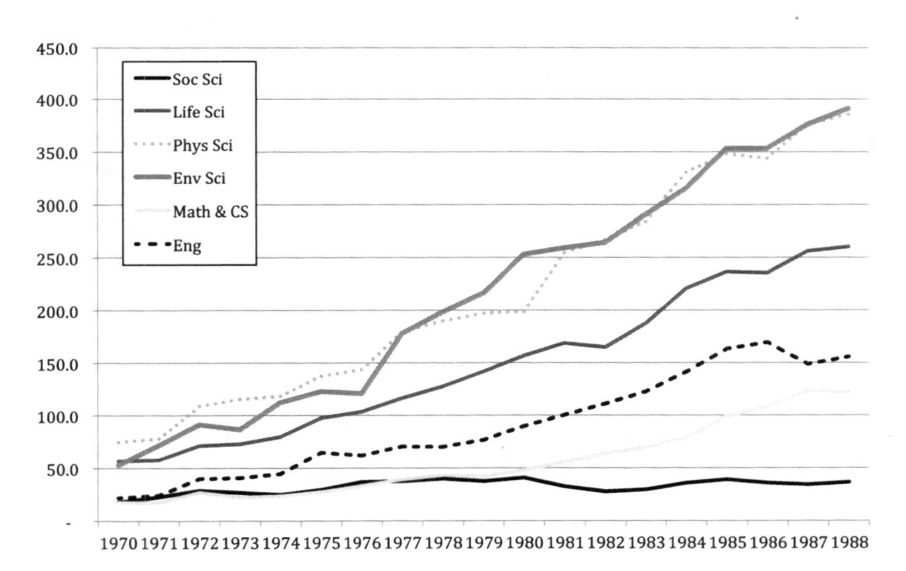

Figure 8.3
Federal obligations for basic research, by field of science. National Science Foundation, 1970–1988 (current dollars, millions). Data from table 2L, federal obligations for basic research, by detailed fields of science and engineering: National Science Foundation, fiscal years 1970–2003. In National Science Foundation, Division of Science Resources Statistics, *Federal Funds for Research and Development: Fiscal Years 1970–2003*, NSF 04-335, Project Officer Ronald L. Meeks (Arlington, VA, 2004). Amounts for life science computed by adding separate amounts for biological aspects of psychology; amounts for social sciences computed by adding separate amounts for social aspects of psychology.

While the engineering sciences flourished, the social sciences floundered. The latter's low status, as seen in meager representation among NSF top decision makers and in the subordinate position of the divisions and programs in this area, contributed to the problems. With this in mind, the 1988 NAS report mentioned above proposed that, along with a substantial increase in federal funding, the "place and role of the behavioral and social sciences" inside "relevant federal agencies," including the NSF, needed to be "critically reappraised." Only significant improvements at administrative levels could "ensure continuous high-level representation of the scientific needs and opportunities in these fields."[49]

But, as was the case with federal social science funding and NSF funding in particular, no significant improvements in organizational standing, representation, and status had materialized as the Reagan era came to a close. To borrow a line from Yogi Berra, the ever-quotable New York Yankee baseball player and manager, it was déjà vu all over again.

★★★

Distinct from yet related to the problems discussed above, the social sciences also faced a revival of critical statements from powerful natural science and engineering figures. At the beginning of the Reagan years, social science leaders emphasized the need to develop better relations with their more highly esteemed and more powerful natural science colleagues. Yet, the sociologist and SSRC representative Richard Rockwell observed that "it really does not help us very much to presume that we are a part of the national science system unless physicists and biologists share that presumption."[50] Indeed, that presumption remained dubious in the minds of some leading figures.

Consider the stance of the physicist George Keyworth. When the Reagan administration created the White House Science Council in 1983, the president gave Keyworth responsibility for choosing its thirteen members. Keyworth's selection included twelve physical scientists, some with close ties to high-tech industry, but not a single social scientist.[51] Furthermore, some of his comments suggest that this omission was not merely an oversight. In 1984, Keyworth expressed nothing but admiration for the administration's recent efforts to curtail NSF funding. As seen in this chapter's opening quote from him, he even asserted that Reagan's team had acted with "great courage and wisdom" by "squashing the daylights out of" NSF's social science programs, adding that "in terms of sheer quality," those programs "rated phenomenally low."

Figure 8.4
Chief science adviser George A. Keyworth II meets with President Reagan in the White House Oval Office, July 2, 1981. Courtesy of Ronald Reagan Presidential Library.

One could not reasonably dismiss Keyworth as an aberrant case either. The social scientist's role in the national science community remained, in Roberta Miller's words, "modest" and "tempered by the belief—held mainly by physical scientists ... that the social sciences were not quite legitimate as science." SSRC president Kenneth Prewitt and the sociologist David Sills agreed, noting that in the battles for acceptance and support, social scientists "generally suffered from the indifference and at times even the hostility of natural scientists."[52]

Inside the NSF, tough conditions added salt to the wound. According to science reporter Constance Holden, social scientists were "not happy" with what they took to be "half-hearted attempts" by the agency to defend them. After working there for nearly five years in two different positions, Otto Larsen offered this dismal assessment in 1985:

> Fundamentally, there has been no change in the way in which social science is understood by the dominant figures from physics, engineering, chemistry, and mathematics. ... It is not just that social sciences are deemed 'soft': rather, it is a genuine skepticism about whether they are sciences at all.[53]

A few years later, Director Bloch repeated the idea, which still passed as official wisdom within the agency, that the social sciences were lagging and thus would benefit from closer relations with more mature sciences. On one occasion, in 1989, North Carolina Democratic Representative David Price, who had a Yale PhD in political science and had been a political science professor at Duke University, wrote to Bloch complaining about the "apparent low regard in which these disciplines are held at the NSF." Bloch responded by asserting that he was "as much a champion of the behavioral and social sciences" as he was of any other scientific discipline.[54] However, we have seen that the social sciences were far from the center of his concern. Moreover, Bloch was not confident about their value. During an invited presentation at a COSSA meeting, Bloch reported that the social sciences still needed to "gain greater recognition as a legitimate scientific activity." He did not suggest that they should be "modeled on physics or astronomy in any general sense." Nevertheless, to help disprove the notion that "anything with 'science' in the title probably isn't," he urged social researchers to make greater use of "experimentation and quantitative methods."[55] The notion that this broad field needed to become harder also came through in another passage from Bloch's reply to Price where he stated that "real opportunities for scientific advances" lay at the "interfaces with other sciences and engineering."[56]

Bloch's view about the post–World War II trajectory of American science policy added a historical perspective that reinforced the social sciences' marginal importance. As the NSF director saw things, during the late 1950s and 1960s, the federal government had responded admirably to the shock of Sputnik by "accepting responsibility for the [nation's] 'science and engineering base,' which also served the goal of 'economic competitiveness.'" However, after 1968 and the Daddario amendment, "the momentum of Sputnik was spent." Subsequently, national attention shifted to "social problems," such as "housing, energy, crime." With this emphasis on social "relevance," federal policies shifted as well, to direct research "toward these efforts." Unfortunately, there was "little concern for economic competitiveness." As the science studies scholar Janet Abbate has explained, Bloch had "disdain for the goal of addressing social problems with applied research." Moreover, in Bloch's telling, the story of federal science policy took a happy turn during the early Reagan years through "an increased recognition of the need to support the science and engineering base."[57]

One should not overlook the fact that at certain moments, some figures from the natural science and engineering communities, such as NAS president Philip Handler, had rallied in support of their lower-status colleagues. Harvey Brooks, a Harvard physicist with extensive science policy experience, also said that the proposed budget cuts in 1981 were "foolhardy" and would all but destroy "a whole broad area of scholarship for the sake of a budgetary saving that is almost 'lost in the noise.'"[58] According to Roberta Miller, one factor lying behind such support was the recognition that if ideological considerations threatened social science budgets, natural science budgets could become vulnerable as well.[59]

However, social scientists, as we have seen, could not count on consistent and strong support from top federal science policy figures. Notwithstanding a few supportive statements from them, and regardless of the particular mix of motivations involved in each case, the problem of winning the respect of high-ranking individuals from the physical science and engineering communities, such as Keyworth and Bloch, reemerged with a vengeance during the 1980s.

★★★

In sum, substantial problems in the areas of funding, organizational status, and respect continued throughout the decade. The notion that the social sciences experienced serious challenges during the early years of the first Reagan administration but then managed to overcome them must be discarded.

It is also crucial to understand that the new politics of science and the White House did not reject social science per se. Remember that soon after Reagan had taken office, his administration established a "social science hour." As the sociologist Martin Bulmer observed (and rightly so), given the administration's framework for "viewing economic and social phenomena," the critical question was not necessarily "social science or no social science?" but, instead, "what kind of social science?"[60]

THE EXCEPTIONAL STATUS OF ECONOMICS

The last point raises an important question: which social sciences did best at the NSF during the Reagan era, and why? This section does not attempt to provide a comprehensive answer. Rather, we will zoom in on economics, an interesting case because during those years, national economic problems and the efforts of economists to address those problems received so much attention in American political culture, in the national science policy arena, and inside the White House.[61] In a number of respects, and due to a variety of factors discussed below, economic science also enjoyed a relatively favored position among the social sciences at the NSF.

To begin with, the favored standing of economics must be seen in the context of diminished NSF social science support during Reagan's two presidential terms, from 4.6 to 3.3 percent of the agency's total budget. After adjustments for inflation, none of the SES division's disciplinary-based programs enjoyed budgetary growth from 1981 to 1989. The budget for Economics rose in current dollars from $9.4 million to $12.9 million. But in constant 1982 dollars, it fell, albeit only minimally, from $10.2 million to 10.0 million, or about 2 percent. The budget for Political Science increased in current dollars from $2.9 million to $3.8 million. Similar to the case of Economics, however, it decreased slightly in constant 1982 dollars, from $3.1 million to $3.0 million, or about 3 percent. More notable in its decline was Sociology's budget, which rose in current dollars from $3 million to $3.7 million but dropped in constant 1982 dollars from $3.3 million to $2.9 million, or about 12 percent.[62]

While none of the programs experienced growth, the figures also show that Economics enjoyed a favored position. Indeed, at the level of funding, this is quite striking. Both at the beginning and end of the two Reagan terms, the budget for economics was roughly three times larger than the budgets for sociology and political science.

The proportion of total SES funding that went to each of the division's disciplinary and nondisciplinary programs also shows that Economics did strikingly well, receiving 39 percent of all such funds during the 1980s. Other disciplinary programs in this division received considerably less: Political Science, Sociology, Geography, and the History and Philosophy of Science received, respectively, 13, 12, 6, and 3 percent. Four nondisciplinary programs—Measurement Methods and Data Desources; Law and Social Science; Decision, Risk, and Management Science; and Regulation and Policy Analysis—received, respectively, 13, 6, 5, and 3 percent.[63] Note as well that these last two programs—Decision, Risk, and Management Science and Regulation and Policy Analysis—would have funded work in economics, although exactly how much is not clear.

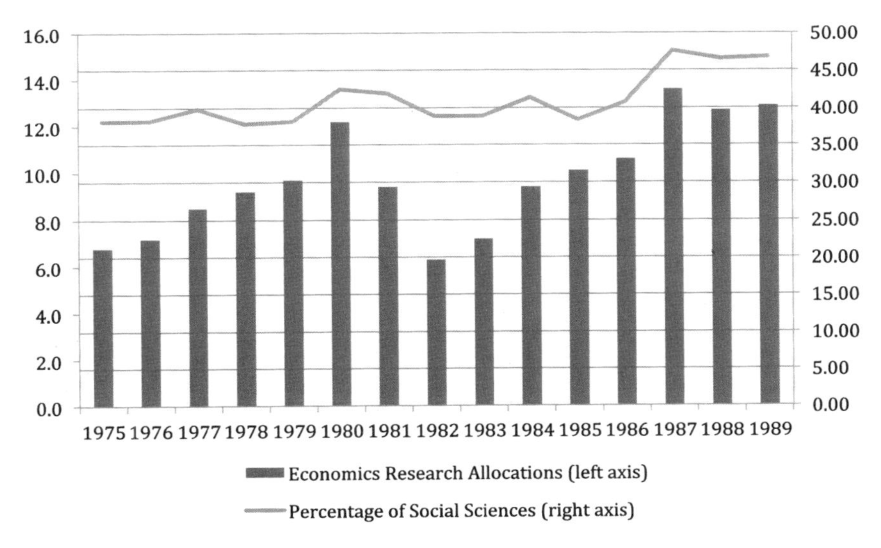

Figure 8.5
NSF research allocations for Economics, 1975–1989: Total (Current Dollars, Millions) and Percentage of Social and Economic Sciences Division Research Allocations. Data compiled from tables 5.3 and 6.2 of Otto N. Larsen, *Milestones and Millstones: Social Science at the National Science Foundation, 1945–1991* (New Brunswick, NJ: Transaction Publishers, 1992).

Figures 8.5 and 8.6 make it easy to see how well the Economics Program did, and over the course of many years, even though it was just one of many such programs in the social sciences division.

We have already seen that the crisis of the early 1980s provided economic stars such as Zvi Griliches, Lawrence Klein, and Robert Lucas with opportunities to argue for the value of economic reasoning in science policy discussions and for the special importance of the NSF's economics program. To understand why this program did so well compared to the other disciplinary programs, we also need to consider four more factors: (1) previous support for economics, (2) the relevance of the agency's work to the "new economics," (3) the development of a favorable environment for economics inside the NSF during the 1980s, and (4) the contributions of economics to big social science.

First, starting well before and leading all the way up to the Reagan years, economics had already acquired a privileged position in financial terms. During the 1960s, when the economics program was part of the original social

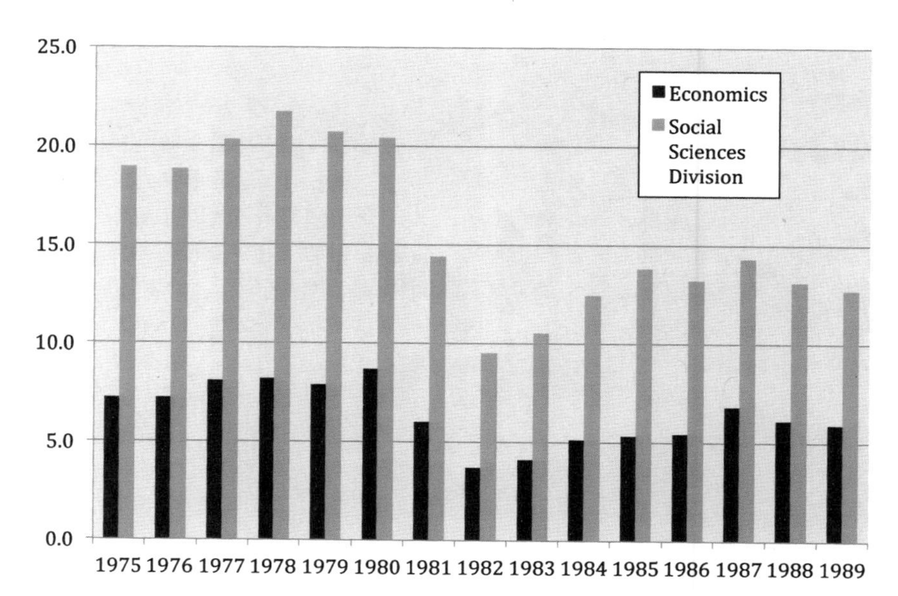

Figure 8.6
NSF research allocations for Economics and for the Social and Economic Sciences Division, 1975–1989 (1976 Dollars, Millions). Data compiled from tables 5.3 and 6.2 of Otto N. Larsen, *Milestones and Millstones: Social Science at the National Science Foundation, 1945–1991* (New Brunswick, NJ: Transaction Publishers, 1992). Current dollars converted to 1976 dollars using cpi calculator from https://data.bls.gov/cgi-bin/cpicalc.pl.

science division, economics received 24 percent of all social science funding. Midway through the 1970s, this program was moved into the newly reconfigured social sciences division—located in the DBBSS, which also had a division for behavioral and neural sciences. During the 1970s, Economics received 26 percent of all SBS funding.[64] Up through 1979, more than 650 different economists had been principal investigators on one or more NSF-sponsored projects.[65] In addition, economics, as chapter 6 noted, was added to the name of the social sciences division in 1979, making it the social and economic sciences division from that point forward.

The elevated position of economics rested on a widespread perception that the discipline had achieved an especially high degree of scientific maturity and rigor. This perception, in turn, supported and reflected the discipline's practical uptake in many vital areas (i.e., national security matters, including operations analysis, nuclear strategy, and defense budgeting, as well as domestic policy issues, such as capital accumulation, economic growth, and poverty reduction). NSF social science leaders themselves regularly emphasized that economics had achieved a commendable level of mathematical sophistication and made extensive use of computers. In fact, the very first grant in economics, awarded through the sociophysical sciences program in 1957, went to the National Bureau of Economic Research to support a project that used computers to analyze economic statistics and business cycle data.[66] One decade later, the 1967 NSF annual report stated that the "increasing use of mathematics and computer-based technology" promised to strengthen theoretical findings in the social sciences, which, when translated for practical matters, would be valuable to policy makers. The report added that of all the social sciences, economics in particular, had found productive ways to use "a large battery of such [mathematical] techniques."[67]

The NSF, in turn, provided generous support for mathematical economics, as seen not only in its grants to hundreds of individual scholars but also in its support for particular projects that involved numerous scholars. For instance, the agency gave many grants to support Stanford's annual summer program at the Institute for Mathematical Studies in the Social Sciences, with the economist Mordecai Kurz as principal investigator. This program brought together economists from around the world for two months of research and seminars. The NSF also gave multiple grants for work in econometrics carried out by a group at the Cowles Foundation, with Tjalling Koopmans and William Brainard as principal investigators. These are but two of many such examples.[68]

Economics also did well because the NSF and its supporters could point to an impressive track record of funding high-quality proposals and star researchers. As we saw in chapter 5, the 1976 Simon Report argued that the agency had been successful in funding high-caliber basic social science and thus should be given additional resources. That argument obviously included economics but did not mention it specifically. However, elsewhere the Simon Report emphasized the agency's support for outstanding economists. Between 1965 and 1985, all eleven winners of the highly coveted John Bates Clark Award had held NSF grants before receiving this award. During these same years, four of the eleven US-based winners of the Nobel Memorial Prize in Economic Sciences had previously received NSF grants.[69]

Experienced, devoted, and well-connected leaders provided additional strength to the economics program. James H. Blackman, a specialist in Soviet economics and former professor at Duke University, headed this program from 1967 to 1980. His successor, Daniel Newlon, had completed a PhD at the University of Virginia and worked as a professor at the State University of New York–Binghamton. In 1974, he began working at the NSF, first under Blackman as associate director and later, after Blackman left, as director, a position he held throughout the Reagan years and beyond. Newlon's expertise in energy economics and public finance cemented his scholarly reputation and facilitated his ties to government and business through numerous consulting positions, for the Navy's Alleghany Ballistics Laboratory, the Center for Naval Analysis, the Federal Energy Administration, the U.S. Army, and IBM. Newlon also cultivated good relations with the American Economic Association and COSSA.[70]

Hence, economics already benefited from a strong base of support when the crisis of the early 1980s hit, which brings us to the second factor

Figure 8.7
Dan Newlon participates in a COSSA-sponsored congressional briefing on "Better Living through Economics," March 15, 2010. Left to right: Nancy Lutz, NSF economics program officer; Alvin Roth, Harvard University; Lawrence Ausubel, University of Maryland; Brigitte Madrian, Harvard University; John L. Siegfried, Vanderbilt University; and Dan Newlon, former NSF economics program officer (1980–2009) and director for government relations of the American Economic Association. Courtesy of COSSA.

that gave it a privileged status among the otherwise beleaguered social sciences. Following the Reagan administration's proposed budget cuts, leading economists mounted a vigorous defense of NSF social science funding and for economics in particular. Moreover, the agency received widespread positive attention for its contributions to the changing landscape of academic economics and economic policymaking. To appreciate the NSF's importance here, a few words about the rise and fall of Keynesianism are necessary.

Back in the mid-1960s, Walter W. Heller, chairman of the Council of Economic Advisers under Presidents Kennedy and Johnson, asserted that remarkable advances in economic science had brought about the "completion of the Keynesian Revolution" and "put the political economist at the President's elbow." This statement was supported by the taken-for-granted belief among liberals as well as moderate conservatives that "the government must step in to provide the essential stability at high levels of employment and growth that the market mechanism, left alone, cannot deliver."[71] However, proponents of the free market from the business and financial communities opposed Keynesianism and the liberal, managerial, engineering outlook associated with it.

Over the next decade, the latter viewpoint gained strength, aided by developments within the economics profession together with the failure of the Keynesian paradigm to account for the twin problems of rising unemployment and rising inflation—stagflation. As the intellectual historian Angus Burgin notes, the new wisdom presented "Keynesianism, rather than laissez-faire ... [as] a relic of a rapidly receding economic world."[72] Similarly, the historian of economics Roger Backhouse has observed that "an ideology favorable to management of the economy gave way to one in which state action was seen as raising more problems than it solved."[73]

This critical shift in economic thinking reflected the impact of many factors. According to Backhouse, these include lessons learned from past mistakes about the difficulties and harms associated with too much governmental intervention; the development of rational choice ideology; strong advocacy of free-market thinking by conservative think tanks such as the American Enterprise Institute and the Heritage Foundation; contributions from universities, including some with exceptional influence (i.e., the University of Chicago), and from powerful international organizations, such as the World Bank; and funding for people, projects, and institutions provided by private organizations, including the two conservative think tanks mentioned above as well as the Lynde and Harry Bradley Foundation and the John M. Olin Foundation.[74]

Although Backhouse did not mention so, the NSF also played a significant role by sponsoring work that contributed to the shifting terrain in economic science and policy. During the 1970s, the agency supported studies on what economics program director James Blackman referred to as the "unemployment-inflation dilemma."[75] The agency also awarded grants to economists who were leading the charge against Keynesian orthodoxy. This group included researchers associated with monetarist theory, such as MIT's Robert Solow, Princeton's Alan Blinder, Chicago's Robert Lucas, and Carnegie-Mellon's Allan Meltzer. As the 1975 NSF annual report explained, this body of work

> questioned a basic economic principle, widely accepted since the work of Keynes, that the federal budget can influence the aggregate level of income and employment. In its simplest form the so-called "monetarist" thesis asserts that each dollar of additional Government spending, no matter how it is financed, simply "crowds out" the same amount of private spending.

The upshot was that "fiscal policy is powerless to alter the overall level of aggregate income."[76]

The agency financed research by Martin Feldstein on the deleterious effects of government programs as well. A Harvard economist and long-time president of the National Bureau of Economic Research (1978–2008), Feldstein's specialties included macroeconomics, finance, and public pension systems. According to the 1977 NSF annual report, his work showed that the enormous size and explosive growth of the U.S. Social Security program had "major effects," including a long-term and "substantial reduction" in savings. The implications seemed grave: such a reduction would result in slower capital accumulation, lower productivity, and less national income.[77] As Daniel Newlon explained, by "documenting the significance of the disincentives caused by the U.S. tax system," NSF-supported projects by Feldstein and his collaborators made a major contribution to the study of public finance.[78]

This body of NSF-funded research acquired greater importance because it lent support to the Reagan administration's economic policy agenda, which had been heavily shaped by free-market scholars such as Martin Anderson and Milton Friedman. In 1981, soon after the administration announced its plans for severe social science funding reductions, the NSF lost no time in identifying the policy relevance of its economics program. In June (and thus around the same time that leading social scientists and their supporters gathered in New York for a strategy meeting to confront the emerging crisis), an internal report emphasized that many NSF-funded economists, including Lucas, Meltzer, and Feldstein, had made crucial contributions to the "new economics."[79]

Discussion of such work extended well beyond the walls of the agency. Writing in the *Washington Post*, reporter Philip Hilts noted that regarding the proposed NSF funding cuts, Feldstein himself criticized the administration for acting on a mistaken belief that NSF social science had a liberal bias: many researchers "critical of the Great Society programs" had, in fact, received NSF support for scientific investigations of pressing issues to conservatives, including Feldstein's own work on the Social Security program's "adverse impact on savings."[80] Other economists hammered home the point for the benefit of Congress, with Feldstein's Harvard colleague Zvi Griliches pointing out that the results of NSF-funded research had supported many "conservative ideas in economics," including "the importance of rational expectations and the impotency of conventional macro-economic policy, the disincentive effects of various income-support programs, the magnitude of the regulatory burden, and the argument for deregulation."[81]

In this context, economics program director Daniel Newlon saw a unique opportunity to rally its supporters. He had to be careful, for he had been told not to lobby against the administration's proposed cuts. Nevertheless, Newlon worked effectively behind the scenes, with help from COSSA and the American Economic Association. He also enlisted support from individual economists, including Feldstein—who in 1981 received his eighth NSF grant, this one for $121,000 to study the macroeconomic effects of fiscal programs.[82] The following year, Feldstein became chairman of the Council of Economic Advisers and thus Reagan's chief economic adviser.

Third, throughout the 1980s, economics benefited from a relatively favorable institutional environment inside the NSF. Even though its directors and board members generally had little knowledge of the social sciences and little interest in them, economics was a modest exception. As noted before, Director Slaughter emphasized in congressional testimony that the empirical basis for supply-side economic policy still needed to be assessed. Doing so required scientific research of the sort that the NSF was especially good at supporting. In addition, the next two directors, Knapp and Bloch, were fully behind the Reagan administration's emphasis on federal science funding in areas that could contribute to economic revitalization and technological innovation, thus giving economic science itself a glow that the other social sciences lacked. Meanwhile, "individually and collectively," economists "developed close relationships with NSF officials, and engaged in lobbying campaigns in Congress and government to assert their membership to the Endless Frontier," as Mata and Scheiding have shown.[83]

Economics also stood out as the only social science that enjoyed regular, albeit modest representation on the governing board. This included two scholars appointed by President Reagan. The first, Annelise Anderson, was married to the prominent libertarian economist and conservative policy adviser Martin Anderson. After completing her PhD in business administration at Columbia University in 1974, Anderson became a professor in the School of Business and Economics at California State University–Hayward (1975–1980). Similar to her husband, she established close ties to the Reagan White House. She served as a senior policy adviser for Reagan's 1980 presidential campaign and subsequently held a series of positions in the new administration working on various policy issues, including the deregulation of financial institutions.[84] John Moore was the second Reagan appointee. He did his PhD at the University of Virginia and later worked there as a faculty member (1966–1977). In addition, at Emory University and the University of Miami, Moore was an associate director for centers dedicated to law and economics, a rapidly growing field generously supported by the John M. Olin Foundation and other conservative organizations. Moore and Anderson also knew one another from the Hoover Institution, where Moore was an associate director and both were senior research fellows.[85]

Furthermore, in 1985, Reagan appointed Moore as NSF deputy director. Moore, who replaced the physicist Donald Langenberg, thus became Director Bloch's main assistant. Remember, it was also in 1985 that DBBSS leader David Kingsbury mentioned that getting NSF leaders to focus on the social and behavioral sciences was difficult. But at least they had "some understanding" of economics, he added.[86]

As the cases of Anderson and Moore already suggest, the development of good relations between the NSF and the conservative Hoover Institution gave economics yet another layer of support. According to an internal NSF report, as of 1986, Hoover had twenty-two resident senior scholars, including four sociologists, six political scientists, and twelve economists. Fourteen of the twenty-two, including six of the twelve economists, had served as reviewers for NSF research proposals. The six economists were W. Glenn Campbell, Milton Friedman, Edward Lazear, Charles E. McClure Jr., Thomas G. Moore, and Thomas Sowell.[87]

Equally significant, inside the NSF, the influence of economics extended beyond the Economics Program itself. Remember that the official reason for the administration's proposed cuts was that the social sciences had little relevance to the nation's economic troubles. In responding to this charge, the agency and its supporters argued that, in fact, the social sciences contributed in a myriad of ways to technological innovation and economic productivity. First presented in an NSF document called "Emerging Issues in Science and Technology, 1981," this argument became the basis for an extended article in the *American Psychologist* by a group of authors including three NSF officials: the sociologist Otto Larsen and two psychologists, Louis Tornatzky and Trudy Solomon.[88] According to this article, it had become "increasingly clear that human, social, and institutional factors" were "particularly important" in the "process of innovation." Social factors influenced everything from "the conduct and management of research" to "the dissemination and marketing of new technical products" as well as "the implementation of new manufacturing processes at the shop floor level." Sound understanding of these processes thus depended on social science research.[89]

The authors even claimed that "virtually all" of the policy issues under debate regarding "national innovation and productivity" rested on "knowledge derived from social science research." Unfortunately, however, this fact was "seldom acknowledged."[90]

Throughout the decade, social science contributions to economic growth and technological innovation became a recurring theme. To be sure, NSF leaders devoted much greater attention to the economic role of the natural and engineering sciences. Still, the notion that the social sciences would make significant contributions now received regular attention. Director Bloch, who according to one NSF official possessed the "conventional views of engineers" when it came to the social sciences, asserted that they could help the nation to address the present "economic challenge of unprecedented magnitude." He then singled out three areas of study that deserved special attention: "world market competition," "research competition," and "the complexity of technology."[91]

The fourth and last factor that gave economics favored status concerns its contributions to big social science. As explained in chapter 6, NSF support in this area had been growing since the mid-1960s. By the late 1970s, big social science, including large-scale econometric studies, enjoyed unusually strong support from the agency's natural science leaders. During the first Reagan administration, big social science stood out again, this time as one type of social science research that nearly everybody rallied to defend in the midst of the administration's budget-cutting efforts. The 1982 NSF annual report pointed out that the agency had become "responsible for creating, maintaining, and making accessible major collections of high-quality data, along with the development of increasingly powerful quantitative techniques to analyze those data."[92] As noted in the previous chapter, such work convinced the agency's leaders that similar to the natural sciences, the social sciences needed long-term funding for large-scale projects, thus bringing them into the realm of big science at the NSF, albeit on a relatively smaller scale. Such work also fit well with the general strategy of promoting the social sciences as part of a unified enterprise led by the natural sciences.[93] The following year, William Mishler, from the political science program, reported that high regard for big social science had helped to ward off the harshest budget-cutting scenario in 1981. Moreover, such positive recognition had contributed to partial restoration of the agency's social science budget.[94]

How economics benefited from participating in big social science can be seen in the NSF-funded Panel Study on Income Dynamics (PSID). In the early 1980s, the agency decentralized its support for big social science projects. Responsibility for supporting the PSID was thus transferred from the measurement methods and data resources program to the economics program.[95] In a 1983 memorandum to the board, Director Knapp explained the unique value of this particular project:

> Every year from 1968 to the present, the PSID staff has collected, processed, and disseminated information about the behavior and economic status of a nationally representative sample of about 5,000 American households. The cumulative database created from fifteen years of PSID interviews provides the foundation for much of the empirical research on family dynamics.

Starting out as "an instrument for assessing the nature and extent of poverty in the U.S.," this study had become "the only survey of income, occupation, education, family composition, and other social and economic family characteristics that combines a long (fifteen year) time period and a nationally representative sample of families and individuals." The data collected and analyzed gave this massive study considerable policy relevance as well. Federal agencies regularly used its results.[96]

The broader trajectory of funding for this economics project sheds further light on its status as big science at the NSF. Although the PSID began in the late 1960s, the NSF did not support it until 1979. And the agency's contribution was modest at first, with the bulk of funding coming from other agencies, including the Department of Health and Human Services (DHHS). But in the early 1980s, Reagan administration cuts in federal social research programs brought an end to HHS funding. By this point, as one commentator noted, the project had already collected "fourteen years of longitudinal data on the same five thousand families and the split-off from those families." If one wanted to create "that kind of longitudinal data base again," you would "have to start from square one and it would have taken fourteen more years."[97] With budget chopping endangering the project's continuity, private patrons, including the Sloan, Ford, and Rockefeller foundations, stepped in by providing a total of $725,000 in 1982 and $450,000 in 1983. However, they provided only temporary relief, based on an understanding that after the NSF had completed an extensive evaluation, private funding would end.[98]

Toward the end of 1983, advice from some twenty-five external reviewers and from the NSF Economics Advisory Panel suggested that this project had become too big to fail. In the panel's assessment:

> The data being collected by the PSID is unique, virtually irreplaceable, and essential for empirical research in important subfields of economics. Ending the PSID now just when the cumulative body of information from past interviews is starting to capture full lifecycle decision making would represent a major setback for empirical research in economics.

Hence, PSID funding for three more years deserved "highest priority."[99]

Following that advice, the agency assumed the great bulk of the funding burden. In the preceding five-year period, from 1979 to 1983, it had provided a total of $1.618 million.[100] For the next three years, from 1984 to 1986, it awarded a large $4.4 million grant, under the direction of James Morgan and Greg Duncan at the University of Michigan. Of that amount, the NSF portion was $3.1 million, with the rest provided by interagency transfers from the NIH and the DHHS.[101] In 1986, the NSF awarded an even bigger grant, this time for $10 million spread over a five-year period. The NSF's contribution was now $7.5 million, with the rest again coming from interagency transfers.[102]

During the 1980s, the NSF supported at least two other big social science projects in economics as well. These were the U.S. Manufacturing Establishment Data Base at Yale's Department of Economics and the Computer Research Center for Economic and Management Science at MIT.[103]

In sum, despite the deep and persistent difficulties facing the social sciences during the Reagan years, economics, due to a number of factors considered above, stood out as an exception. The relatively privileged status of economics during the two Reagan terms should not be overstated, however. After all, the economics program's budget actually decreased, albeit by only about 2 percent. Still, the new politics of science, the Reagan administration's policy agenda, and conditions inside the NSF gave economics a comparatively privileged position. This analysis also reveals how the NSF fits into the larger story since the 1970s wherein, as Dorothy Ross has summarized it, "the political shifts that battered the other social sciences served to benefit economics," particularly that segment of the discipline associated with "conservative and libertarian politics."[104]

At this point, one might suspect that the budgetary drama, policy pressures, and constrained opportunities of the Reagan years produced a deep alignment between NSF social science and the Republican administration's policy goals. Perhaps alterations at the level of policies, priorities, programs, and practices were so deep that they even produced a fundamental rupture. That would leave us with two somewhat distinctive periods in this study: one period, running from the Harry Alpert years through the end of Richard Atkinson's tenure as director, when the agency promoted the social (and behavioral) sciences in a nominally nonpartisan and nonideological fashion (even though critics did not always accept this characterization as accurate), and a second period, from the crisis of the early 1980s and continuing throughout Reagan's two presidential terms, when the agency sought to direct these sciences in ways supportive of the administration's policy agenda. But is this a fair assessment?

PARTIAL (BUT NOT THOROUGH) ALIGNMENT

To address this question, it will be helpful once again to focus on the case of economics. No doubt, this discipline enjoyed a relatively favored position compared to the others because common wisdom inside and outside the agency said that NSF support for economics was especially relevant to key policy issues associated with the new economics and the new politics of science. In addition, certain lines of NSF-funded research carried out by prominent scholars such as Harvard's Martin Feldstein received considerable attention at least in part because it was understood that this research supported the Reagan administration's agenda. However, taking a broader view reveals that the NSF's engagement with economics was not characterized by a conservative bent in a more general or profound sense. Thus, it is best to understand this storyline not as a case of a pervasive and deep alignment but as a limited and partial one. Three lines of evidence support this interpretation.

First, if there had been a deep and pervasive alignment, the agency would have been discouraged from supporting work that did not fit well with the administration's agenda. However, the agency did fund such work, at least in some cases. In fact, the PSID, one of the largest and most visible NSF-funded projects in economics during those years, challenged certain views about poverty that were prominent in conservative circles.

Remember that two decades before, poverty had been considered the nation's enemy. Inspired by the spirit of 1960s' reform liberalism, the Johnson administration had launched a war on poverty, supported by a raft of new federal policies, programs, and commitments. But in the 1980s, right-wing policy intellectuals and the Reagan administration itself portrayed the government-supported welfare system as the new enemy. According to them, welfare programs, whatever their designers' intentions may have been, did not really alleviate poverty. Rather, they actually made things worse, by spawning a massive, ineffective, and self-serving bureaucracy, by sapping individual initiative, and by encouraging unhealthy dependency, which, in turn, kept certain individuals and groups mired in a culture of poverty, thereby creating a relatively permanent underclass.[105]

PSID results challenged some of these claims. According to a 1983 document prepared for the NSF governing board, the study had provided "evidence" that contradicted "theories" that assumed there were relatively separate labor markets for the permanently poor and for the rest of the population. Specifically, the findings undermined the view that women and blacks were blocked from moving out of many low-paying and precarious jobs. Furthermore, related research found that poverty caused attitudes of dependency and lack of achievement, but these attitudes did not appear to cause poverty. In the broadest terms, PSID results raised sharp doubts about the claim that a large group of people were in danger of remaining poor because they were caught in a culture of poverty.[106] Four years later, Robert Haveman, an economist and one of the leading academic experts in this area, confirmed that many PSID findings had "challenged conventional wisdom" by revealing, among other things, that "poverty is largely a transitory problem and not a permanent state of affairs" and that "few workers are trapped in certain types of jobs."[107]

Furthermore, the NSF continued to provide PSID with substantial support. As we saw before, when major funding from private patrons ended in late 1983, the NSF director and governing board agreed to assume the lion's share of the financial burden. Accordingly, NSF funding rose considerably: from $1.6 million between 1979 and 1983 to $3.1 million between 1984 and 1986, and then $7.5 million for the next five years.

More generally, as far as NSF leaders were concerned, claims about economic matters put forth by the discipline or by the Reagan administration remained open to scrutiny and could thus be challenged by further research. We have already seen that in 1981, Director Slaughter made this point forcefully, by telling Congress that the agency had a key role to play in funding the basic research needed to assess the validity of the "new economics." The story of the PSID reveals that throughout the 1980s, the notion that the agency should support economic research in a partisan manner, by promoting conservative ideas and policies, never took hold, despite the fact that the next two directors, Knapp and Bloch, were strongly supportive of the White House's broad policy aims.

The second line of evidence concerns continuity in the criteria used to evaluate NSF grant applications. The notion that the agency funded research projects that first and foremost met its criteria for scientific excellence had remained in place ever since its founding. With the 1968 Daddario amendment, consideration about practical relevance became important in many cases as well. However, the agency never suggested that this should include relevance to one or another partisan agenda. To be sure, a good deal of

NSF-funded research in economics addressed issues that had direct bearing on policy issues at the heart of partisan struggles. But recognition of this fact does not undermine the key point that even within the highly charged partisan context of economic policymaking during the Reagan years, the agency never indicated that it would give preference to research projects because they supported the administration's views.

Hence, one should not infer that NSF support for Martin Feldstein, for example, meant that the agency had jettisoned its commitment to scientific merit. Nor did it mean that in light of his work's practical relevance, the agency was acting out of a special interest in research with an explicitly conservative orientation. Feldstein himself emphasized the NSF's distinctive role in providing support for "pure, nonpartisan studies."[108]

Third, unflagging devotion to putatively nonpartisan and nonideological research was reinforced by the NSF's multilayered review process. As the economics program director, Daniel Newlon summarized this process as follows:

> Researchers prepare and submit proposals to the National Science Foundation. My two colleagues and I select six or more specialist reviews for each proposal. Twice a year pending proposals, with any written reviews received, are evaluated by a panel of fourteen distinguished economists. The panel makes recommendations. The staff uses the information in the written reviews and from the panel discussion and recommendations to make its own decisions, subject to approval of the Division Director and the Grants and Contracts Office. Funds are then given to the winning projects.[109]

Nowhere did the agency ever suggest that review of a particular proposal should include its likely contribution to partisan causes of any stripe.

This doesn't mean that individual reviewers or NSF personnel never allowed considerations of partisan relevance to influence their personal preferences for one research project or another. But there is no reason to believe that this sort of consideration significantly influenced the overall process of evaluation on a regular basis. By relying on assessments from many different reviewers and discussions involving a large panel of scholars, all of whom were instructed to give top priority to the scientific merits of individual proposals, the economics program continued to focus first and foremost on promoting work with strong scientific promise.

Newlon himself also looked to give proposals based on controversial ideas a fair hearing. Accordingly, for the peer-review process, he made a point of selecting many young and midcareer scholars, based on the assumption that they would be more open to controversial ideas than senior scholars, who seemed more likely to be set in their thinking. In addition, when Newlon thought the peer-review process had been too harsh in judging a proposal that he considered sufficiently promising, he could request support for that proposal from a special fund.[110] It would be helpful to have more information about how often Newlon used this special fund and what the circumstances of each case were, but he did not provide such specifics.

In any case, the three issues discussed above suggest that there remained a big difference between the NSF's engagement with economics, on one hand, and the support provided by partisan organizations, on the other. Although the NSF was responsive to the Reagan administration's priorities, including its economic concerns, the agency did not become the public equivalent of a conservative research or policy institute, such as the Heritage Foundation, the American Enterprise Institute, or the Hoover Institution. Nor did the NSF become the public equivalent of a conservative funding organization supportive of free-market economics, such as the J. M. Olin Foundation, the Smith Richardson Foundation, or the Bradley Foundation. In contrast to the more academically oriented NSF, with its commitments to the unity of the sciences and scientific objectivity, these private organizations sought to promote projects, people, and ideas with an explicitly conservative orientation.[111]

Equally noteworthy, NSF funding did not become susceptible to partisan interference to the extent as some mission-oriented federal agencies. Consider, for example, what happened at the Department of Health and Human Services. In the early 1980s, budget cutting brought an end to DHHS funding for the PSID, though not at the NSF. Under strong conservative pressure, the DHHS declared that it would no longer support "studies of large scale social conditions or problems." Furthermore, a DHHS call for proposals to study the merits of public- and private-sector social service programs mentioned that the department would evaluate proposals according to the researcher's understanding of the administration's preference for private-sector programs. Specifically, the agency asked scholars to consider if "privately funded programs" operated "more efficiently" than "publicly funded programs" and if the former "are more productive according to commonly accepted measures of service performance."[112]

Not surprisingly, this call for proposals provoked a worry among social scientists that partisan pressure was corrupting the review process, such that proposals most likely to produce results aligned with the administration's partisan goals would have an advantage. COSSA's Roberta Miller thus asked the agency to withdraw its call for proposals because it seemed "strongly political in tone": its phrasing suggested that "the political orientation of the proposal will influence the decision as to who receives the contract."[113]

But the DHHS rejected Miller's request. According to the department's assistant secretary, nothing unseemly was going on: "Obviously, the subjects we study are initiatives that the Administration is interested in."[114]

Of course, one might wonder if this position was—and is—a sound one for a mission-oriented agency. Should an agency such as the DHHS seek to support social science research whose orientation is shaped by partisan pressures coming from the White House—or the Congress? The main point above, however, concerns the contrast with the NSF, where, even in the case of economics, the agency's program and the review process for evaluating proposals continued to reflect a steadfast commitment to scientific excellence rather than partisan relevance.

CONCLUSION

Previous chapters examined an array of challenges to NSF social science during the 1970s, followed by an unprecedented budget-cutting threat and widespread sense of crisis in the wider social science community during the early Reagan years. This chapter has revealed that as the 1980s wore on, serious problems at the levels of funding, organizational position, high-level representation, and status persisted at the natural science–oriented NSF. During the decade, the social science share of funding dropped steeply, from 6 percent in 1980 to 3.3 percent in 1989. Although the notion of giving the social sciences a directorate of their own had received some high-level attention back in 1980, this notion never received serious consideration during the Reagan era, and social science programs at the NSF remained inside the DBBSS, where leadership always rested in the hands of biologists. Against this background, the case of the sociologist Otto Larsen is telling. From 1983 to 1986, he occupied a new position established to address difficulties associated with the lowly status of the social sciences, both at the NSF and in the national science policy arena more generally. But his time there proved to be frustrating. And when Larsen left the agency in 1986, the position itself was discontinued. Meanwhile, some (but not all) national science leaders, including Reagan's chief science adviser George Keyworth and NSF director Erich Bloch, spoke about the social sciences in unflattering terms.

No doubt, none of these problems was entirely new. But during the 1980s, they were all exacerbated by the conservative counterrevolution's continuing depreciation of the social sciences, by the Reagan administration's science policy agenda (with its emphasis on national security, economic revitalization, the natural sciences, and the engineering sciences but not social reform or the social sciences), and by the NSF's responsiveness to that agenda, especially during the directorships of Knapp and Bloch, both Reagan appointees.

Yet, these years were not inhospitable to all of the social sciences, as the case of economics shows. Although the NSF economics program experienced a slight drop in funding from 1981 to 1989, the agency provided this program with a relatively supportive environment. Before Reagan came to power, economics already had a favored status compared to neighboring programs in Sociology and Political Science—due to widespread regard for its allegedly high level of scientific rigor and its quantitative sophistication; the relevance of economic science to pressing public policy concerns such as economic growth, inflation, and unemployment; a strong track record as seen in NSF support for many of the discipline's top prize winners; and effective leadership under James Blackman and Daniel Newlon. When conservative attacks on social science funding kicked into high gear, economists and others pointed out that the agency sponsored many lines of investigation that had undermined confidence in Keynesianism and contributed to the rise of the new economics. During Reagan's two presidential terms, economics also enjoyed a comparatively favorable institutional environment, as seen in the discipline's regular, albeit modest, representation among top decision makers and in the prominence of economic issues in social science work beyond the economics program itself. Finally, economics participated in and benefited from its involvement with big social science, as indicated in the case of large-scale economic modeling during the 1970s and the story of the PSID during the 1980s.

Equally important, we have seen that conservative pressures did not alter the fundamental commitment to scientific advance within the agency and within the economics program in particular. As a strategy for defending against conservative attacks, economists such as Martin Feldstein and Zvi Griliches as well as agency leaders drew attention to NSF-funded research that supported the administration's views. Still, nobody ever suggested that the process of reviewing research proposals and making awards should favor conservative viewpoints per se. Instead, supporters of the economics program, both inside and outside the agency, valued the fact that the agency, both in policy and practice, maintained its longstanding commitment to first-rate scientific inquiry and scholarship at the hard-core end of the social research continuum. Thus, at a time when incentives for pursuing economic inquiry with a decidedly conservative bent had grown strong in the world of conservative think tanks and policy institutes, and while conservative pressures were reshaping the scope and directions of social research programs at mission-oriented agencies such as the DHHS, the NSF's insistence on funding science-driven research—in economics but also more generally in the social sciences—continued to give this agency enormous importance within the national science funding landscape.

Throughout the Reagan years, a commitment to the unity of the sciences, along with the notion that the social sciences were relatively immature compared to the natural sciences and thus the former would do well to take after the latter, remained firmly entrenched at the NSF. Yet, within American society and academia, that scientistic viewpoint received considerable attention and criticism. In fact, whereas during the 1950s and early 1960s, serious challenges to scientism sometimes seemed like relics from a less-advanced or even prescientific epoch, such challenges had gained considerable power since the mid-1960s. What were the main thrusts of these challenges? And what if anything did critics of scientism say about the NSF?

9

ALTERNATIVE VISIONS: FRAGMENTATION BEHIND THE SCIENTISTIC FRONT

> The main inclination ... was to break away somehow from the prevailing paradigms in the social sciences, poor imitations, mostly, of misunderstood physics, and to adapt those sciences to the immediate peculiarities of their supposed subject matter, the human way of being in the world. The aim was and ... still is, not just to measure, correlate, systematize, and settle, but to formulate, clarify, appraise, and understand.
>
> —Clifford Geertz, anthropologist and long-time leader of the Institute for Advanced Study's School of Social Science, 2001[1]

In a 1983 essay, Henry Riecken, former NSF social science division leader, commented on the persistent power of scientistic commitments:

> The policies guiding the selection of research to be supported reflect a view of social science that is epistemologically and methodologically congruent with the position of the physical and biological sciences. In this sense, the influence of NSF has nurtured a science that is positivistic, empirical (as social scientists use this term), quantitative, analytic, value-neutral, and fundamental or basic in orientation. ... By and large, for successful NSF grantees, "social science research" means experiments, field studies, or the quantitative analysis of archival data to test an hypothesis about some basic social, economic, political, or other behavioral process.[2]

Toward the decade's end, yet another NSF-commissioned report from the National Academy of Sciences reaffirmed the dominance of that viewpoint, starting on the very first page: the "proximate goal of the behavioral and social sciences is to discover, describe, and explain behavioral and social phenomena in accord with the canons of scientific logic and methods."[3]

Many social scientists could be happy, for they embraced that position as their own, or at least found it useful when it came to matters of status, resources, and influence. Riecken himself observed that the unity-of-science viewpoint, in which the natural sciences were often taken as the gold standard, remained the "dominant form of contemporary, academic social science."[4]

Yet by the 1980s, opponents of scientism had made substantial headway. From the late 1940s to early 1960s, opponents associated with humanistic, left-liberal, and conservative perspectives engaged in an uphill battle vis-à-vis the scholarly mainstream and its institutional pillars of support, including the federal science establishment and more specific nodes of influence from the NAS to the NSF. But in light of the intertwined political, social, and intellectual upheavals associated most often (although not to say exclusively) with the 1960s, challenges to scientism gained considerable ground in the following decades, within the social sciences, within academia more generally, and within American political culture. Any effort to understand and assess the NSF's importance needs to consider this larger context and the agency's positioning with respect to the core issues under debate.

This chapter begins, in the first section, by examining four sources of ferment that nurtured opposition during the 1970s and 1980s: the legacy of the New Left, the resurgence of conservatism, the Kuhnian vision, and the interpretivist tradition. By no means do these four cover all of the developments and ideas relevant to the ongoing debate over scientific identify. However, the challenges they raised were among the most important ones—other major challenges that would merit discussion in a more comprehensive analysis include certain currents in feminist scholarship, critical race theory, postmodernism, and the work of the French philosopher Michel Foucault. The second section discusses the special force of the interpretivist tradition in challenging scientism's dominant position within federal science policy circles and at the NSF in particular. The third and last section examines how such challenges led to some specific proposals for reform.

SOURCES OF INTELLECTUAL FERMENT

Let's start with the legacy of the New Left. During the 1960s and continuing into the post-1960s era, some proponents of the New Left and its followers—including many sociologists, political scientists, anthropologists, and scholars in related fields—advanced biting critiques of the scientistic outlook. As we have noted before, they raised doubts about the value of supposedly objective and apolitical expertise in the policy arena and wider society. In doing so, they argued that the scholarly mainstream, including prominent areas of study such as the behavioral sciences, structural-functional analysis, and social systems research, gave short shrift to important value-laden research and studies explicitly critical of the status quo. Informed by the ideals of justice, democracy, and human dignity, the latter work seemed especially useful for identifying, scrutinizing, and unmasking oppressive forces in American society and the world more generally (i.e., racism, militarism, colonialism, capitalism, sexism, etc.).

Remember, too, that in the late 1960s, Senator Harris's legislation to establish a new social science agency supported the call for critical scholarship. This was clear in his suggestion that the proposed agency should fund "controversial" studies that could interrogate the unjust status quo and advance constructive alternatives. Although Harris's initiative failed and the New Left itself lost much of its power and identity by the late 1970s, critical lines of analysis associated with the New Left remained prominent in American society and academia, especially in the humanities and social sciences but also in some new interdisciplinary fields, including women's studies.

In addition, critical modes of inquiry that could inform progressive change in domestic and international affairs gained support from the period's most influential left-leaning think tank, the Institute for Policy Studies (IPS; f. 1963). As James A. Smith noted in an important book on the history of think tanks in America, the IPS's founders and intellectual leaders worked to "recover the radical activist spirit of American pragmatism in which ends are discovered and refined in action." These scholar-activists were thus "suspicious of the claims of a 'value-free' social science that could direct policy."[5] Marcus Raskin, arguably the most influential thinker in shaping the institute's orientation, was himself a prominent New Left intellectual and activist in the 1960s and later decades. During the middle of the Reagan years, Raskin spearheaded an ambitious project to establish "new ways of knowing" that critiqued scientistic and associated technocratic modes of analysis, hoping to replace them with what he called "reconstructive knowledge," about nature, society, and culture.[6]

A second source of anti-scientistic ferment came from the resurgence of conservative thought and political culture. We have seen that during the 1970s and 1980s, the push to develop social research, political and cultural analysis, and policy proposals from an explicitly conservative orientation received support from an influential group of right-wing foundations and

Figure 9.1
Marcus Raskin takes part in 25-Year Reunion of the Boston Five, who in 1968 were indicted for conspiracy to aid resistance to the draft. Left to right, Dr. Benjamin Spock, Rev. William Sloan Coffin, Michael Ferber, Marcus Raskin, Mitchell Goodman. Courtesy of Erika Raskin.

think tanks. In earlier decades, concern about embracing an explicitly partisan or ideological orientation had been widespread in these quarters, for such an orientation seemed to leave researchers, policy intellectuals, and their patrons vulnerable to damaging charges of bias. However, in the post-1960s era, a number of organizations, including the John M. Olin Foundation, the Heritage Foundation, and the American Enterprise Institute, became vigorous advocates of explicitly conservative positions on economics, politics, education, foreign policy, legal issues, and other matters. As explained by Jason Stahl, a U.S. political and intellectual historian, these organizations worked to "theorize and 'sell' conservative public policy and ideologies to both lawmakers and the public as large."[7]

As seen in chapter 6's discussion of the attack on MACOS for allegedly promulgating such things as moral neutrality and cultural relativism, right-wing foundations, think tanks, and journalists sometimes targeted the scientistic project as well. Similar to the charges put forth during the McCarthy era (and at other times) by conservative figures, they found the social sciences guilty of contributing to a host of ills during the 1970s and 1980s, including atheism, moral relativism, liberal social engineering, and socialism. At the same time, conservative funding sources advocated lines of political, economic, and social inquiry supportive of the "right" sort of values. This context also informed discussions during the early 1980s' funding crisis, when defenders of the NSF hailed its commitment to putatively objective and nonpartisan research, in contrast to organizations that supported research with an explicitly partisan bent.

The evolving debate over scientific identify also reflected the impact of revisionist accounts of science whose roots were, on the whole, more scholarly and less political in character. Such accounts included a third source of anti-scientistic ferment based on new studies in the history, philosophy, and social studies of science. Here, I will only mention the enormously influential work of the physicist-turned-historian-and-philosopher-of-science Thomas Kuhn. In 1962, Kuhn, then a professor at the University of California–Berkeley, although he soon moved to Princeton and later to MIT, published *The Structure of Scientific Revolutions*, containing his famous ideas about normal science, scientific paradigms, paradigm shifts, and scientific revolutions. Based mainly on investigations of the physical sciences, especially astronomy, physics, and chemistry, Kuhn argued that in science, an established paradigm channels the intellectual activity of a given research community in an orderly manner such that it leads to successful puzzle-solving, advances in knowledge, and cumulative scientific progress. However, from time to time, a period of revolutionary change culminates in the replacement of an established paradigm with a new one.[8]

What, if anything, did this account mean for the social sciences and their scientific standing? Kuhn himself did not say much, although he did suggest that they were in a preparadigmatic state of development. This reinforced the notion that they were lagging behind the natural sciences and especially the physical sciences. Furthermore, if to be a science depended on having a paradigm, it might seem that the social sciences were not real sciences, at least not yet.[9]

Beyond Kuhn's limited comments, a cottage industry of scholarly studies revealed divergent viewpoints regarding implications for the social sciences. Not surprisingly, one common response from social scientists focused on strengthening the scientific status of their work by claiming that specific disciplines or particular research communities operated in a manner similar to Kuhn's physical science communities. As the philosopher of science Steve Fuller has observed, a good number of sociologists, political scientists, and the like found Kuhn's account attractive because "it seemed to provide a blueprint for how a community of inquirers can constitute themselves as a science, regardless of subject matter."[10]

But another and quite different response held that Kuhn's work showed why the social sciences should not take physics (or any other science) as a model. Accounts in this vein pointed out that in Kuhn's view, the establishment of a paradigm in a particular field depended not on following general or abstract rules about the process of scientific inquiry, but on developing a community-wide consensus about the field's foundations. From this viewpoint, it seemed that social scientists should proceed not by embracing some set of abstract notions about the proper conduct of scientific inquiry associated with physics or chemistry, but by deciding for themselves what sort of inquiry, what type of investigative framework, and what set of methods would be most appropriate in light of their field's specific character, needs, and opportunities. In a 1980 survey of the lively discussion about Kuhn's influence, Gary Gutting, another philosopher of science, captured this position nicely with the following suggestion: if the social sciences were "lacking the consensus on a single paradigm," then perhaps the most valuable take-away message for the practicing scholar was not to insist on one or another paradigm but, rather, to "cultivate" one's "own garden."[11] As we saw in chapter 4, leftist scholars wanting to reform the discipline of political science also invoked Kuhn's ideas as justification for a paradigm shift.

Growing opposition to scientism also gained support from a resurgence of the interpretivist tradition, whose roots lie deep in the history and philosophy of modern social science. In broad brushstrokes, proponents of this tradition held that knowledge of social phenomena depends on understanding the meaning of social actions, and understanding requires interpretation on the part of the investigator. Consequently, the social sciences differ in fundamental respects from the natural sciences, which don't have to deal with meaning, understanding, and interpretation in the specific ways needed in the social sciences. We have encountered key tenets of this fourth source of intellectual ferment before, including in Harry Alpert's view that the study of human actions in social context depends on *verstehen* analysis, which produces rich qualitative understanding of human meanings, intentions, and purposes. Arguably, such research was quite different from work in the physical sciences at the levels of ontology, methodology, epistemology, and practical value. During the 1970s and 1980s, interpretivism acquired renewed energy as it was embraced by leading scholars in various disciplines.[12]

The interpretivists found a powerful advocate in the figure of Clifford Geertz. An anthropologist trained in the Boasian tradition of fieldwork and cultural interpretation, which emphasized the need to understand the uniqueness of each culture, its particular constellation of values and standards, and its singular historical trajectory, Geertz received his PhD from Harvard's Department of Social Relations in 1956, then worked as a professor at the University of Chicago. In 1970, he moved to the Institute for Advanced Study in Princeton, where he helped establish the School of Social Science. As the school's founding figure and long-time leader, Geertz nurtured an interdisciplinary milieu conducive to interpretivist study. This meant welcoming contributions from scholars in the humanities, such as history and literature, and from humanistically oriented social scientists.[13]

The main goal, as noted in this chapter's opening quote from Geertz, was to push beyond "the prevailing paradigms in the social sciences," which seemed to be "poor imitations, mostly, of misunderstood physics," with the aim of adapting these sciences "to the immediate peculiarities of their supposed subject matter, the human way of being in the world." Scholars at the

Figure 9.2
Clifford Geertz, participating in a political theory panel, celebrating twenty-five years of the IAS School of Social Science. Left to right, William Sewell from the University of Chicago, Clifford Geertz, and Laura Englestein from Princeton University. May 1997. Photo by Randall Hagadorn. From the Shelby White and Leon Levy Archives Center, Institute for Advanced Study, Princeton, NJ, USA.

School of Social Science were thus encouraged to free themselves from "the narrowed confines of a fixed and schematized scientific method." Liberation would enable them to address "moral, political, and spiritual concerns," which, in turn, meant reaching out to "the humanities—to philosophy, literature, history, art, religion."[14]

In the context of this multipronged attack on scientism described above, what, if anything, did critics say about federal science policy and the NSF?

THE SPECIAL FORCE OF THE INTERPRETIVIST CHALLENGE

One cannot fail to be impressed by the persistence of the NSF's scientistic commitment throughout the 1970s and 1980s, as we have seen in agency policies and programs as well as in efforts to defend those by agency leaders and other supporters in the scholarly and political arenas. Yet the growth of anti-scientistic ferment meant that the conception of the social sciences advanced by the NSF was increasingly suspect from a number of different viewpoints. The interpretivist critique received the most attention.

A 1977 congressional background report on NSF programs—mentioned in the previous chapter—pointed out that some social scientists were unhappy with the agency's hard-core emphasis. They questioned its effort to "impose the criteria of the physical and natural sciences on subject matter which does not lend itself to quantification."[15] That same year, discussion of this issue appeared in *Science*, in its coverage of a dispute over the criteria used to review NSF proposals in social and cultural anthropology.

The inciting incident occurred in February, when an article by journalist Gina Kolata reported that the NSF anthropology program officer, Nancie Gonzalez, thought that scholars in the social and cultural wing of this discipline received relatively few research grants because they often failed to present their work in rigorous scientific terms. According to Kolata, Gonzalez "was shocked when she first saw some of the 'mushy' proposals submitted by social anthropologists." Thus, to have a better chance of receiving a grant, these scholars would need to satisfy the agency's scientific criteria better. In a follow-up letter, Gonzalez—who, interestingly, was a social anthropologist herself—added that archeologists were more successful in getting grants because they did a better job in their applications of stating a specific problem and explaining why it was interesting and important.[16]

Although Kolata's article said little more about what Gonzalez meant by thinking scientifically, the history of NSF funding for anthropology suggests why her remarks provoked critics of scientism. During the period of convergent research in the mid-1950s, the NSF had privileged work that overlapped with the natural sciences, particularly physical anthropology and archaeology. As early as 1956, it began to support social and cultural anthropology as well, but mainly for work on the hard-core end of the social research continuum. Over the years, this included studies on the functional integration of social systems and on the general processes of cultural evolution. Meanwhile, scholars from the discipline's humanistic wing, who sought to understand the uniqueness of particular cultures in their specific historical contexts, had a tough time getting funding.[17]

Against this background, Gonzalez's position did not sit well with anthropologists who believed its implications for scholarship were stifling. In the same issue of *Science* that published Gonzalez's letter, another letter from Richard Adams, a Latin American specialist from the University of Texas, complained that the NSF continued to insist on narrowly construed scientific criteria, as seen in its favoring of research that used "the hypothetical-deductive method." As Adams saw it, that outlook reflected the more general "behavioristic fad" so prevalent in American social science during the last quarter century. Consequently, important types of anthropological inquiry, including "general ethnography" and studies employing an "interpretative approach," would have difficulty acquiring funding. If this situation persisted, the overall impact would, Adams predicted, be "simply disastrous" for the discipline.[18]

To underscore the depth of this problem, Adams speculated that if the NSF continued on this path, even such a luminary as Clifford Geertz would be excluded. Although Geertz was "certainly among the most distinguished anthropologists of his generation," an application from this champion of interpretive studies would probably not satisfy "the 'scientific' criteria" used by the agency's anthropology program.[19]

Placed immediately below Adams's letter in *Science* was another letter of complaint, this one signed by eight scholars. Many had positions at top universities, including Harvard, Yale, Princeton, and the University of California–Berkeley. And all of them were anthropologists, including Geertz himself. Their letter charged that the anthropology program officer's approach was misguided because "social science is neither biology nor physics." In an exasperated tone, they asked,

> Are we to assume that NSF has an official policy subscribing to a simplistic and rigid view of social science harkening back to 19th century positivism or, even more disturbing, to an authoritarian insistence that those dispensing funds may dictate to scientists what science is or is not?

Hoping to correct such a view, this group of eight urged the agency to abandon its one-shoe-fits-all outlook, because the "study of society involves factors of value and history radically different from research in those sciences apparently endorsed by Gonzalez as models we must emulate."[20]

Gonzalez responded with her own letter to the editor, claiming that those criticisms were unfair. She also claimed that one reason the NSF gave less support to social anthropology and more to archaeology was that "much respectable social anthropological research addresses humanistic, rather than scientific questions."[21] Thus, in the end, neither side conceded much ground in the ongoing debate over scientific identity and the agency's steadfast position.

During the Reagan years, the interpretivist challenge received favorable consideration from a small but well-informed group of scholars who were concerned about the mounting political attacks on the social sciences and associated threats to their funding. Among this group was SSRC president Kenneth Prewitt. Regarding the June 1981 strategy meeting in New York (discussed in chapter 7), Prewitt observed that many social science leaders and supporting organizations wanted to establish better ties with the natural sciences. Elsewhere, however, Prewitt acknowledged the seriousness of the interpretivist challenge to the unity-of-science standpoint associated with that strategy. According to what Prewitt called the "dominant tradition of American social science," the basic data of inquiry consisted of "systematic observations or measures of individuals or groups, extracted for analytical purposes from their larger context." Scholars treated these observations as"discrete variables," "hypothesized causal relationships between them," and tested the hypotheses "by examining the impact of changes in one variable on changes in the others." "Modeled on the physical sciences," this type of investigation emphasized "the use of quantitative data, especially survey data, and to a lesser extent, experimental data." The underlying expectation was that "an understanding of relationships among significant variables will enable society to develop the means for alterations and improvements." Prewitt believed that critics of this tradition deserved serious consideration.[22]

Foremost among the critics were the advocates of "interpretive social science." Their humanistic approach, Prewitt observed, "seeks a detailed understanding of the meaning of actions, customs, events, and institutions to the individuals and groups that perform and participate in them." To obtain this special type of understanding, interpretivists employed distinctive methods, including the "techniques of literary analysis." They employed "metaphor, narrative, and scripts as tools for understanding social behavior." Summing up, the SSRC president emphasized that the interpretivist challenge provoked deep controversy "because it redefines the objectives of the [social science] enterprise, the kind of knowledge desired, and the appropriate ways of obtaining that knowledge."[23]

Although the previously mentioned 1982 NAS report defended the unity-of-science viewpoint (as NAS reports always had done and would continue to do), this report also acknowledged that interpretivism offered an alternative. As presented here, the stakes in terms of understanding the nature of the social sciences and promoting them in a healthy way could hardly have been bigger. Similar to other accounts that presented the matter as a contest between irreconcilable positions, the report identified "two largely competitive visions" of the social sciences. Their existence lent credence to the idea that these sciences comprised an "intermediate" branch of inquiry between the natural sciences and the humanities—an idea suggested three decades earlier by the British cultural commentator C. P. Snow. While the "dominant vision" upheld the unity-of-science position and sought to establish broad generalizations or laws of social behavior, the minority vision, held by "a sizable fraction" of scholars in several disciplines, didn't aim at "generalization—many would question whether the sort of generalization that characterizes the physical and biological sciences is possible—but interpretation." These scholars viewed human behaviors and social arrangements as arising from particular concrete historical circumstances best studied by paying careful attention to those circumstances. They emphasized the goal of "understanding what is distinctive rather than what is general," a difference, as the NAS report also pointed out, that Clifford Geertz had referred to as the "cases-and-interpretations" versus the "laws-and-instances ideal" of explanation.[24]

NEW CALLS FOR REFORM

In light of the gathering challenges, some scholars called for major reforms in federal social science research policies. It would be a mistake to believe that this push for reform received widespread attention within the federal science establishment or within the NSF. And the agency had no program to support qualitative social research, philosophical social inquiry, or critical scholarship that could have provided a complement to its many disciplinary and nondisciplinary programs that supported research grounded in quantitative methods and putatively objective, value-neutral analysis. I have found no evidence, either in published documents or archival records, that suggests the agency even considered the idea of creating a program dedicated to those alternatives, never mind giving its existing programs responsibility for supporting them. If somebody did suggest as much, the idea certainly did not gain traction. Nevertheless, as we will see below, the reformist agenda that challenged the established NSF framework was advanced by a small group of sociologists, political scientists, and, interestingly, a mathematician who had substantial knowledge about funding policies and their impact on the production and uses of knowledge.

In the early 1970s, the sociologist Harold Orlans reported on the deep divide over Senator Harris's NSSF proposal within the social sciences, as was noted in chapter 4. Orlans himself also advocated a new approach to funding, which he described as the exact "opposite of that which governs scientistic research." More specifically, he said federal policies should encourage research where people are "described as people—as children, citizens, unemployed blacks or busy congressmen—not as abstract and timeless 'subjects'"; where institutions are described as "specific organizations with stated characteristics, not as an abstract form which exists only in the academic mind"; and where data are understood to be "the residue of certain procedures employed by designated persons, not as elementary particles of an unchanging universe."[25]

Other scholars argued that the established funding regime with its scientistic commitment had contributed not only to troubling limitations on the pursuit of knowledge but also to widespread disillusionment with applied social research. How the project of advancing the social sciences in the image of the natural sciences could have such a disconcerting effect was explained by the Princeton political scientist Richard Nathan, in his book *Social Science in Government: Uses and Misuses* (1988). After studying a series of cases involving the public-policy relevance of applied social science, Nathan concluded that the "tendency to emulate the natural sciences" had become a bad habit. As a result, social scientists generally preferred "quantitative research designs and techniques," while they unfairly dismissed or simply overlooked the value of "qualitative research methods and data." These were serious mistakes, asserted Nathan, as the quest to make social research scientific through the application of quantitative methods often resulted in a "spurious precision." Furthermore, although the effort to construct scientific models that made accurate predictions had proven effective in certain natural science fields, in the social sciences this effort tended to produce poor results. Was it possible to "predict human behavior using the 'objective method'" allegedly found in the natural sciences? The answer, said Nathan, was definitely not, because the data needed to construct such models did not exist or could never be collected. The natural science model of inquiry had turned out to be the real problem, then; work based on this model had failed to produce the promised policy-relevant results, which, in turn, fueled disillusionment with the social sciences more generally.[26]

As part of this wider rethinking about intellectual foundations and practical value, the impact of the national funding system on the social sciences received critical scrutiny from the Yale mathematician Neil Koblitz. As he saw it, the political economy for scholarly research had unfortunately privileged work of a supposedly rigorous scientific nature at the expense of humanistic studies, including humanistic social research. Not only did "prestige and money go disproportionately to the sciences, especially applied science and engineering," but "the administrations at most large universities" also tended to favor "those departments that are most successful in attracting outside funding." Professors "in social and behavioral fields" thus had strong incentives to portray their work "as being closer to the sciences than to the humanities." On the other hand, "professors in those fields who prefer the honestly subjective approaches of earlier years are often regarded as old-fashioned and out of step." A fierce critic of the misuses of mathematics and quantitative methods in social research, Koblitz issued a stern warning:

> As long as our society continues to undervalue humanistic and cultural pursuits, we can expect to see a proliferation of mathematical jargon and pseudoscience in fields far removed from what one would have thought to be readily quantifiable—fields in which mathematical methods are rarely appropriate and are often misused.[27]

Against this background, the NSF itself came under scrutiny as well. In a 1969 book, the sociologist Gene Lyons had presented a history of the "uneasy partnership" between the social sciences and the federal government. Two decades later, in a 1986 essay aptly called "The Many Faces of Social Science," Lyons pointed out that previous efforts to reform the federal science establishment in ways that would have encouraged social scientists to pursue a wider range of work had ended without much change. Not only had Senator Harris's NSSF proposal failed, but a series of national science policy reports published at the end of the 1960s stood firmly behind the scientistic position. With dismay, Lyons added that two decades later, NSF funding still "tended toward a narrow range of work."[28]

In Lyons's view, the basic problem arose from a mistaken unity-of-science notion and the oft-invoked corollary that said the social sciences could and should try to emulate the natural sciences. Yet he emphasized that he was not against calling the social sciences scientific. Similar to the position taken by the economist Carl Kaysen and some others during the NSSF debate, Lyons wanted people to recognize that a great diversity existed among the sciences, even though popular English-language usage of the term science obscured this fact. He proposed that one might reasonably say that the social sciences were scientific, just as one might justifiably assert that they should employ "rigorous methods of research." However, this did not mean that the social sciences were just like the "other sciences." Similar to what Senator Harris had suggested, and thus contrary to the stance taken by a Herbert Simon or by any of the NSF directors up through Erich Bloch in the 1980s, Lyons proposed that the social sciences should be encouraged to push forward on multiple and diverse fronts, maintaining "strong lines" to the natural sciences for certain purposes but simultaneously pushing hard to "break out on their own" for other purpose.[29]

The call for reform went one step further in the hands of the sociologists Samuel Klausner and Victor Lidz, who questioned whether one could realistically expect the social sciences to achieve a breakout on their own. In 1986, they published an edited volume of essays called *Nationalization of the Social Sciences*, which included the aforementioned pieces by Lyons and Riecken as well as the first published version of Talcott Parson's stillborn report for the SSRC in the late 1940s. According to Klausner and Lidz, increased federal funding for the social sciences since World War II had proven to be "a mixed blessing." Early in the postwar NSF debate, Parsons and a clutch of other SSRC scholars had claimed, perhaps too optimistically, that these sciences could be part of a new national science agency without relinquishing scholarly autonomy. In the ensuing decades, however, the undeniable power of the federal purse together with selective funding policies had given the NSF considerable "leverage ... to impose certain standards" on social science research and training.[30]

Of particular concern, the social sciences had suffered from "fiduciary overlordship" exercised by the natural sciences. Lidz and Klausner suggested that an entire generation of social scientists had been influenced in questionable ways by NSF policies and programs together with a broader climate of opinion shaped by "figures who are neither social scientists nor profoundly schooled in the nature of social scientific knowledge." The resulting problems included an unhealthy split between quantitative scholarship and more richly descriptive qualitative scholarship.[31]

Looking for a way forward during a period of lean budgets and conservative displeasure, Klausner and Lidz reasoned that the NSF was not about to reform its ways. An examination of the agency's track record in combination with political, institutional, and ideological pressures during the Reagan years rendered the prospects for a major reorientation in social science funding unlikely. The two sociologists thus called for reviving the proposal for a separate social science agency. To be sure, the road ahead would not be easy. Common worries about the relationship of social science to social ideology and social reform would surely cause headaches for the agency they envisioned. The possibility that its mandate to promote the social sciences would stir up conservative worries that it was supporting a "form of socialism" could not be dismissed either. For social scientists, letting go of natural scientists' coattails would also mean losing some supporters, certainly in the short term.[32] Nevertheless, if they had an agency of their own, the pressures to mimic the natural sciences would, presumably, be reduced. The influence of natural scientists on funding policies and priorities would decrease dramatically as well. Presumably, the proposed agency would be able to support a much broader range of social research than the established NSF could.

CONCLUSION

Appreciating the significance of NSF social science policies, programs, and practices requires that we consider them in light of evolving challenges to the scientistic project. Two of the main challenges considered here, the legacy of the New Left and the resurgence of conservatism, were inspired by political and intellectual movements associated with left-leaning and right-leaning agendas. The other two challenges were rooted mainly—although not to say exclusively—in academic currents in the history and philosophy of the sciences and were associated with the figures Thomas Kuhn and Clifford Geertz. All four challenges had gained considerable momentum since the late 1960s. Throughout the 1970s and 1980s, they informed the broader discourse concerning the social sciences in American politics, science, and higher education.

Anti-scientistic criticism thus produced a palpable degree of fragmentation behind the scientistic front. The particular challenge posed by the interpretivists received special attention from a group of figures with substantial knowledge about the politics and funding of social science. SSRC president Kenneth Prewitt and a number of anthropologists underscored the power of this challenge and the central contributions of Geertz, who by this time was the head of the School of Social Science at Princeton's Institute for Advanced Study. In addition, passages from NSF and NAS documents acknowledged the importance of the interpretivist challenge in a respectful tone. Meanwhile, some well-informed social scientists, such as the sociologist Harold Orlans and the political scientist Richard Nathan, argued that the dominance of a natural science model of inquiry had actually done significant harm to the social sciences and their standing in the wider society.

Yet the multiple sources of anti-scientistic ferment had little impact at the NSF and within the federal science establishment more generally. During the 1980s, no legislative proposal comparable to Senator Harris's NSSF initiative emerged, even though the sociologists Klausner and Lidz called for a revival of that initiative. As the 1980s and the Reagan era came to an end, any hope that the flourishing of alternative views about scientific identity and social relevance would inspire major reforms at the NSF, never mind the creation of a separate social science agency, seemed nothing more than that—just a hope.

10

THE SOCIAL SCIENCES AT THE NSF: PAST, PRESENT, AND FUTURE

> Mark Twain once wrote that history does not repeat itself, but it rhymes. One good reason for studying the past is to make ourselves more sensitive to these rhyming patterns.
> —Robert M. Collins, historian, 2000[1]

> The point is not to belabor the aphorism that "those who fail to study the past are condemned to repeat it," but rather to recognize that both continuity and change in history need to be understood to deal effectively with the present.
> —George Mazuan, former NSF historian, 1994[2]

So far, we have examined the NSF's engagements with the social sciences from the mid-1940s to late 1980s, by focusing on four somewhat distinctive eras. The first one encompassed the agency's legislative origins and the debate about whether the social sciences should be included, followed by the initial development of NSF social science policies, programs, and practices during the Alpert years. Next came an expansionary era, from the time of Sputnik through the heyday of 1960s' reform liberalism. The third era saw mounting criticisms during the increasingly conservative 1970s and a loss of momentum and retreat. The fourth period spanned the two Reagan administrations and was marked by even deeper troubles.

This final chapter begins by taking stock of what we have learned, by reviewing some of the key findings. The second section shows how understanding what happened from the end of World War II through the end of the Reagan years provides a valuable foundation for interpreting subsequent developments and the continuing controversy about NSF social science, leading up to present day. In the third and last section, I offer an assessment of what the agency has done well, where it has fallen short, culminating with a call to reconsider the proposal to create a national social science foundation as a basis for moving forward in the future.

TAKING STOCK

THE POLITICS–PATRONAGE–SOCIAL SCIENCE NEXUS

At the broadest level of historical significance, we have seen that the NSF was a patron of crucial importance in the evolution of the politics–patronage–social science nexus in Cold War America. To appreciate this point, we have situated NSF social science in the context of major changes in private and public funding. World War II marked a crucial divide. Previously, the large private foundations had taken responsibility for promoting the development of academic social science on a broad front. But after the war and continuing into the Cold War era, the federal government assumed a much greater role. Furthermore, although the Ford, Rockefeller, and Carnegie foundations provided considerable support for a wide range of social science research, especially investigations related to practical concerns such as population growth or development in the Third World, over time the large foundations relinquished any responsibility for looking after the health of social science in a general sense. This was especially clear after the Ford Foundation decided in the late 1950s to close its behavioral sciences program. So, even though the social sciences received only a thin slice of the dramatically enlarged federal science budget during the Cold War years, support from federal agencies became increasingly important. Moreover, following the 1960s' debate over Project Camelot and protests against military-funded social research units on university campuses, the political and academic communities recognized the need to ensure substantial funding from civilian agencies.

Those developments opened up a critical space within the politics–patronage–social science nexus for NSF social science to grow in size and significance. The agency emerged as a major patron, especially for academically oriented scholarship carried out at the nation's research universities and social science research centers closely affiliated with the academic world. As this happened, the agency's commitment to basic science, its dedication to first-rate research, and its special role in looking after the overall health of the sciences made it increasingly important within the national social science funding system.

Yet none of this seemed likely in light of the agency's legislative origins and early development. In fact, during the postwar NSF debate, considerable opposition to the social sciences had left it unclear whether they would be included in the new agency at all. The 1950 charter also failed to mention them directly. And when the agency decided in the early to mid-1950s to test the social science waters, it proceeded cautiously, starting with two programs in "convergent research" that were located in its natural science divisions and considered "experimental."

SCIENTISM

If the future importance of NSF social science was not readily apparent, that formative period nevertheless had a great impact, because it established the agency as an intensive site for scientific boundary work and a vigorous advocate of a scientistic strategy for promoting the social sciences. In retrospect, it almost seems as if fate had decreed that the NSF would have a special role. Right from the outset, the question of whether the social sciences should be included or not prompted interested parties in the scientific and political communities to worry about the matter of scientific credentials. Although questions about the scientific status of the social sciences along with related questions about their social relevance and their involvement with political affairs were certainly not new, these issues obtained heightened significance when a loose alliance of conservative skeptics in the natural science and political communities managed to place the social sciences on the margins. Under those conditions, leading scholars from the Social Science Research Council (SSRC) agreed that gaining a place in the NSF was a top priority and advocated inclusion based on a unity-of-science stance that recognized the social sciences as junior partners to the allegedly more advanced natural sciences. The course of that debate together with subsequent developments during the McCarthy era led to a common understanding that securing public funding would require concerted efforts to distinguish the social sciences from such things as socialism, social reform, and social philosophy and to convince powerful natural science leaders that they were maturing as legitimate sciences.

In that context, the NSF hired Harry Alpert for the sensitive task of studying and making recommendations regarding social science funding, which he did by crafting a carefully circumscribed framework that focused exclusively on promoting work at the hard-core end of the social research continuum. This viewpoint, which was readily accepted by NSF leaders as a basis for moving forward in a cautious manner, assumed a fundamental unity of the sciences, often accompanied by an understanding that the natural sciences were the gold standard.

Moreover, that framework remained in place long after the Alpert years. In fact, the agency's dedication to funding the hard-core, as defined by allegedly universal criteria and associated inside the agency with scientific rigor most commonly found in the natural sciences, became firmly entrenched over the long run. Remarkably, at no point did anyone inside the NSF ever mount a serious challenge. Recall how quickly Father Hesburgh's call for including representatives with a wide range of viewpoints on the agency's social science advisory panel was shut down in 1958, a time when the agency's scientistic strategy was still rather new and thus perhaps could have been reformed before hardening and becoming part of the bedrock.

THE NSF

Beyond its rhetorical importance, the scientistic strategy became deeply consequential in the evolution of NSF policies, programs, and practices. This took place through the elaboration of funding criteria, including the original trio of principles proposed by Alpert (i.e., objectivity, generalizability, and verifiability), which were then implemented in policies, programs, and practices through organizational units dedicated to the social sciences, the behavioral sciences, and more specialized programs corresponding to the major disciplines or another meaningful rubric, such as social indicators. Throughout this study, a number of cases have revealed how the agency's programs implemented the general commitment to the unity of the sciences, the underlying imperative to make the social sciences more rigorous, and the oft-repeated aim that these sciences would follow in the footsteps of the natural sciences. These cases include the development of convergent research programs during the early to mid-1950s; the process by which political science gained inclusion during the 1960s, as well as the agency's support, from the mid-1960s through the mid-1970s, for social science curriculum building through MACOS; the development, during the 1970s and continuing into the 1980s, of programs for social indicators and for big social science, including large quantitative databases, the National Election Studies, social survey research, and econometrics; and the conditions that gave economics a relatively privileged standing and stronger funding compared to the other well-established social science disciplines.

The scientistic strategy also acquired importance because it was closely associated with the social sciences' second-class status. Abundant evidence for their lowly status includes the fact that from the early 1950s to late 1980s, they received only about 2.9 percent of total NSF research support—about $580 million out of nearly $20 billion.[3] In terms of their organizational standing, the social sciences first got a toehold in the convergent research programs under the auspices of the agency's natural science divisions. A major advance came in the early 1960s with the creation of a separate social science division. But, then, as part of the mid-1970s' agency-wide reorganization, the social and behavioral sciences were split into separate units, and both were placed in a new directorate that was always led by a biologist and in which the biological sciences enjoyed higher status and greater funding. Meanwhile, natural scientists and, in the later years, engineers had a dominant presence in the top leadership positions. Of the nine men appointed NSF director by seven different presidents, only one could be considered a social scientist. Furthermore, this was Richard Atkinson, a mathematical psychologist whose scholarly interests concerned psychology at the level of the individual, rather than any type of social analysis, and whose intellectual outlook and professional trajectory made him comfortable in an environment where the natural sciences led the way. At most points, the social sciences also had only one or two representatives on the agency's twenty-four-member governing board.

During the postwar NSF debate, Vannevar Bush had suggested that a future "partnership" between the social sciences and the natural sciences could be used as a basis for inclusion. But neither he nor the agency's leaders from the 1950s to the 1980s ever tried to make this a partnership of equals.

To be sure, a belief that the social sciences were part of a unified scientific enterprise did not necessarily imply second-class status. Indeed, one could argue the opposite. And the reasons why the social sciences were subordinate to the natural sciences rested not only on the question of scientific identify but also on a widespread belief that compared to, say, political science or sociology, physics, chemistry, biology, and engineering had much greater practical importance. Still, at the natural science–oriented agency, doubts about whether the social sciences were really scientific gave rise to a seemingly unending struggle to demonstrate that at least some types of social research were really scientific. The scientistic strategy thus reflected and contributed to the subordinate status of the social sciences. As Alpert's successor Henry Riecken put it, "essentially" the social sciences embraced "a strategy of protective coloration, of allying one's cause with stronger others, a strategy that has been used by countless minorities and other underdogs to secure a share of power and position."[4]

THE FEDERAL SCIENCE ESTABLISHMENT

The story of NSF social science cannot be understood without seeing the agency as an important part of a much larger and dynamic federal science establishment. The social sciences' low status reflected the fact that in the immediate postwar years and throughout the Cold War era, they had much lower status and considerably less influence than the natural sciences, especially the physical sciences, in the rest of the federal science establishment (i.e., in the massive science programs in the Defense Department, in the prestigious and influential National Academy of Sciences, and in the top echelon of national science policy advisers, including the President's Science Advisory Committee). However, as the NSF grew in size and importance, it emerged as a crucial site for thinking about the nature of the social sciences, the path to progress, and the ways in which the federal science establishment could contribute to that progress. Here, the agency's scientistic strategy came to exert a powerful influence in its own right.

As a civilian science agency dedicated to first-class research and as a major node within the federal government for scientific boundary work, the NSF exercised its influence as an unwavering and vigorous advocate of scientism through a range of channels. These included regular announcements about its social science programs and funding criteria, more detailed discussions in publications by program leaders, and coverage of the social sciences in annual reports. In annual appropriations hearings and other congressional hearings concerning the social sciences and government programs, NSF leaders explained and defended its social science policies and activities. The agency also commissioned studies from the NAS, which resulted in a series of published reports, including the 1976 Simon Report. When it came to promoting the social sciences as junior partners in a unified scientific enterprise and dismissing any significant alternatives to the scientific strategy, the NAS was a consistent and strong ally.

Within the federal science establishment, the NSF also became an important site for working out the social relevance of the social sciences. In its early days, the agency's basic science mission plus the pressure to keep its social science efforts far from social problem solving and public policy matters suggested that the best way of understanding the relevance of NSF-funded research involved its long-term payoffs: presumably, hard-core social research would yield advances in knowledge, some of which would have practical applications, while others would at least contribute to the general process of enlightenment in society. The events of the 1960s, however, placed increasing pressure on the agency to pursue relevant research, culminating in the passage of the Daddario amendment. This new context stimulated extensive efforts to link social science to social action in a more deliberate fashion. With this aim in mind, the NSF created a new program called RANN—Research Applied to National Needs—that provided funds for scientific studies, often interdisciplinary, designed to address specific issues and serve identifiable user communities.

Yet, within a short time, a number of developments, including widespread criticism of social science–informed social programs associated with the 1960s' Great Society, produced considerable disillusionment. RANN also inspired anxiety among defenders of the agency's original and still dominant commitment to basic research. Moreover, the agency itself pulled back from the high hopes associated with applied social science and the social engineering viewpoint in particular. In the 1970s, NSF publications incorporated criticisms of such work put forth by a growing chorus of scholars, including Carol Weiss and others from an emerging field called the sociology of knowledge utilization. Subsequently, with the help of the 1976 Simon Report, which roundly criticized RANN, the agency reasserted its commitment to basic social science and disavowed any commitment to support social research directly concerned with social problem solving and social action.

As revealed by the difficulties associated with Proxmire's Golden Fleece Awards and then by the new politics of science during the Reagan era, making the case that the NSF wasn't wasting taxpayer money on esoteric research projects remained a pressing issue as well. Indeed, the problem of establishing relevance hardly subsided, as seen in the efforts by the NSF and its supporters to show that agency-funded social research had great importance for understanding and addressing urgent national priorities, such as economic revitalization. Furthermore, it wasn't only economists who argued vociferously for their relevance in this way. So, too, did a broader range of scholars from the social and psychological sciences, including a number who were closely associated with the NSF. Equally significant, however, defenders of the NSF insisted that funding for basic research in particular had great value, because it provided scientific knowledge of a nonpartisan, value-neutral, and objective sort that could serve as the basis for effective public policy making, especially in light of supply-side economics or some other policy-relevant doctrine whose validity had not yet been examined in a rigorously scientific manner. Shortly after the Reagan years, David Featherman, a sociologist and the SSRC president, contributed to the ongoing discussion by suggesting it would be useful to think about "mission-oriented basic research."[5]

AMERICAN POLITICAL CULTURE AND PARTISAN POLITICS

The previous chapters have also highlighted the significance of the NSF as a frequent focal point for discussions in America political culture about the nature and meaning of the social sciences. We have paid special attention to congressional deliberations and controversy, which occurred with regularity as part of the annual budget-making cycle. Because of the NSF's status as an Executive Branch agency, presidential administrations also weighed in at certain moments, and sometimes with major consequences, as seen, for instance, in the case of the early Reagan administration's efforts to slash social science funding. Furthermore, whereas the intensity of political and academic discussions about other social science patrons waxed and waned considerably over time (i.e., the 1960s' debate over military patronage reached a highpoint in the mid-to-late 1960s but diminished dramatically shortly thereafter), discussion of the NSF was more or less constant.

Furthermore, this part of the story was strongly marked by partisan differences, with the most consistent and strongest support for the social sciences and federal funding coming from liberal quarters, while conservatives were much more likely to raise doubts and propose cutbacks. This dynamic surfaced right at the beginning, in the postwar NSF debate, when the effort to include the social sciences was associated with the liberal policy agenda of President Truman and Senator Kilgore, while a series of conservative legislators raised sharp doubts about their scientific credentials and charged them with being political and ideological in character. The general pattern of partisan support and criticism reappeared once the agency was up and running. Thus, more favorable conditions prevailed during the liberal 1960s, which encouraged expansion and deepening of NSF social science, giving rise to important legislative initiatives from Representative Daddario and Senator Harris that aimed to strengthen federal funding, and resulted in passage of the Daddario amendment (although Harris's proposal for a new social science agency failed). By contrast, during the increasingly conservative 1970s and 1980s, American political culture supported mounting criticisms and efforts to slash social science funding, as seen in the stories of MACOS and the crisis of the early Reagan years.

THE SOCIAL SCIENCES

All of the developments discussed above contributed to the broader story of the social sciences in recent American history. They shaped the intertwined stories of social science patronage, the place of the social sciences in the federal science establishment, and their status in American political culture.

The NSF also encouraged the development of the social sciences along lines that meshed well with its scientistic strategy. With the help of a multilayered evaluation process and heavy reliance on peer review of proposals submitted by the scholarly community, the agency sought to fund first-rate research that had a rigorous scientific basis according to NSF funding criteria. Successful applicants were thus rewarded with valuable funding and a stamp of high scientific status that such funding conferred.

At the level of individual programs and specific research projects, peer reviewers and NSF staff surely had some wiggle room for deciding how various scientific criteria should be applied. Nevertheless, I have found no evidence suggesting that established funding criteria were regularly ignored. Meanwhile, abundant evidence, gathered from many programs, specific lines of research, individual projects, and various initiatives considered throughout this study, shows that there was, in fact, a close concordance between the rhetoric of scientific rigor and agency practices that encouraged scholarship along certain lines but not others. Furthermore, we have encountered many cases where individuals with firsthand knowledge observed that the agency really did stick to its guns by supporting what Harry Alpert had initially identified as hard-core research.

The NSF also became an object of widespread concern and organized activity in the social science community. Going back to the agency's legislative origins, we found that leading social scientists working with the SSRC were at the forefront of developing the case for inclusion. Yet as the debate progressed, the weak condition of organized social science at the level of national science policy affairs also became painfully evident. Two decades later, during the mid-to-late 1960s, major representatives from the social science community had the chance to testify at congressional hearings for the legislative initiatives put forth by Representative Daddario and Senator Harris. But, even under the more favorable climate of those years, the social sciences did not have a strong organized presence in the national science policy arena.

During the early 1980s, however, the crisis sparked by the Reagan administration's proposed cuts galvanized the social science community into action, resulting in the transformation and dramatic expansion of the Consortium of Social Science Associations.[6] COSSA first demonstrated its value by helping to defeat the legislative measure to scale back NSF social science funding put forth by Republican Representative Winn in the name of loyalty to the administration's plans. Thereafter, COSSA distinguished itself through lobbying, public relations, and educational efforts carried out on behalf of the social science community, especially at the levels of national politics and federal science policy. The consortium's importance also became unmistakable when, starting in 1984, Roberta Miller moved from her position as its executive director to become head of the NSF Social and Economic Sciences Division.

Last but not least, the NSF had considerable importance for the social sciences because the agency itself had become a public patron of major significance. The bare numbers are revealing. In 1989, the final year of the Reagan presidency, the overall NSF budget passed the $2 billion mark for the first time. That year the agency received more than 37,000 proposals, and the number of awards it made exceeded 16,000.

CRITICISMS

The final point in this recap concerns the views presented by figures who took a deep interest in the social sciences and generally supported increased funding for them but found the agency's efforts to do so woefully inadequate. One main group of criticisms focused on the scientistic strategy. As early as the postwar science debate, the sociologist Louis Wirth had warned that it would be a grave mistake to include the social sciences in a framework that presented them as immature versions of the natural sciences and would thus encourage the former to ape the later. Subsequently, those familiar with the agency's development, such as the political scientist James Robinson—who served in the 1960s on the first political science advisory committee—noted that it was not by accident but design that the agency ruled out various types of social research that in the view of many scholars had substantial value and deserved support. Such research included historical studies; investigations grounded in qualitative data and descriptive analysis; research that engaged directly with normative questions about the individual, the social order, politics, and so on; and scholarship with an explicitly critical dimension. In addition, by the 1970s and 1980s partisan think tanks and policy institutes on the left and right presented alternatives to the viewpoint that said social inquiry should be objective, value neutral, and nonpartisan. Gaining steam in those same decades, research associated with "interpretive" social science and some perspectives inspired by Thomas Kuhn's work presented important challenges to the scientistic outlook.

Another set of criticisms focused on the agency's institutional environment, where the social sciences were second-class citizens and depended on the good graces of natural science leaders for support and respect. As the sociologist Kingsley Davis observed at one point, natural scientists, including physical scientists, "inevitably have a 'layman's view' of the social sciences." Furthermore, physical scientists were often "quite unconsciously ... patronizing toward their poor relations in the social sciences."[7] Limited understanding among natural science leaders, including many directors and board members, was an ongoing problem and forced social scientists to engage in what Henry Riecken called an "endless need for justification." These conditions also led to the suggestion, put forth in the mid-1980s by the sociologists Samuel Klausner and Victor Lidz, that increased federal funding for the social sciences since World War II had been a "mixed blessing," as these sciences had suffered from "fiduciary over-lordship" by the natural sciences.

In short, the NSF deserves our attention because, in the many ways noted above, it became a central patron in the evolution of the politics–patronage–social science nexus from the postwar NSF debate through the end of the Reagan era. It's now time to see how our understanding of that period provides a foundation for examining more recent developments, enabling us to appreciate how continuities and changes leading to the present day have ensured that the NSF has remained both important and controversial.

RHYMING PATTERNS

If Rip Van Winkle fell asleep in the late 1980s until the mid-2010s, what would he conclude about NSF social science during the intervening quarter century or so? Although a full examination of the post-Reagan era lies beyond the scope of this book, I suggest that Van Winkle would have little reason to be surprised.[8] During his long snooze, two milestones marked notable changes in the organizational standing and representation of the social sciences at the agency. At the same time, both milestones had clear precedents. Moreover, in many other basic respects, little had changed.

The first milestone was the establishment of the NSF Directorate for Social, Behavioral, and Economic Sciences in 1992. After the mid-1970s' reorganization created the Directorate for Biological, Behavioral, and Social Sciences, grumblings about the subordinate position of the social sciences arose. But for the next decade and a half, such complaints had little effect. However, in March 1989, Herbert Simon proposed, during congressional testimony, that the time had come to give the social and behavioral sciences their own directorate. One year later, an NSF advisory committee observed that the agency continued to treat "these disciplines with 'benign neglect' based on a lack of knowledge."[9]

In June 1991, "Looking to the 21st Century," a draft report from the Biological, Behavioral, and Social Sciences Task Force, provided additional impetus for change:

> Although the current leadership of the BBS directorate has worked very hard to include the SEPS [social, economic, and political sciences] in the reports and funding priorities for the directorate, funding remains at pre-1980 levels in most of these disciplines (even in actual dollar terms) and they are often not included in major Foundation programs and initiatives ... SEPS ... need representation at the highest levels of decision-making in the Foundation through an Assistant Director, [which would enable them to] participate in decisions on NSF resources allocation as equal partners with colleagues in the biological, geological, and physical sciences and engineering.[10]

Since 1980, overall NSF science funding, when measured in constant dollars, had increased by 27 percent. Meanwhile, funding for the psychological and social sciences had fallen 38 percent.[11]

The big news arrived in October 1991, when NSF director Walter E. Massey—a physicist who had taken up this position in March—announced the creation of a new Directorate for Social, Behavioral, and Economic Sciences (SBE). Soon thereafter, the University of Wisconsin sociologist Cora Marrett became the first SBE leader. Marrett had completed her PhD in sociology at the University of Wisconsin in 1968, before holding academic appointments at the University of North Carolina–Chapel Hill and Western Michigan University, and then returning to Wisconsin in 1974 as a faculty member. Marrett's qualifications also included her experience as chair of the NSF Committee on Equal Opportunities in Science, Engineering, and Technology; her membership on the Board of Trustees at the Center for Advanced Study in the Behavioral Sciences; and her expertise in energy policy. In addition, Director Massey knew Marrett from a few previous occasions. Among other things, they had both been members of a presidential commission studying the partial nuclear reactor meltdown at Three Mile Island in 1979.[12] Marrett's tenure as SBE leader lasted four years, until 1996.

The other milestone occurred nearly two decades later, when Marrett became the first sociologist to lead the agency. After finishing her position as SBE leader in 1996, she had returned to academia, first as provost and professor of Afro-American studies at the University of Massachusetts–Amherst, and then back to the Midwest as the University of Wisconsin system's senior vice-president for academic affairs. In 2007, she returned to the NSF, this time as assistant director and leader of its Education and Human Resources Directorate. Three years later, she moved into the agency's top position as acting director for five months, from June to October 2010. When the agency had a new permanent director, Marrett stayed on as deputy director. A couple years later, she became acting director once again, this time for a year, from March 2013 to March 2014. After that, she left the agency for good, retired, and moved back to Wisconsin.

The fact that a sociologist rose to the position of acting director twice gave the social sciences some additional status and visibility at the agency, as did the presence of the SBE directorate. These developments also had precedents. Recall that from 1961 to 1975, the social sciences had their own division, on an organizational par (although certainly not a financial par) with the physical science and biological science divisions. And in the second half of the 1970s, the psychologist Richard Atkinson was NSF director. Remember also that during the Atkinson years, the social sciences had not fared well, which contributed to mounting discontent inside and outside the agency. So, placing a social or behavioral scientist at the helm of an agency still focused predominantly on the natural sciences did not necessarily translate into broader benefits. In addition, in the case of Marrett, I have not seen any evidence that suggests her expertise as a sociologist per se was a significant factor in her two appointments as acting director. Nor does it seem that during those two appointments she reshaped and/or strengthened

Figure 10.1
NSF acting director Cora Marrett, speaking at a COSSA meeting about the agency's social and behavioral sciences programs. November 12, 2013. Photo by Chris Flynn. Courtesy of COSSA.

the agency's social science activities in any notable respects. The fact that she never became permanent director is also telling.

Furthermore, if we take the period from 1990 to 2019 as a whole, it is clear that the social sciences remained marginal in NSF leadership positions. During these years, the directorship lay in the hands of a social scientist only during Marrett's two temporary terms, which together account for merely 1.5 of the total 30 years. The full list of directors shows that physics, engineering, and biology enjoyed much stronger representation: from nuclear chemistry and physics, Frederick M. Bernthal (acting director, September 1990–March 1991 and April 1993–October 1993); from physics, Walter E. Massey (March 1991–April 1993); from physics, Neal F. Lane (October 1993–August 1998); from microbiology, Rita R. Colwell (August 1998–February 2004); from engineering, Arden L. Bement Jr. (acting director, February 2004–November 2004, and permanent director, November 2004–May 2010); from sociology, Cora B. Marrett (June 2010–October 2010 and March 2013–March 2014);

from engineering, Subra Suresh (October 2010–March 2013); and from astrophysics, France A. Cordova (March 2014–).

As for the governing board, recent data confirm that the social and behavioral sciences have remained marginal here as well, while the physical and engineering sciences continued to enjoy a much stronger presence. As of April 2018, the board's chairman, Maria Zuber, was a geophysicist, while the vice-chairman, Diane Souvain, was a computer scientist. The twenty-four board members included ten from the mathematical and physical sciences broadly speaking (i.e., astronomy, physics, chemistry, atmospheric science, computer science), seven from the engineering sciences (i.e., chemical, electrical, and mechanical engineering), three from the biological sciences, and one with a specialty in mathematics education. The remaining three included a social psychologist, James Jackson; a social statistician from sociology, Robert Groves; and Emilio Moran, whose work in the human environmental sciences bridged the natural and social sciences. Thus, the NSB—as of April 2018—had nobody from economics, political science, or anthropology.[13]

In the past couple decades, questions about the nature of the social sciences, their relationship to the natural sciences, and their connections to the humanities have also remained contentious ones for the NSF. For good reasons, scholars continued to wonder about the agency's power to shape the directions of social science research, to elevate certain types of scholarship above others, and to use the power of the purse and its influential position within the federal science system to promote a particular vision of social science inquiry based on a presumed unity of the sciences.

One might suppose that after 1992, when the SBE directorate was created, social scientists would have finally had the freedom to be themselves. Perhaps with their own directorate, they would no longer need to worry much about what natural scientists thought of them or what other people who viewed the agency primarily in terms of its natural science activities thought. Writing shortly after the SBE's establishment, the psychologist David Johnson suggested that after residing for the past seventeen years "as tenants in someone else's house," social scientists might be "free to hang our own pictures, paint the walls the colors of our choosing, and make this house our home." However, realizing such aspirations would be hard, as Johnson also emphasized: "The house might be new, but it is located in an old, established neighborhood whose ways we would do well to recall and reexamine in light of our aspirations for this new house. Changing some of the neighborhood's old ways may be essential for the realization of those aspirations."[14]

Persistent consternation over this matter surfaced in a controversy in the early 2000s about limited NSF support for qualitative social research. In 2003, following criticisms of the sociology program by advocates of qualitative inquiry, this program sponsored a conference with thirty-two scholarly participants, which culminated in a 2004 report called *Workshop on Scientific Foundation of Qualitative Research*. Subsequently, the NSF sponsored a second workshop about funding for qualitative inquiry across a broader range of the social sciences. This workshop had twenty-nine scholarly participants and resulted in a 2009 report called *Workshop on Interdisciplinary Standards for Systematic Qualitative Research*.[15]

In a critical commentary, the sociologist Howard Becker took this second report to task for embracing scientific rigor inappropriately. As he presented it, the report's message to scholars was basically this: "Quit whining and learn to do real science by stating theoretically derived, testable hypotheses, with methods of data gathering and analysis specified before entering the field. Then you'll get NSF grants like the real scientists do."[16]

More recently, an accomplished Ivy League sociologist told me that incorporation into the NSF was "one of the worst things" that had happened to his discipline since the middle of the twentieth century. When I asked him why he was so critical, his answer was straightforward: the agency had strongly encouraged sociologists to pursue a model of scientific investigation that he believed was rather narrow and, in fact, downright unhealthy.[17]

Meanwhile, the main challenges facing the social sciences in the federal science policy arena had changed little since the early Reagan years. In 2014, COSSA's executive director, Wendy Naus, characterized the predicament as a continuous struggle: "having to justify how or why federally funded social and behavioral science research is in our 'national interest,' fending off attacks on individual grants simply because their titles lure additional scrutiny, or beating back attempts to pit fields of research against one another, especially in times of scarce resources."[18]

Funding levels also remained paltry, at least compared to natural science funding. As of 2017, the budget for the NSF social science division was impressive in the sense that it "amounted to 55 percent of all such federal funding." That budget was also impressive in a different and unflattering sense, however, as it amounted to "less than 4 percent of the agency's total research budget."[19]

The challenges mentioned by Naus still had a strongly partisan character as well, with the great bulk of disapproval expressed by conservative Republicans—the Democratic Senator William Proxmire's highly publicized attacks on certain NSF grants back in the mid-to-late 1970s stands out as the main exception. In 1995, Republican Representative Robert Walker, at the time chairman of the House Science Committee, commended the NSF's dedication to basic science and recommended growth in this area. But he excluded the social sciences. In fact, just three years after the creation of the SBE directorate, Walker proposed eliminating social science funding wholesale, which amounted to $110 million at the time:

> In large part, we think that's an area where the National Science Foundation has largely wandered into [and] that was kind of a politically correct decision in recent years. And that is a place where the science budgets can be rescoped. We think that the concentration ought to be in those areas of the physical sciences.[20]

A decade later, Senate Republicans attacked again. Kay B. Hutchison from Texas introduced a bill—S. 2802, The American Innovation and Competitiveness Act—that would have ended NSF funding for social and behavioral sciences.[21]

In 2009, Oklahoma Republican Senator Tom A. Coburn returned to the cause, with an amendment to an appropriations bill that, as was noted in this book's introduction, would have terminated NSF funding for political science. Coburn's press secretary, Don Tatro, explained the rationale: "Federal research dollars should go to scientists who work on finding solutions for people with severe disabilities, or the next generation of biofuels, or engineering breakthroughs." Coburn himself added that the federal government should not be spending taxpayer dollars on political science research, such as the NSF-funded National Election Studies, still being carried out under the auspices of the University of Michigan. Although he recognized that such work might involve "interesting theories about recent elections," he added that "Americans who have an interest in electoral politics" already had good sources of information. They could "turn to CNN, Fox News, MSNBC, the print media, and a seemingly endless number of political commentators on the Internet."[22]

Coburn also questioned NSF social science funding more broadly. During the previous decade, the agency had provided $91.3 million. The full amount, suggested the Oklahoma Republican, would have been better spent on real sciences such as biology and chemistry.[23]

Of course, not all voices on the right took such a dismissive view. To note just one interesting example, in 2011, the best-selling author and *New York Times* journalist David Brooks, a moderate conservative whose writings often referred to social science studies, opposed a congressional bill that called for closing the SBE directorate. "This is exactly how budgets should not be balanced—by cutting cheap things that produce enormous future benefits," wrote Brooks. Furthermore, he suggested that the current generation was living in "the middle of a golden age of behavioral research." The nation would be wise to "design policies around that knowledge." Eliminating the directorate for the purpose of cost saving would thus be a colossal mistake, "like cutting off navigation financing just as Christopher Columbus hit the shoreline of the New World."[24]

Still, Brooks's message had little if any impact in conservative circles. Republican proposals to curb social science funding in the name of prudent fiscal management persisted.

Nevertheless, within the contexts of the nation's science funding system and the federal science establishment, a good case can be made that the NSF's special importance vis-à-vis the social sciences has remained intact. As of 2007, the SBE directorate provided "61 percent of federal support for basic research in anthropology, social psychology, and the social sciences at U.S. academic institutions." For a number of fields, "including archaeology, political science, linguistics, and non-medical aspects of anthropology, psychology, and sociology," the directorate was "the predominant or exclusive source of federal basic research support."[25] A decade later, in 2017, when the NSF provided 24 percent of all federal support for basic research in the sciences at the nation's universities and colleges, the directorate provided 55 percent of all federal funding for such research in the social sciences.[26]

NSF support remained important in many specific areas of investigation as well. These include long-term, large-scale studies that continued to enjoy high regard. For example, the agency consistently supported the National Election Studies, which, according to the journalist Dylan Matthews, stood out as "the single best source of survey data on American voters' opinions, going back as far as 1948." The agency also provided ongoing funding for the Panel Study on Income Dynamics, the "single best data source on economic mobility, among the most hotly contested topics of political debate at the time."[27] The economics program retained its special significance as well, as "the only program in the federal government with a broad mandate to strengthen basic economic science." As of November 2018, the agency provided more than half of all federal support for such work.[28]

Moreover, the agency has remained valuable as a balance wheel. As COSSA has put it, NSF funding for "basic scientific discovery, workforce training, and state-of-the-art facilities" is crucial in helping to keep "the U.S. ahead of its global competitors." Its unique value reflects the fact that it continues to be "the only U.S. federal agency tasked with **supporting scientific research across all fields of science.**"[29]

Against this background, COSSA has continued to cultivate bipartisan support for the social sciences and NSF funding among a wide range of stakeholders. According to Wendy Naus, these include the NSF, the NAS, and other nodes in the federal science system; elected officials and staff in the Executive and Legislative Branches; the American Association for the Advancement of Science, the Association of Public and Land-Grant Universities, and other "national associations and societies representing broad fields of science, higher education associations, university presidents, [and] corporate heads." "Advocating for social science" has thus become "a team sport."[30]

The charge that social science research is esoteric and has little practical value has received regular attention. Following the 2016 national elections, COSSA prepared a report for the new Trump administration whose basic message was suggested in the title: *Social and Behavioral Science Research: Essential to Keeping America Competitive, Prosperous and Safe*. According to this report,

> Federally-supported scientific research—**including social and behavioral science research**—provides an evidence base that the President and Executive Branch agencies can use to produce science-backed strategies for addressing issues of national importance, such as crime prevention, health care for the underserved, the safety of our troops, early childhood education, and improved efficiency of American businesses, to name a few.[31]

★★★

Very recently, events closely associated with the nation's newest Republican president have given traction to the notion that we are now living in a "post-truth" era. I believe that these peculiar conditions should prompt us to think carefully about what the NSF has done well vis-à-vis the social sciences and where it has come up short. Here is a second area where the insights from previous chapters can be fruitfully extended.

ASSESSMENT AND PROPOSAL

"Do you know where the truth is?" This question, raised by Sheila Jasanoff, a professor of science and technology studies at Harvard's Kennedy School of Government, is just as relevant today as it was when her article first appeared in the summer of 2017. As Jasanoff notes,

> Hardly a day passes without some major accusation in the media that the nation's highest office has become a source of unfounded stories, claims without evidence, even outright lies. As the charges ... pile up, the White House counters that institutions long seen as standing above partisan wrangling can no longer be trusted. ... Even scientific consensus can be dismissed as politics by other means.

In an age when the president's spokeswoman Kellyanne Conway suggested the value of considering "alternative facts," it's clear that the claim that a certain type of human endeavor— science—has succeeded reasonably well in finding out the truth and judging various truth claims needs an urgent and robust defense.[32]

The challenge, Jasanoff explains, is to find a way of addressing this "retreat from reason," in order to "restore confidence that 'facts' and 'truth' can be reclaimed in the public sphere." This will not be easy, however, for "truth in the public domain is not simply out there, ready to be pulled into service like the magician's rabbit from a hat." What is needed is not a defense that simply asserts "science knows best" but a defense that focuses on convincing the public that "science itself has been subjected to norms of good government."[33]

Not only are these reflections timely and compelling. In addition, I believe that Jasanoff's analysis can help us to appreciate anew the NSF's longstanding importance in promoting good governance of science and public trust in ways that have purposefully included the social sciences.

In the post–World War II and early Cold War years, American natural science and social science leaders, from the electrical engineer Vannevar Bush to the economist Wesley Mitchell, feared that dramatic expansion in federal funding would lead to the corruption of scientific inquiry. At the time, the infamous examples of Soviet genetics and Nazi racial science revealed just how much damage could be done by the politicization of science. But the problems were not confined to that moment in time. Nor were they limited to foreign contexts. Of special relevance to the present study, American science has also faced repeated threats to its integrity due to the power of various patrons in the public and private sectors (i.e., the military, the CIA, the corporation, the partisan think tank, and the policy institute). Such threats challenged public confidence in science as a source of nonbiased and reliable expert knowledge about a broad array of essential matters: medicine, nutrition, national security, public safety, crime, immigration, racism, sexuality, gender roles, environmental degradation, climate change, and many others.

Over the course of nearly seven decades, the NSF has established an impressive (although certainly not perfect) record when it comes to managing threats to scientific integrity posed by federal funding. Right from the outset, the legislative proposals for the NSF and then the young agency itself focused on establishing institutional arrangements, practices, and norms to ensure that "good science" got funded. Among other things, this meant that the agency provided public funding within a framework that limited the influence of political, ideological, and social pressures on the course and conduct of scientific inquiry, leaving scientists themselves largely responsible for such matters. The agency also placed great emphasis on methodological rigor, often accompanied by quantification, which, as the historian of science Theodore Porter observes, has been "part of the drive to make a science of society that can hold its head up in the company of physics, chemistry, and engineering."[34]

Many social scientists have been greatly appreciative of this point. This makes perfect sense because warding off the danger that patron interests might bend research in worrisome ways has been central to the ongoing project of establishing the scientific legitimacy and public value of the social sciences. Moreover, containing this danger became an ongoing challenge in the specific case of what the original 1950 NSF charter vaguely referred to as the "other sciences." On a few occasions, conservative critics claimed to have found a disturbing left-wing bias in NSF-funded social research, as seen, for example, in the uproar over MACOS. Nevertheless, as the years passed, the agency's system of peer review, its commitment to funding research of the highest quality, and its steadfast position that social science is part of a unified scientific enterprise largely succeeded in placing NSF-funded work above such criticism.

Indeed, it is striking that the long line of critics have, by and large, not claimed that NSF-funded social research is biased in a partisan direction. Over the decades, many Republicans along with many conservative scientists, scholars, and commentators have made this charge against the social sciences and various funding sources. Meanwhile, the worry that patrons associated with the national security state exert a nefarious conservative influence over social research has been particularly widespread among liberals and those with stronger leftist commitments. This can be seen, for example, in the controversy about the role of scholars and academic institutions in the War on Terror.[35] Still, critics of the NSF have more commonly suggested that NSF-sponsored social research does not warrant federal funding because it has little practical value and distracts the agency from focusing on more important work in the natural sciences.

In short, when it comes to addressing longstanding worries about the corruption of research due to the agendas of funding bodies, the NSF's approach has largely succeeded. A recent COSSA document puts it this way: "NSF, through its merit review process, allows the demands of scientific discovery to dictate how best to spend basic research dollars, leaving political and individual ideologies at the door." Moreover, "this process has been emulated the world over."[36] For these reasons, the agency's past and ongoing efforts to promote good science deserve public admiration and strong support within the national science policy arena. The view that the federal government and the NSF in particular have an important role to play in providing broad-based support for the social sciences, not just the natural sciences, has considerable merit. This is especially so for social research with a strong scholarly orientation that does not duplicate the work of mission-oriented agencies and that is distinct from the sort of ideologically driven work sponsored by private patrons with partisan agendas. Just as in the case of the physical sciences, life sciences, earth sciences, and medical sciences, providing healthy federal support for work that is not narrowly focused on practical payoffs or on matters of partisan conflict and ideological warfare has been, and still is, vital in maintaining a strong social science enterprise.

The present study has also made it clear, however, that NSF social science activities have had considerable limitations over the long term. Interestingly, some of these were hinted at by the psychologist and former NSF director Richard Atkinson in a 2006 essay, coauthored by William Blanpied, a physicist, former NSF staff member, and historian of American science policy. Their essay presented the extensive involvement of social scientists in federal science policy making during the New Deal as a telling contrast to the very different situation that has prevailed since World War II. Atkinson and Blanpied wrote that although "the Bush report is justifiably regarded as the cornerstone of post-war U.S. science policy," Bush "did not have a high regard for the social sciences." Furthermore, "the bias against the social sciences on the part of the 'hard' sciences and engineering persisted for many years." "Arguably," the authors concluded, "the nation has continued to be ill served by its neglect of the social sciences."[37]

Although Atkinson and Blanpied did not elaborate, the unhappy consequences of that neglect include, as we have seen, all of the following: relegating the social sciences to second-class status within the NSF, within the federal science establishment, and within the nation's scientific enterprise most generally; giving preferential treatment to certain "hard-core" investigative approaches to social inquiry while minimizing support for other "softer" approaches that have their own particular value; reinforcing a hierarchy among the sciences that suggests scientific inquiry about nature is more valuable to national well-being and human welfare than scientific inquiry about human beings and society; and contributing to a science policy environment that has made it difficult to develop national policies and funding programs based on the plausible notion that at the levels of scientific ontology, scientific epistemology, scientific methodology, and social relevance, there are some crucial differences, at least in many cases, between research in the social sciences and research in the natural sciences.

I am not suggesting that the NSF should embrace an anything-goes approach. However, the early establishment and subsequent entrenchment of a scientistic outlook, in principle and in practice, has been problematic because that made it very difficult to obtain support for various types of research, however important, simply because they did not seem to be "scientific" enough based on a narrow understanding of the term—even if the agency's carefully delimited approach also helped to ward off charges of political or ideological bias. Ironically, the first NSF social science policy architect, Harry Alpert, understood that the broad field of inquiry known as social science had roots in humanistic scholarship, philosophical inquiry, and social criticism. Moreover, work along those lines deserved continued encouragement, respect, and nourishment.

Even though Alpert himself was unable to promote this broader understanding during his tenure at the NSF, he did propose—in a 1958 article aptly titled "The Knowledge We Need Most"—that the social sciences should be placed on an equal basis with the better-established natural sciences: "For the long run, steps must be initiated to develop among natural scientists and social scientists a sense of mutual respect and understanding and a community of interest." Looking forward, Alpert wrote that "today's invidious hierarchical distinctions among specialists of the various [social and natural science] disciplines must disappear."[38]

But this did not happen in Alpert's lifetime. Nor was there any noticeable decline in the power of those hierarchical distinctions at the NSF through the end of the Reagan era and beyond. Moreover, there is no good reason to believe that major reform in this direction is about to occur any time soon under current arrangements.

In light of this situation, it seems useful to think about possible alternatives. Across the ocean at the European Research Council, one finds three divisions, one for the physical sciences, another for the life sciences, and a third for the social sciences and humanities. The Canadian federal funding system has three major branches as well, although with a slightly different configuration: one for the natural sciences, another for the medical sciences, and a third for the social sciences and humanities—SSHRC, the Social Sciences and Humanities Research Council. It is important to recognize that other funding systems have different structures and offer different opportunities for working out the nature and meaning of the social sciences. Otherwise, it can easily seem like the established arrangement in the U.S. is the only feasible one. And it's possible to develop alternative funding structures that place the social sciences in closer proximity to the humanities, rather than positioning the social sciences as a minor adjunct to the natural sciences and engineering.

With the above points in mind, I suggest that the idea of creating a national social science foundation deserves serious reconsideration. This possibility should interest a number of participants and stakeholders: the scientific community, including the federal science establishment; the social sciences especially, including the SSRC and COSSA; organizations and leaders in the humanities who can promote stronger ties with the social sciences; and of course politicians and staff in the White House and Congress who are responsible for making sure the federal science system is working well. To revisit the idea for a new agency intelligently, however, we first need to understand that the character and fate of Senator Fred Harris's NSSF proposal in the late 1960s has been badly misunderstood. Historical commentaries over the past half-century have claimed that his proposal failed because it was redundant and thus basically did not offer anything beyond the changes in the successful proposal put forth by Senator Emilio Daddario. But, this is simply not the case, as we have seen.

Their two proposals differed in crucial respects, as those paying careful attention to the NSSF debate in the nation's political, scientific, and academic communities would have known at the time. In a 1967 *Science* article explaining the pro-NSSF case, Harris pointed out that the established NSF had failed the social sciences in many ways. Social scientists thus no longer needed "the cover of the natural-science umbrella."[39] Daddario proposed nothing like that, however. Thus, after the Daddario amendment became law, it made sense for someone well informed to point out, as the author Michael Reagan did in his 1969 book *Science and the Federal Patron*, that "the best argument" for a new social science agency was that, unlike at the existing NSF, "the full range of social science approaches and subdisciplines would be supported and given a chance to flower, without being held down by natural scientists' skewed pictures of what the social sciences are or can become."[40] What is more, the Daddario amendment never aimed to strengthen the independence, status, and influence of the social sciences within the federal science system to nearly the extent that Harris's mature NSSF proposal did.

Admittedly, in the present moment, it seems unlikely that a proposal for a separate social science agency could gain much traction. Under the current Trump administration, and with conservative Republicans still exercising considerable influence in Congress (even though the 2018 elections gave Democrats a majority in the House of Representatives), the political environment does not seem welcoming—and there are too many other more pressing items on the national agenda.

If history is a useful guide, it also seems unlikely that the NSF itself would support such a proposal. Remember that during the debate over Harris's proposal, the NSF used the occasion to strengthen its own efforts in this area while rejecting all of the more ambitious aspects of Harris's proposal as unworkable or unnecessary. Of course, that happened some time ago. Still, half a century later, any good argument for a new social science agency will need to include criticism of the NSF. The chances that this well-established agency would support such a proposal thus seem slim. In addition, because of the NSF's special position within the federal science system, the agency's track record of engaging with the social sciences would probably receive strong support from other major nodes in this system, including the NAS.

Yet, however formidable such obstacles may appear, they should not be used as an excuse for resigned acceptance of the status quo. Any concerted effort to address the difficulties will depend heavily on Congress. As the biologist Anne Fausto-Sterling once observed, "The most commonplace influence our society exerts on scientific activity is the direct political authority by which Congress can determine what kinds of research and how much of it will be supported."[41]

Indeed, Congress has exercised great influence by establishing agencies with particular mandates and policies for promoting scientific activities in certain ways and not in others, by providing these agencies with budgetary appropriations to finance their activities, and by regularly reviewing their effectiveness. Of special relevance here, Congress debated and passed the legislation creating the NSF. And it has continued to support the agency's engagements with the social sciences, sometimes a bit more generously than at other times, but nearly always without challenging their marginal position and a raft of other related difficulties. Senator Harris's NSSF initiative stands out as the only serious exception.

Although this is not the place to provide more specific suggestions for crafting a new NSSF proposal, such an initiative would provide an excellent occasion for reassessing the position of the social sciences at the NSF and within the federal science establishment. This would include taking a deep look at the agency's special importance in shaping the social sciences and their influence within the broader society. Consideration of such a proposal should thus stimulate widespread discussion over a suite of fundamental questions: about the past, present, and future of the social sciences; about the changing landscape of public and private patronage for them; about their evolving relationships with the natural sciences and the humanities; about their contributions to the national interest and human welfare; and about their status and influence in American politics, science, and society.

Such a proposal and such a reassessment are long overdue. Efforts in this direction, informed by historical scholarship, should be encouraged.

NOTES

INTRODUCTION

1. Smith quoted in Dylan Matthews, "Why Congressional Republicans Want to Cut Social Science Research Funding," *Vox*, Nov. 12, 2014, https://www.vox.com/2014/11/12/7201487/congress-social-science-nsf-funding, accessed Dec. 15, 2018.

2. Roberta B. Miller, "The Social Science Lobby in the United States," in *The Human Sciences: Their Contribution to Society and Future Research Needs*, ed. Baha Abu-Laban and Brendan G. Rule (Edmonton: University of Alberta Press, 1988), 241–252, at 245.

3. Harry Alpert, "The Government's Growing Recognition of Social Science," *Annals of the American Academy of Political and Social Science* 327 (1960): 59–67, at 64.

4. Boehner and Cantor proposal mentioned in Dan Berett, "Picking on Social Science," *Inside Higher Ed*, Dec. 21, 2010, https://www.insidehighered.com/news/2010/12/21/picking-social-science, accessed Dec. 15, 2018.

5. Tom Coburn, *The National Science Foundation: Under the Microscope*, April 2011, p. 53, http://lcweb2.loc.gov/service/gdc/coburn/2014500020.pdf, accessed Nov. 15, 2019.

6. Flake quoted in John Sides, "Congressman Flake's Remarks," *The Monkey Cage blog*, May 10, 2012, http://themonkeycage.org/blog/2012/05/10/congressman-flakes-remarks, accessed Dec. 15, 2018.

7. See Matthews, "Why Congressional Republicans."

8. Smith quoted in Matthews, "Why Congressional Republicans."

9. A total of 10.4 percent was noted on p. 37 in COSSA, *Analysis of the President's FY 2018 Budget Request for Social and Behavioral Science*, May 2017, http://www.cossa.org/wp-content/uploads/2017/05/COSSA-FY-2018-Budget-Analysis.pdf, accessed Dec. 15, 2018.

10. Five percent and two-thirds were noted on p. 4 in COSSA, *Social and Behavioral Science Research: Essential to Keeping America Competitive, Prosperous, and Safe: Ten Recommendations for the 45th President of the United States,* Dec. 2016, http://www.cossa.org/wp-content/uploads/2017/01/COSSA-Recommendations-for-Trump-Administration-Dec-2016.pdf, accessed Dec. 15, 2018.

11. COSSA, *Social and Behavioral Science Research*, 3.

12. COSSA, *Social and Behavioral Science Research*, 2, 4.

13. Vannevar Bush, *Science—The Endless Frontier: A Report to the President on a Program for Postwar Scientific Research* (Washington, D.C.: GPO, July 1945), hereafter, *SEF*. J. Merton England, *A Patron for Pure Science: The National Science Foundation's Formative Years, 1945–57* (Washington, D.C.: NSF, 1982).

14. Public Law 507, 81st Congress, S. 247, the "National Science Foundation Act of 1950," https://www.nsf.gov/about/history/legislation.pdf, accessed Dec. 15, 2018.

15. See Roger L. Geiger, *Research and Relevant Knowledge: American Research Universities since World War II* (New York: Oxford University Press, 1993), 159.

16. George Mazuzan, "Good Science Gets Funded: The Historical Evolution of Grant Making at the National Science Foundation," *Science Communication* 14 (1992): 63–90.

17. See my book Mark Solovey, *Shaky Foundations: The Politics–Patronage–Social Science Nexus in Cold War America* (New Brunswick, NJ: Rutgers University Press, 2013), esp. chaps. 1 and 4, and my recent essay "The Impossible Dream: Scientism as Strategy Against Distrust of Social Science at the U.S. National Science Foundation, 1945–1980," *International Journal for History, Culture, and Modernity* 7 (2019): 209–238.

18. Miller, "The Social Science Lobby in the United States."

19. Many Americans are accustomed to viewing the years from the late 1940s through the mid-to-late 1980s as the era of Cold War. However, there has been considerable scholarly debate about how to conceptualize and periodize the Cold War. See Duncan Bell and Joel Isaac, eds., *Uncertain Empire: American History and the Idea of Cold War* (New York: Oxford University Press, 2012).

20. Thomas Gieryn, *Cultural Boundaries of Science: Credibility on the Line* (Chicago: University of Chicago Press, 1999). See also Charles A. Taylor, *Defining Science: A Rhetoric of Demarcation* (Madison: University of Wisconsin Press, 1996); Michele Lamont and Virag Molnar, "The Study of Boundaries in the Social Sciences," *Annual Review of Sociology* 28 (2002): 167–195, esp. 178–181 on "Science, Disciplines, and Knowledge"; Gerhard Sonnert, "Social Science and Sozialwissenschaft: Categorical and Institutional Boundaries of Knowledge," *Journal of the History of the Behavioral Sciences* 54 (2018): 178–197.

21. Major historical studies include Mary O. Furner, *Advocacy and Objectivity: A Crisis in the Professionalization of American Social Science, 1965–1905* (Lexington: University Press of Kentucky, 1975); Thomas I. Haskell, *The Emergence of Professional Social Science: The American Social Science Association and the Nineteenth-Century Crisis of Authority* (Urbana: University of Illinois Press, 1977); Donald T. Crithlow, *The Brooking Institution, 1916–1952: Expertise and the Public Interest in a Democratic Society* (DeKalb: Northern Illinois University Press, 1985); Robert C. Bannister, *Sociology*

and Scientism: The American Quest for Objectivity, 1880–1940 (Chapel Hill: University of North Carolina Press, 1987); Peter Novick, *That Noble Dream: The 'Objectivity Question' and the American Historical Profession* (New York: Cambridge University Press, 1988); Dorothy Ross, *Origins of American Social Science* (New York: Cambridge University Press, 1991); Donald Fisher, *Fundamental Development of the Social Sciences: Rockefeller Philanthropy and the United States Social Science Research Council* (Ann Arbor: University of Michigan Press, 1993); Mark C. Smith, *Social Science in the Crucible: The American Debate over Objectivity and Purpose* (Durham: University of North Carolina Press, 1994); John M. Jordan, *Machine-Age Ideology: Social Engineering and American Liberalism, 1911–1939* (Chapel Hill: University of North Carolina Press, 1994); Julie A. Reuben, *The Making of the Modern University: Intellectual Transformation and the Marginalization of Morality* (Chicago: University of Chicago Press, 1996); Sarah Igo, *The Averaged American: Surveys, Citizens, and the Making of a Mass Public* (Cambridge, MA: Harvard University Press, 2007); Alice O'Connor, *Social Science for What? Philanthropy and the Social Question in a World Turned Rightside Up* (New York: Russell Sage Foundation, 2007); Andrew Jewett, *Science, Democracy, and the American University: From the Civil War to the Cold War* (New York: Cambridge University Press, 2012); Joel Isaac, *Working Knowledge: Making the Human Sciences from Parsons to Kuhn* (Cambridge, MA: Harvard University Press, 2012).

22. Recent contributions that survey a wide terrain when considered together and that include plentiful references to earlier studies: Audra J. Wolfe, *Competing with the Soviets: Science, Technology, and the State in Cold War America* (Baltimore: Johns Hopkins University Press, 2013); Naomi Oreskes and John Krige, eds., *Science and Technology in the Global Cold War* (Cambridge, MA: MIT Press, 2014); Jeroen van Dongen, ed., *Cold War Science and the Transatlantic Circulation of Knowledge* (Leiden: Brill, 2015). On the importance of following the money in the history of science more broadly, see Casper Andersen, Jakob Bek-Thomsen, and Peter C. Kjaergaard, "The Money Trail: A New Historiography for Networks, Patronage, and Scientific Careers," *Isis*, 103 (2012): 310–315; Noortje Jacobs and Pieter Huistra, "Funding Bodies and Late Modern Science," *International Journal for History, Culture and Modernity*, 7 (2019): 887–898.

23. For historiographic discussions, see Hunter Crowther-Heyck, "Patrons of the Revolution: Ideals and Institutions in Postwar Behavioral Science," *Isis* 97 (2006): 420–446; Joel Isaac, "The Human Sciences in Cold War America," *Historical Journal* 50 (2007): 725–746; Mark Solovey, "Cold War Social Science: Specter, Reality, or Useful Concept?," in *Cold War Social Science: Knowledge Production, Liberal Democracy, and Human Nature,* ed. Mark Solovey and Hamilton Cravens (New York: Palgrave Macmillan, 2012), 1–22. Important studies include Christopher Simpson, *Science of Coercion: Communication Research and Psychological Warfare, 1945–1960* (New York: Oxford University Press, 1994); Ellen Herman, *The Romance of American Psychology: Political Culture in the Age of Experts, 1940–1970* (Berkeley: University California Press, 1995); Christopher Simpson, ed., *Universities and Empire: Money and Politics in*

the Social Sciences during the Cold War (New York: New Press, 1998); James H. Capshew, *Psychologists on the March: Science, Practice, and Professional Identity in America, 1929–1969* (New York: Cambridge University Press, 1999); Alice O'Connor, *Poverty Knowledge: Social Science, Social Policy, and the Poor in Twentieth-Century U.S. History* (Princeton, NJ: Princeton University Press, 2001); Ron Robin, *The Making of the Cold War Enemy: Culture and Politics in the Military-Industrial Complex* (Princeton, NJ: Princeton University Press, 2001); Michael A. Bernstein, *A Perilous Progress: Economists and Public Purpose in Twentieth-Century America* (Princeton, NJ: Princeton University Press, 2001); Philip Mirowski, *Machine Dreams: Economics Becomes a Cyborg Science* (New York: Cambridge University Press, 2002); Nils Gilman, *Mandarins of the Future: Modernization Theory in Cold War America* (Baltimore: Johns Hopkins University Press, 2003); S. M. Amadae, *Rationalizing Capitalist Democracy: The Cold War Origins of Rational Choice Liberalism* (Chicago: University of Chicago Press, 2003); David Engerman, *Know Your Enemy: The Rise and Fall of America's Soviet Experts* (New York: Oxford University Press, 2009); Matthew Farish, *The Contours of America's Cold War* (Minneapolis: University of Minnesota Press, 2010); Solovey and Cravens, eds., *Cold War Social Science*; Solovey, *Shaky Foundations*; Jamie Cohen-Cole, *The Open Mind: Cold War Politics and the Sciences of Human Nature* (Chicago: University of Chicago Press, 2014); Joy Rohde, *Armed with Expertise: The Militarization of American Social Research during the Cold War* (Ithaca, NY: Cornell University Press, 2013); Hunter Heyck, *Age of System: Understanding the Development of Modern Social Science* (Baltimore: Johns Hopkins University Press, 2015); David H. Price, *Cold War Anthropology: The CIA, The Pentagon, and the Growth of Dual-Use Anthropology* (Durham, NC: Duke University Press, 2016); Susan Lindee and Joanna Radin, "Patrons of the Human Experience: A History of the Wenner-Gren Foundation for Anthropological Research, 1941–2016," *Current Anthropology* 57, supplement 14 (2016), S218–S301.

24. For example, O'Connor, *Poverty Knowledge*.

25. Solovey, *Shaky Foundations*.

26. See Rik Peel, "Ten Reasons to Embrace Scientism," *Studies in History and Philosophy of Science Part A* 63 (2017): 11–21.

27. Friedrich A. Hayek: "Scientism and the Study of Society," *Economica* 9 (1942): 267–291, 10 (1943): 34–63, 11 (1944): 27–39; *The Counter-Revolution of Science: Studies in the Abuse of Reason* (Glencoe, IL: Free Press, 1952).

CHAPTER 1: TO BE OR NOT TO BE INCLUDED

1. President Roosevelt, "Let Us Move Forward with Strong and Active Faith," undelivered address prepared for Jefferson Day, Apr. 13, 1945, in Samuel I. Rosenman, ed., *The Public Papers and Addresses of Franklin D. Roosevelt, 1944–1945* (New York: Harper, 1950), 613–616, at 615.

2. Fulbright quoted on pp. 397–398 in George A. Lundberg, "The Senate Ponders Social Science," *Scientific Monthly* 64 (May 1947): 397–411.

3. Paul K. Hoch, "The Crystallization of a Strategic Alliance: The American Physics Elite and the Military in the 1940's," in *Science, Technology and the Military*, ed. Everett Mendelsohn, Merritt Roe Smith, and Peter Weingart (Boston: Kluwer, 1988), 87–116; Daniel J. Kevles, *The Physicists: The History of a Scientific Community in Modern America* (Cambridge, MA: Harvard University Press, rev. ed. 1995); Gregg Herken, *Cardinal Choices: Presidential Science Advising from the Atomic Bomb to SDI* (Stanford, CA: Stanford University Press, rev. ed. 2000); Audra J. Wolfe, *Competing with the Soviets: Science, Technology, and the State in Cold War America* (Baltimore: Johns Hopkins University Press, 2013).

4. Roger E. Backhouse and Philippe Fontaine, "Introduction," in Roger E. Backhouse and Philippe Fontaine, eds., *The History of the Social Sciences since 1945* (New York: Cambridge University Press, 2010), 1–15, at 8.

5. Joel Isaac, "The Human Sciences in Cold War America," *The Historical Journal* 50 (2007): 725–746; Backhouse and Fontaine, *History of the Social Sciences since 1945*; Mark Solovey and Hamilton Cravens, eds., *Cold War Social Science: Knowledge Production, Liberal Democracy, and Human Nature* (New York: Palgrave Macmillan, 2013).

6. Richard S. Kirkendall, *Social Scientists and Farm Politics in the Age of Roosevelt* (Columbia: University of Missouri Press, 1966); Patrick D. Reagan, *Designing a New America: The Origins of New Deal Planning, 1890–1943* (Amherst: University of Massachusetts Press, 2000); Allan M. Winkler, *The Politics of Propaganda: The Office of War Information, 1942–1945* (New Haven, CT: Yale University Press, 1978).

7. David H. Price, *Threatening Anthropology: McCarthyism and the FBI's Surveillance of Activist Anthropologists* (Durham, NC: Duke University Press, 2004); Mike F. Keen, *Stalking the Sociological Imagination: J. Edgar Hoover's FBI Surveillance of American Sociology* (Westport, CT: Greenwood, 1999); Alice O'Connor, "The Politics of Rich and Rich: Postwar Investigations of the Foundations and the Rise of the Philanthropic Right," in *American Capitalism: Social Thought and Political Economy in the Twentieth Century*, ed. Nelson Lichtenstein (Philadelphia: University of Pennsylvania Press, 2006), 228–248.

8. Teller's comment on p. 40 in Theda Skocpol, "Governmental Structures, Social Science, and the Development of Economic and Social Policies," in *Social Science Research and Government: Comparative Essays on Britain and the United States*, ed. Martin Bulmer (New York: Cambridge University Press, 1987), 40–50.

9. For a useful but brief account, see J. Merton England, *A Patron for Pure Science: The National Science Foundation's Formative Years, 1945–57* (Washington, D.C.: NSF, 1982), esp. 266–273.

10. See, for example, James H. Capshew, *Psychologists on the March: Science, Practice, and Professional Identity in America, 1929–1969* (New York: Cambridge University Press, 1999), 176–179; Michael A. Bernstein, *A Perilous Progress: Economists and Public Purpose in Twentieth-Century America* (Princeton, NJ: Princeton University

Press, 2001), 100–101; David Haney, *The Americanization of Social Science: Intellectuals and Public Responsibility in the Postwar United States* (Philadelphia: Temple University Press, 2008), 30–38, which concentrates on sociology.

A handful of studies by sociologists have undertaken a deeper examination of social science and the NSF inspired by their discipline's interest in the sociology of knowledge, science, and the professions. These works also emphasize issues that figure prominently in my analysis, especially scientific legitimacy, boundary work, natural science–social science relations, and scientific patronage. See Gene M. Lyons, *The Uneasy Partnership: Social Science and the Federal Government in the Twentieth Century* (New York: Russell Sage Foundation, 1969), 126–136; Samuel Z. Klausner and Victor M. Lidz, eds., *The Nationalization of the Social Sciences* (Philadelphia: University of Pennsylvania Press, 1986); Thomas F. Gieryn, "The U.S. Congress Demarcates Natural Science and Social Science (Twice)," in Gieryn's book *Cultural Boundaries of Science: Credibility on the Line* (Chicago: University of Chicago Press, 1999), 65–114; Otto N. Larsen, *Milestones and Millstones: Social Science at the National Science Foundation, 1945–1991* (New Brunswick, NJ: Transaction, 1992), 1–18, 33–58.

11. The present chapter pursues lines of analysis that I have developed elsewhere: Mark Solovey, "Riding Natural Scientists' Coattails onto the Endless Frontier: The SSRC and the Quest for Scientific Legitimacy," *Journal of the History for the Behavioral Sciences* 40 (2004): 393–424; Mark Solovey, *Shaky Foundations: The Politics–Patronage–Social Science Nexus in Cold War America* (New Brunswick, NJ: Rutgers University Press, 2013), 20–55, chap. 1, "Social Science on the Endless (and End-less?) Frontier: The Postwar NSF Debate."

12. Vannevar Bush, *Science—The Endless Frontier* (Washington, D.C.: NSF, 1995, 50th anniversary reissue of 1945 edition), hereafter, *SEF*. G. Pascal Zachary, *Endless Frontier: Vannevar Bush, Engineer of the American Century* (New York: Free Press, 1997), on *SEF* and NSF's origins, 327–334. Daniel J. Kevles, "The National Science Foundation and the Debate over Postwar Research Policy, 1942–1945: A Political Interpretation of Science—The Endless Frontier," *Isis* 68 (1977): 5–26. Shelby Scates, *Warren G. Magnuson and the Shaping of Twentieth-Century America* (Seattle: University of Washington Press, 1997).

13. Robert F. Maddox, *The Senatorial Career of Harley Martin Kilgore* (New York: Garland, 1981), on the NSF legislation, 162–173, 329–330. U.S. Congress, Senate, Committee on Military Affairs, Subcommittee on War Mobilization, *Hearings on Science Legislation (S. 1297 and Related Bills)*, hereafter, *1945 Senate Hearings*, 79th Cong., 1st sess. (Washington, D.C.: GPO, 1945). On the contending science policy visions, see Daniel L. Kleinman, *Politics on the Endless Frontier: Postwar Research Policy in the United States* (Durham, NC: Duke University Press, 1995); Jessica Wang, "Liberals, the Progressive Left, and the Political Economy of Postwar American Science: The National Science Foundation Debate Revisited," *Historical Studies in the Physical and Biological Sciences* 26 (1995): 139–166.

14. Bush, *SEF*, 9.

15. Bush, *SEF*, 23; Bush to Truman, July 5, 1945, included in *SEF*, 1.

16. S. 1297 in *1945 Senate Hearings*.

17. Letter from Vannevar Bush to D. C. Josephs, Oct. 10, 1946, Folder 12, Box 329, Series III.A Grants, Carnegie Corporation of New York Records, Rare Book and Manuscript Library, Columbia University Libraries, New York.

18. Kevles, "The National Science Foundation and the Debate over Postwar Research Policy, 1942–1945," 24.

19. Bush to Josephs, Oct. 10, 1946, 3.

20. Vannevar Bush to D. C. Josephs, Sept. 19, 1946, Folder Bush Papers, Box 14, Series Historian Files, hereafter, NSF HF, NSF Collection Record Group 307, hereafter, RG 307, National Archives–College Park, MD.

21. Nathan Reingold, "Vannevar Bush's New Deal for Research; or, The Triumph of the Old Order," in Nathan Reingold, *Science, American Style* (New Brunswick, NJ: Rutgers University Press, 1991), 284–333, at 288.

22. Reagan, *Designing a New America*.

23. On the board's vision for the postwar era, see U.S., NRPB, *Post-War Agenda: Full Employment, Security, Building America* (Washington, D.C.: GPO, 1942). Other liberal scholars involved with the NRPB include the economists Paul Samuelson, Alvin Hansen, and John Kenneth Galbraith.

24. Donald Fisher, *Fundamental Development of the Social Sciences: Rockefeller Philanthropy and the United States Social Science Research Council* (Ann Arbor: University of Michigan Press, 1993); Kenton W. Worcester, *Social Science Research Council, 1923–1998* (New York: SSRC, 2001); Denis Bryson, "Personality and Culture, The Social Science Research Council, and Liberal Social Engineering: The Advisory Committee on Personality and Culture, 1930–1934," *Journal of the History of the Behavioral Sciences* 45 (2009): 355–386.

25. On Mitchell, see Mark C. Smith, *Social Science in the Crucible: The American Debate over Objectivity and Purpose, 1918–1941* (Durham, NC: Duke University Press, 1994), 49–63; on Yerkes, see Capshew, *Psychologists on the March*, 42–51; on Ogburn, see Robert C. Bannister, *Sociology and Scientism: The American Quest for Objectivity, 1880–1940* (Chapel Hill: University of North Carolina Press, 1987), 161–187; on Nourse, see Joseph G. Knapp, *Edwin G. Nourse—Economist for the People* (Danville, IL: Interstate, 1979).

26. Committee on Problems and Policy, Minutes, Apr. 14, 1945, p. 1, Folder 1785, Box 315, Sub-series 1, Series II, RG 1, Social Science Research Council Archives, hereafter, SSRC Archives, Rockefeller Archive Center, hereafter, RAC, Sleepy Hollow, NY. All other quotes in Appendix 1, "The Federal Government and Research," attached to the minutes.

27. Fisher, *Fundamental Development of the Social Sciences*, 191.

28. Committee on Problems and Policy, Minutes, July 28–29, 1945, Folder 1785, SSRC Archives.

29. Harry S. Truman, "Special Message to the Congress Presenting a 21-Point Program for the Reconversion Period, September 6, 1945," in *Public Papers of the Presidents, Truman, 1945* (Washington, D.C.: GPO, 1962), 292–294.

30. Smith quoted on p. 311 in Reingold, "Vannevar Bush's New Deal for Research."

31. Board of Directors Meeting, Minutes, Sept. 10–13, 1945, pp. 8, 17, Folder 2098, Box 357, Series 9, RG 1, SSRC Archives. Klausner, "The Bid to Nationalize American Social Science," in Klausner and Lidz, ed., *Nationalization of the Social Sciences*, 3–40, at 7.

32. Distribution of views noted on p. 113 in U.S. Congress, House, *Technical Information for Congress*, Report to the Subcommittee on Science, Research, and Development of the Committee on Science and Astronautics, 92d Cong., 1st sess., Serial A (Washington, D.C.: GPO, Apr. 25, 1969, rev. May 1, 1971).

33. Bush in *1945 Senate Hearings*, 200.

34. *1945 Senate Hearings*, Compton 631, Rabi 998–999, Dewey 818, Adams 826–827, Bakhmeteff 715, Bowman 23, Simms 1170, Fishbein 496.

35. Adams in *1945 Senate Hearings*, 827.

36. Rabi in *1945 Senate Hearings*, 999, 998.

37. England, *Patron for Pure Science*, 36.

38. Simms in *1945 Senate Hearings*, 1170.

39. Union statement in *1945 Senate Hearings*, 1153.

40. Bush, *SEF*, 18, 19, 5. For background, see Ronald Kline, "Construing 'Technology' as 'Applied Science': Public Rhetoric of Scientists and Engineers in the United States, 1880–1945," *Isis* 86 (1995): 194–221.

41. Dewey in *1945 Senate Hearings*, 818.

42. Warren G. Magnuson to the Social Science Research Council, Sept. 19, 1945, Folder 1894, Box 168, Sub-series 37, Series 1, RG 2, SSRC Archives.

43. William F. Ogburn, "The Folkways of a Scientific Sociology," *Scientific Monthly* 30 (1930), 300–306, at 301.

44. Robert M. Yerkes, "The Scope of Science," *Science* 105 (May 2, 1947): 461–463, at 462.

45. "Memorandum of the Social Science Research Council, on the Federal Government and Research in the Social Sciences, October 1945," hereafter, *1945 SSRC Memorandum*, submitted by Mitchell in *1945 Senate Hearings*, 740–743, at 743. Mitchell's testimony runs just a bit longer, from 738–744, which suggests that he

believed the main points he wanted to make were already included in the memorandum. Presumably, he also wrote the memorandum albeit with input from others.

46. *1945 SSRC Memorandum*, 743.

47. Nourse in *1945 Senate Hearings*, 759.

48. *1945 SSRC Memorandum*, 741.

49. On this point, the sociologist of science Thomas Gieryn's analysis of social science testimony in this episode is largely compatible with mine. In "The U.S. Congress Demarcates Natural Science and Social Sciences (Twice)," Gieryn notes three main points favoring inclusion (pp. 74–84): that social science is useful in the same way as natural science; that the social and natural sciences use similar methods, including observation and quantitative analysis; and that since reality is one package, any division among the sciences impedes understanding.

50. Nourse in *1945 Senate Hearings*, 757.

51. *1945 SSRC Memorandum*, 743.

52. Yerkes in *1945 Senate Hearings*, 755.

53. *1945 SSRC Memorandum*, 741.

54. Nourse in *1945 Senate Hearings*, 757.

55. Gaus in *1945 Senate Hearings*, 747.

56. Mitchell in *1945 Senate Hearings*, 739.

57. Ogburn in *1945 Senate Hearings*, 769.

58. Cooper in *1945 Senate Hearings*, 778.

59. On Mitchell's vote for Thomas, see Smith, *Social Science in the Crucible*, 50–51.

60. On Ogburn's politics, see Bannister, *Sociology and Scientism*, 175.

61. Nourse's views discussed in Knapp, *Edwin G. Nourse*.

62. Heather Douglas, *Science, Policy, and the Value-Free Ideal* (Pittsburgh: University of Pittsburgh Press, 2009), chap. 3.

63. Isaiah Bowman, *Geography in Relation to the Social Sciences* (New York: Charles Scribner's Sons, 1934), 224. For more on Bowman, see Neil Smith, *American Empire: Roosevelt's Geographer and the Prelude to Globalization* (Los Angeles: University of California Press, 2004).

64. Bowman Committee letter included in "Truman Aid Asked for Magnuson Bill," *New York Times*, Nov. 27, 1945, reprinted in *1945 Senate Hearings*, 1126–1129; "Pending Legislation for Federal Aid to Science," *Science* 102 (Nov. 30, 1945): 545–548.

65. Bowman in *1946 House Hearings*, 13, 14. For full citation, see p. 326, n. 68.

66. "Original Members of the Committee [for a National Science Foundation]," attached to Harlow Shapley & Harold C. Urey to Dear Member, July 18, 1946, Folder 8, Box 30, Wirth Papers, Special Collections Research Center, University of Chicago Library, Chicago.

67. Harlow Shapley, *Through Rugged Ways to the Stars* (New York: Charles Scribner's Sons, 1969), 145, 149–157.

68. U.S. Congress, House, Committee on Military Affairs, Subcommittee on War Mobilization, *National Science Foundation, Report on Science Legislation*, hereafter, *1946 House Hearings*, 79th Cong., 2nd sess. (Washington, D.C.: GPO, Feb. 27, 1946), 6, 16.

69. See Kilgore in *Congressional Record—Senate,* July 3, 1946, 8231–8232.

70. Brown in *1946 House Hearings*, 13.

71. Hart in *Congressional Record-Senate*, July 3, 1946, 8230.

72. Smith in *Congressional Record-Senate*, July 3, 1946, 8231.

73. The amendment, discussion, and vote in *Congressional Record-Senate*, July 3, 1946, 8230–8232.

74. "Science Bill Gets Senate Approval," *New York Times*, July 3, 1946.

75. See William J. Barber, *Designs within Disorder: Franklin D. Roosevelt, the Economists, and the Shaping of American Economic Policy, 1933–1945* (New York: Cambridge University Press, 1996), 153–169.

76. Fulbright in *Congressional Record-Senate*, May 4, 1948, 5251.

77. President's Scientific Research Board (John R. Steelman, chair), *Science and Public Policy: A Report to the President, vol. 1* (Washington, D.C.: GPO, 1947), viii. On Steelman, see Richard Pearson, "Leading Truman Aide John R. Steelman Dies at 99," *Washington Post*, July 21, 1999; on his graduate studies in sociology, see Oral History Interview with John R. Steelman by Niel M. Johnson, Feb. 28, 1996, pp. 6–11, http://www.trumanlibrary.org/oralhist/steelm2a.htm#transcript, accessed Apr. 21, 2016.

78. John Dewey, "Liberating the Social Scientist," *Commentary* 4 (1947): 378–385, at 378, 379, 380. Charles A. Beard, "Neglected Aspects of Political Science," *American Political Science Review* 42 (1948): 211–222. Robert S. Lynd, "The Science of Inhuman Relations," *New Republic* 121 (Aug. 1949): 22–25. Louis Wirth, "Preface" to *Karl Mannheim, Ideology and Utopia: An Introduction to the Sociology of Knowledge* (New York: Harcourt, Brace & Co., 1936), xxvii–xxviii. Gunnar Myrdal, with the assistance of Richard Sterner and Arnold Rose, *An American Dilemma: The Negro Problem and American Democracy* (New York: Harper, 1944), Appendix 2 "A Methodological Note on Facts and Valuations in Social Science," 1035–1064.

79. Roger A. Salerno, *Louis Wirth: A Bio-Bibliography* (New York: Greenwood, 1987).

80. Louis Wirth, "The Unfinished Business of American Democracy," *Annals of the American Academy of Political and Social Science* 244 (1946): 1–9, at 2, 9.

81. Louis Wirth, "Report of the History, Activities and Policies of the Social Science Research Council," Aug. 1937, esp. 151, Folder 2, Box 32, Wirth Papers.

82. Louis Wirth to E. W. Burgess, May 29, 1946, p. 2, Folder 1895, Box 168, Sub-series 37, Series 1, RG 2, SSRC Archives.

83. Louis Wirth, "Responsibility of Social Science," *Annals of the American Academy of Political and Social Science* 249 (1947): 143–151, at 147, 148.

84. Minutes, Feb. 9, 1949, Sub-Committee on Social Sciences and Values, Committee on Commissions of Inquiry, Division of the Social Sciences, University of Chicago, pp. 1, 2, Folder 1369, Box 226, Sub-series 19, Series 1, RG 1, SSRC Archives.

85. Louis Wirth, "Comments on Social Science and Values [rough notes on remarks made at SSRC meeting, Sept. 49]," p. 1, Folder 1370, Box 226, Sub-series 19, Series 1, RG 1, SSRC Archives.

86. Minutes, P&P Committee, July 15, 1946, p. 4, Folder 865, Box 154, Sub-series 19, Series 1, Record Group 1, SSRC Archives.

87. Five thousand copies noted on p. 3 in Minutes, Board Meeting, July 20–21, 1946, Folder 1787, Box 316, Sub-series 1, Series 2, RG 1, SSRC Archives. Waldemar Kaempffert, *Should the Government Support Science?* (New York: Public Affairs Committee, 1946), 21.

88. Minutes, CFGR, Nov. 15, 1946, p. 3, Folder 864, Box 154, Sub-series 19, Series 1, RG 1, SSRC Archives.

89. Parsons's best-known books were *The Structure of Social Action: A Study in Social Theory with Special Reference to a Group of Recent European Writers* (New York: McGraw Hill, 1937) and *The Social System* (Glencoe, IL: The Free Press, 1951). Regarding interpretative analysis, see Talcott Parsons, "Weber's Methodology of Social Science," in *Max Weber, The Theory of Social and Economic Organization,* trans. A. M. Henderson and T. Parsons (New York: Oxford University Press, 1947), 8–29. On Parsons's philosophy of social science in the context of his Harvard milieu, see Joel Isaac, *Working Knowledge: Making the Human Sciences from Parsons to Kuhn* (Cambridge, MA: Harvard University Press, 2012), 179–190. On the development of Parsons's social thought, see Howard Brick, *Transcending Capitalism: Visions of a New Society in Modern American Thought* (Ithaca, NY: Cornell University Press, 2006), esp. 121–151.

90. Talcott Parsons, "The Science Legislation and the Role of the Social Sciences," *American Sociological Review* 11 (1946): 653–666, at 662–663.

91. Talcott Parsons, "A Basic National Resource," 41–111 in *Nationalization of the Social Sciences*, eds. Klausner and Lidz, 106, 42.

92. Talcott Parsons, "National Science Legislation, Part 2: The Case for the Social Sciences," *Bulletin of the Atomic Scientists* 3 (1947): 3–5, at 3.

93. For criticisms of Parsons's draft, see Minutes, Board Meeting, Sept. 13–16, 1948, Folder 2100, Box 358, Series 9, RG 1, SSRC Archives. Parsons, "A Basic National Resource," 107.

94. See documents in Folder SSRC Project, Box 19, Talcott Parsons Papers HUGFP 42.8.4, Harvard University Archives, Cambridge, MA.

95. Wang, "Liberalism, the Progressive Left, and the Political Economy of Postwar American Science," 140. Also see Jessica Wang, *American Science in an Age of Anxiety: Scientists, Anticommunism, and the Cold War* (Chapel Hill: University of North Carolina Press, 1999).

96. Hadley Cantril, "Polls and the 1948 U.S. Presidential Election: Some Problems It Poses," *International Journal of Opinion and Attitude Research* 2 (1948): 309–320, at 319.

CHAPTER 2: STAKING OUT THE HARD-CORE

1. Alan T. Waterman to John H. Teeter, Dec. 22, 1955, Folder Social Science Research 1953–56, Box 20, Series Waterman Director Files, hereafter, Waterman File, NSF Collection Record Group 307, hereafter, RG 307, National Archives–College Park, MD.

2. Harry Alpert, "The Knowledge We Need Most," *Saturday Review*, Feb. 1, 1958, 36–38, at 37–38.

3. On the NSF's early development, see J. Merton England, *A Patron for Pure Science: The National Science Foundation's Formative Years, 1945–57* (Washington, D.C.: NSF, 1982); Toby Appel, *Shaping Biology: The National Science Foundation and American Biological Research, 1945–1975* (Baltimore: Johns Hopkins University Press, 2000), chaps. 2–5. From the beginning, the agency also had a number of other responsibilities and activities, some of which involved the social sciences, such as the maintenance of a register of scientific personnel in the U.S., but they have little relevance to the present study.

4. NSF, *1950–51 Annual Report*, vii—from the Foreword by James B. Conant, the first chairman of NSF's governing board.

5. NSF, *1954 Annual Report*, 118. NSF, *1953 Annual Report*, "What Is Basic Research?" 38ff.

6. Daniel J. Kevles, *The Physicists: The History of a Scientific Community in Modern America* (Cambridge, MA: Harvard University Press, rev. ed. 1995), 365.

7. Public Law 507, 81st Cong., the National Science Foundation Act of 1950.

8. Neither American historians nor historians of American science have devoted much attention to the NSF's social science efforts during these early years, although England, *Patron for Pure Science*, 266–273, provides a short and reliable discussion.

Accounts by social scientists include Henry W. Riecken, "Underdogging: The Early Career of the Social Sciences in the NSF," in Samuel Z. Klausner and Victor M. Lidz, eds., *The Nationalization of the Social Sciences* (Philadelphia: University of Pennsylvania Press, 1986), 209–225; Otto N. Larsen, *Milestones and Millstones: Social Science at the National Science Foundation, 1945–1991* (New Brunswick, NJ: Transaction, 1992), 33–58; and Desmond King, "The Politics of Social Research: Institutionalizing Public Funding Regimes in the United States and Britain," *British Journal of Political Science* 28 (1998): 415–444.

Also of interest: Krist Vaesen and Joel Katzav, "The National Science Foundation and Philosophy of Science's Withdrawal from Social Concerns," *Studies in History and Philosophy of Science* (2019, in press).

I have analyzed many events and issues discussed in this chapter elsewhere: Daniel L. Kleinman and Mark Solovey, "Hot Science/Cold War: The National Science Foundation after World War Two," *Radical History Review* 63 (1995): 110–139; Mark Solovey and Jefferson D. Pooley, "The Price of Success: Sociologist Harry Alpert, the NSF's First Social Science Policy Architect," *Annals of Science* 68 (2011): 229–260; and Mark Solovey, *Shaky Foundations: The Politics–Patronage–Social Science Nexus in Cold War America* (New Brunswick, NJ: Rutgers University Press, 2013), 148–187, chap. 4, "Cultivating Hard-Core Social Research at the NSF."

9. Durkheim was the subject of Alpert's PhD dissertation and his book, *Emile Durkheim and His Sociology* (New York: Columbia University Press, 1939).

10. Memo from Harry Alpert to Dr. Waterman, July 1, 1953, Progress Report No. 1, hereafter, Progress Report 1, Folder Role of the Foundation with respect to Social Science Research (Alpert, Mar. 1954), Alpert quoting General Counsel at 1, Box 8, Series Historian Files, hereafter, NSF HF, RG 307.

11. Alpert, "The Knowledge We Need Most," 37.

12. Figures from "NSF Requests and Appropriations By Account: FY 1951–FY 2019," https://dellweb.bfa.nsf.gov/NSFRqstAppropHist/NSFRequestsandAppropriationsHistory.pdf, accessed Dec. 7, 2018. On the development of military support for the social and psychological sciences, see Solovey, *Shaky Foundations*, 56–102, chap. 3, "Defense and Offense in the Military Science Establishment: Toward a Technology of Human Behavior."

13. On Waterman, see England, *Patron for Pure Science*, 130–132.

14. Sophie D. Aberle, *The Pueblo Indians of New Mexico: Their Land, Economy and Civil Organization* (Menasha, WI: American Anthropological Association, 1948).

15. Frederick A. Middlebush and Chesney Hill, *Elements of International Relations* (New York: McGraw-Hill, 1940).

16. England, *Patron for Pure Science*, 127, 169. On American science and anti-communism more generally, see Jessica Wang, *American Science in an Age of Anxiety:*

Scientists, Anticommunism, and the Cold War (Chapel Hill: University of North Carolina Press, 1999), in relation to NSF, esp. 25–37, 254–262.

17. Louis Wirth, "The Social Sciences," in *American Scholarship in the Twentieth Century*, ed. Merle Curti (Cambridge, MA: Harvard University Press, 1953), 32–82, at 37.

18. On the Reece Committee, see Solovey, *Shaky Foundations*, 122–127.

19. On these consultations, see Appendix to Progress Report No. 1, also Appendix to Memorandum, Harry Alpert to Dr. Waterman, Nov. 1, 1953, Progress Report 2, hereafter, Progress Report 2, Folder Role of the Foundation with respect to Social Science Research (Alpert, Mar. 1954), Box 8, NSF HF.

20. Memorandum, Harry Alpert to Dr. Waterman, Nov. 1, 1953, Position Paper No. 2, hereafter, Position Paper 2, Bush quoted at 1, Folder Role of the Foundation with respect to Social Science Research (Alpert, Mar. 1954), Box 8, NSF HF.

21. Barnard's point mentioned on p. 1 in Memo from Harry Alpert to Dr. Waterman, May 22, 1953, Folder Social Science Research 1953–56, Box 20, Waterman File.

22. Chester I. Barnard, "Social Science: Illusion and Reality," *American Scholar* 21 (Summer 1952): 359–361, at 359. William G. Scott, *Chester I. Barnard and the Guardians of the Managerial State* (Lawrence: University of Kansas Press, 1992) claims on p. 83 that after Barnard joined the NSF's governing board in 1951, he underwent a "sudden conversion to orthodox positive science methods." However, I see no evidence that Barnard's views really changed.

23. Position Paper 2, all quotes at 3.

24. Position Paper 2, 2.

25. Position Paper 2, 4.

26. Riecken, "Underdogging," 211.

27. See Appendix C, Recommendations Approved by National Science Board, August, 1954, attached to Harry Alpert, with the assistance of Bertha W. Rubinstein, Progress Report No. 5 (Revised), Feb. 1, 1956, hereafter, Progress Report 5, Folder Role of the Foundation with respect to Social Science Research (Alpert, Mar. 1954).

28. NSF, *1955 Annual Report*, 59.

29. Project noted in "Inventory of Natural Science–Social Science Interdisciplinary Activities of National Science Foundation (Through Fiscal Year 1954)," Folder Social Science Research Program, Box 40, Waterman File.

30. Harry Alpert, "Demographic Research and the National Science Foundation," *Social Forces* 36 (1957): 17–21, project noted on 19.

31. Harry Alpert, "Social Science, Social Psychology, and the National Science Foundation," *American Psychologist* 12 (1957): 95–98, projects noted on 96.

32. Progress Report 5, 3, 4.

33. See George T. Mazuzan, "'Good Science Gets Funded': The Historical Evolution of Grant Making at the National Science Foundation," *Science Communication* 14 (1992): 63–90; Marc Rothenberg, "Making Judgments about Grant Proposals: A Brief History of the Merit Review Criteria at the National Science Foundation," *Technology and Innovation* 12 (2010): 189–195.

34. Bertha W. Rubinstein, "Federal Support of Unclassified, Extramural Social Science Research: 1953–1955," rankings on 1, Folder Social Science Research 1953–56, Box 20, Waterman File.

35. Memorandum, Study Director for Social Science Research to Advisory Panel for Social Science Research, Oct. 1, 1956, figures on 2, Folder Social Science Research 1953–56.

36. Robert Yerkes received a large grant from the psychobiology program, as noted in Appel, *Shaping Biology*, 87.

37. See NSF, *1955 Annual Report*, 61.

38. Harry Alpert, "Congressmen, Social Scientists, and Attitudes toward Federal Support of Social Science Research," *American Sociological Review* 23 (1958): 682–686.

39. NSF annual reports list the agency's advisory panels and their members. Memorandum, Study Director for Social Science Research to Advisory Panel for Social Science Research, Oct. 1, 1956, recommendations on 3.

40. Harry Alpert to Director [Waterman], June 14, 1957, Folder Social Science Research Program, Box 40, Waterman File.

41. NSF, *1958 Annual Report*, 41–42.

42. Harry Alpert to Director [Waterman], June 14, 1957, 1.

43. Harry Alpert to Director [Waterman], June 14, 1957, 19; NSF, *1960 Annual Report*, impact of Ford Foundation decision noted on 56. Also see Solovey, *Shaky Foundations*, 103–147, chap. 3, "Vision, Analysis, or Subversion: The Rocky Story of the Behavioral Sciences at the Ford Foundation"; Jefferson D. Pooley, "A 'Not Particularly Felicitous' Phrase: A History of the Behavioral Sciences Label, *Serendipities* 1 (2016): 38–81.

44. Roger L. Geiger, *Research and Relevant Knowledge: American Research Universities since World War II* (New York: Oxford University Press, 1993), 168.

45. Memo, Director Waterman to Members of the National Science Board, July 29, 1952, p. 1, Board Books, Tab #2a, 15th Meeting, Aug. 7–8, 1952, National Science Board Library, hereafter, NSB Library, NSF headquarters, Alexandria, VA.

46. Executive Order 10521, Mar. 17, 1954, reprinted in England, *Patron for Pure Science*, 353–355.

47. See Dael Wolfle, ed., *Symposium on Basic Research* (Washington, D.C.: American Association for the Advancement of Science, 1959).

48. Phrase taken from Mazuzan's article "Good Science Gets Funded."

49. Dael Wolfle, "Social Science," *Science*, Dec. 16, 1960, 1795.

50. Alpert's articles include "The National Science Foundation and Social Science Research," *American Sociological Review* 19 (1954): 208–211; "Anthropological Research and the National Science Foundation," *Bulletin of the American Anthropological Association* 3 (1955): 1–3; "The Social Sciences and the National Science Foundation," *Proceedings of the American Philosophical Society* 99 (1955): 332–333; "The Social Sciences and the National Science Foundation, 1945–1955," *American Sociological Review* 20 (1955): 653–661; "Demographic Research and the National Science Foundation"; "Geography, Social Science, and the National Science Foundation," *Professional Geographer* 9 (1957): 7–9; "Social Science, Social Psychology, and the National Science Foundation"; "The Social Science Research Program of the National Science Foundation," *American Sociological Review* 22 (1957): 582–585.

51. Alpert, "The National Science Foundation and Social Science Research," 211.

52. Author's interview with Bertha Rubinstein, Feb. 4, 1992.

53. Author's interview with Pendleton Herring, Oct. 3, 1992. Fred I. Greenstein and Austin Ranney, "Pendleton Herring: 27 October 1903–17 August 2004," *Proceedings of the American Philosophical Society* 150 (2006): 487–492.

54. McCann quoted in Board Books, 55th Meeting, Sept. 16–17, 1958, Tab A, at 7, NSB Library. Kevin McCann, *Man from Abilene* (Garden City, NY: Doubleday, 1952). Joan D. Goldhamer, "General Eisenhower in Academe: A Clash of Perspectives and a Study Suppressed," *Journal of the History of the Behavioral Sciences* 33 (1997): 241–259.

55. Riecken, "Underdogging," 217.

56. On juvenile delinquency, Kefauver, and social scientists, see James Gilbert, *A Cycle of Outrage: America's Reaction to the Juvenile Delinquent in the 1950s* (New York: Oxford University Press, 1986). The problem of juvenile delinquency also became relevant to the new specialty of family therapy. See Deborah Weinstein, *The Pathological Family: Postwar America and the Rise of Family Therapy* (Ithaca, NY: Cornell University Press, 2013).

57. The subcommittee's recommendation and its legislative supporters noted in Appendix J to Progress Report 5.

58. NSB Minutes, 39th Meeting, Mar. 12, 1956, closed session, p. 2, NSB Library.

59. "Excerpts from Juvenile Delinquency, Report of Senate Committee on the Judiciary's Subcommittee on Juvenile Delinquency," Senate Report No. 130, 1957, in *Congressional Record—Senate*, June 3, 1957, 8212.

60. Alan T. Waterman to Mr. Clyde C. Hall, July 29, 1957, Folder Social Science Research Program, Box 40, Waterman File.

61. Theodore M. Hesburgh, *God and the World of Man* (Notre Dame, IN: University of Notre Dame Press, 1950), 5. Michael O'Brien, *Hesburgh: A Biography* (Washington, D.C.: Catholic University of America Press, 1998).

62. T. M. Hesburgh, chairman, "Report of Committee on Social Sciences," May 18, 1958, in Board Books, 54th Meeting, June 24, 1958, Tab Q, NSB Library.

63. The quotations are paraphrases of Stratton's comments as reported in NSB Minutes, Closed Session, 54th Meeting, June 28–30, 1958, at 6, 7, NSB Library.

64. Weaver's views and Waterman's reminder in NSB Minutes, Closed Session, 54th Meeting, 6–7. Warren Weaver, *Scene of Change: A Lifetime in American Science* (New York: Scriber, 1970), 154.

65. NSB Minutes, Closed Session, 54th Meeting, McLaughlin's suggestion on 7.

66. "Report of the Social Sciences Committee" in Board Books, 56th Meeting, Oct. 13–14, 1958, 3–4, NSB Library.

67. Theodore Hesburgh Oral History Transcript, interviewed by Marc Rothenberg, Oct. 27, 2008, at 4, NSF Oral Histories, NSF headquarters, Alexandria, VA.

68. Harry Alpert, "The Social Sciences: Problems, Issues, and Suggested Resolutions," April 15, 1958, at 2, Tab R in Board Books, 54th Meeting, June 24, 1958, NSB Library.

69. Alpert, "The Social Sciences: Problems, Issues, and Suggested Resolutions," 3, 2.

70. Memo, from Program Director [Riecken], Social Science Research Program, to Director [Waterman], Sept. 26, 1958, figures on 2, Folder Social Science Research Program, Box 40, Waterman File.

71. Charles O. Porter, "Social Sciences Ignored in National Science Foundation Board Nominations," *Congressional Record-House*, Aug. 19, 1958, figures on 16987. However, the comparable figure given in Larsen, *Milestones and Millstones*, p. 47, is only 2.2 percent.

72. Memo, Riecken to Waterman, Sept. 26, 1958, at 2, 3.

73. Press Release, "National Science Foundation Announces Award of 841 Fellowships in Science for 1958–1959," Mar. 18, 1958, Folder Social Science Research Program, Box 40, Waterman File.

74. See Joel Isaac, "The Human Sciences in Cold War America," *Historical Journal* 50 (2007): 725–746; David Engerman, "Social Science in the Cold War," *Isis* 101 (2010): 393–400; Mark Solovey, "Cold War Social Science: Spectre, Reality, or Useful Concept?" in *Cold War Social Science: Knowledge Production, Liberal Democracy, and Human Nature*, ed. Mark Solovey and Hamilton Cravens (New York: Palgrave Macmillan, 2012), 1–22.

75. These figures are for 1955 to 1957, taken from a NSF report excerpt placed by Senator Wayne Morse in the *Congressional Record—Senate*, June 3, 1957, 8213.

Larsen, *Milestones and Millstones*, p. 24, has somewhat higher figures, with the social sciences' share of overall federal science obligations between 3 percent and 4 percent for the years 1956, 1957, and 1958. Whatever figures one chooses, though, the point that the social sciences' share was rather small is clear.

76. Harry Alpert, "The Government's Growing Recognition of Social Science," *Annals of the American Academy of Political and Social Science* 327 (1960): 59–67, at 62.

77. For further discussion, see Solovey and Pooley, "The Price of Success: Sociologist Harry Alpert, the NSF's First Social Science Policy Architect."

78. Jennifer Platt, *A History of Sociological Research Methods in America, 1920–1960* (New York: Cambridge University Press, 1996), 69. Also Stjepan G. Mestrovic, "Introduction to the Gregg Revivals Edition," 3–12 in Harry Alpert, *Emile Durkheim and His Sociology* (UK: Gregg Revivals, 1993).

79. Alpert, *Emile Durkheim and His Sociology*, on scientific criteria, 111–112; Harry Alpert, "France's First University Course in Sociology," *American Sociological Review* 2 (1937): 311–317, at 315. Also see Alpert's essays: "Emile Durkheim and Sociologismic Psychology," *American Journal of Sociology* 45 (1939): 64–70; "Emile Durkheim: A Perspective and Appreciation," *American Sociological Review* 24 (1959): 462–465.

80. Harry Alpert, review of Robert M. MacIver, *The More Perfect Union* (1948), in *Jewish Social Studies* 11 (1949): 189–191, at 190–191. Also see Alpert's reviews of Ruth G. Weintraub, *How Secure These Rights?* (1949), in *Jewish Social Studies* 11 (1950): 261–263, and of Carey McWilliams, *A Mask for Privilege: Anti-Semitism in America* (1948), in *American Journal of Sociology* 54 (1948): 177.

81. Alpert, *Emile Durkheim and His Sociology*, 108–109.

82. Harry Alpert, "Robert M. MacIver's Contributions to Sociological Theory," in *Freedom and Control in Modern Society*, ed. Morroe Berger, Theodore Abel, and Charles H. Page (New York: D. Van Nostrand, 1954), 286–292, at 290, 291, 292.

83. Alpert, "Robert M. MacIver's Contributions to Sociological Theory," 292.

84. Harry Alpert, "Public Opinion Research as Science," *Public Opinion Quarterly* 20 (1956): 493–500, quoting Bush at 494–495.

85. Harry Alpert, "The Growth of Social Research in the United States," in *The Human Meaning of the Social Sciences*, ed. Daniel Lerner (Cleveland, OH: World Publishing, 1959), 73–86, at 82.

86. Harry Alpert, "Some Observations on the State of Sociology," *Pacific Sociological Review* 6 (1963): 45–58, esp. 48.

87. Harry Alpert, "George Lundberg's Social Philosophy: A Continuing Dialogue," in *The Behavioral Sciences: Essays in Honor of George A. Lundberg*, ed. Alfred de Grazia et al. (Great Barrington, MA: Behavioral Research Council, 1968), 48–62, esp. 58.

88. The two plaques and the doggerel noted in Robert B. Clark, "Harry Alpert Memorial," Nov. 9, 1977, Harry Alpert Collection, UA REF 1, Biographical Collection, Box 1, University of Oregon Library, Division of Special Collections and University Archives, Eugene, Oregon.

89. Ogburn in *1945 Senate Hearings*, 774. For full citation, see p. 322, n. 13.

90. Record of interview with JP and Theodore Hesburgh, Dec. 2, 1958, Folder 8, Box 722, Series III.A Grants, Carnegie Corporation of New York Records, Rare Book and Manuscript Library, Columbia University Libraries.

CHAPTER 3: HELP FROM ABOVE

1. Henry W. Riecken, "Social Sciences and Social Problems," *Social Science Information* 8 (1969): 101–129, at 104.

2. Evron M. Kirkpatrick, "The Impact of the Behavioral Approach on Traditional Political Science," in *Essays on the Behavioral Study of Politics*, ed. Austin Ranney (Urbana: University of Illinois Press, 1962), 1–30, at 29.

3. Daniel J. Kevles, *The Physicists: The History of a Scientific Community in Modern America* (Cambridge, MA: Harvard University Press, rev. ed. 1995), 386.

4. Roger L. Geiger, "What Happened after Sputnik? Shaping University Research in the United States," *Minerva* 35 (1997): 349–367, percentages on 363.

5. George T. Mazuzan, "NSF 88–16, a Brief History," July 15, 1994, at 10, https://www.nsf.gov/about/history/nsf50/nsf8816.jsp, accessed Dec. 7, 2018.

6. Noted in NSF, *1958 Annual Report*, 11–13.

7. Waterman in NSF, *1961 Annual Report*, xvii.

8. Haworth in NSF, *1966 Annual Report*, xix.

9. Henry J. Aaron, *Politics and the Professors: The Great Society in Perspective* (Washington, D.C.: Brookings Institution, 1978); Robert Wood, *Whatever Possessed the President, Academic Experts and Presidential Policy, 1960–1988* (Amherst: University of Massachusetts Press, 1993); Alice O'Connor, *Poverty Knowledge: Social Science, Social Policy, and the Poor in Twentieth-Century U.S. History* (Princeton, NJ: Princeton University Press, 2001).

10. Michael Latham, *Modernization as Ideology: American Social Science and "Nation Building" in the Kennedy Era* (Chapel Hill: University of North Carolina Press, 2000); Nils Gilman, *Mandarins of the Future: Modernization Theory in Cold War America* (Baltimore: Johns Hopkins University Press, 2003).

11. See Clayton Alderfer, "Henry W. Riecken: Present at the Beginning (Many Times)—Biography of an Applied Behavioral Scientist," *Journal of Applied Behavioral Science* 35 (1999): 142–144; Robert F. Boruch, "Henry W. Riecken, Jr.

(1917–2012)," *American Psychologist* 68 (2013): 398; Bart Barnes, "Henry W. Riecken, Jr., Social Scientist," *Washington Post*, Jan. 8, 2013.

12. Henry W. Riecken, *The Volunteer Work Camp: A Psychological Evaluation* (Cambridge, MA: Addison-Wesley, 1952), ix.

13. Henry W. Riecken and Robert F. Boruch, eds., *Social Experimentation: A Method for Planning and Evaluating Social Intervention* (New York: Academic Press, 1974).

14. Henry W. Riecken, "Underdogging: The Early Career of the Social Sciences in the NSF," in *Nationalization of the Social Sciences*, ed. Samuel Z. Klausner and Victor M. Lidz (Philadelphia: University of Pennsylvania Press, 1986), 209–225, at 216.

15. Henry W. Riecken, "Social Change and Social Science," in *Science and the Evolution of Public Policy*, ed. James A. Shannon (New York: Rockefeller University Press, 1973), 135–156, at 140–141.

16. Author's interview with Bertha Rubinstein, Feb. 4, 1992.

17. "Howard H. Hines," Obituary, *Washington Post*, April 1997, https://www.washingtonpost.com/archive/local/1997/04/07/diaconate-head-thomas-knestout-dies/d8aaea55-514b-4cbb-a9b5-e76e7623d66a/?utm_term=.1e732695bb02, accessed Dec. 21, 2018. Otto N. Larsen, *Milestones and Millstones: Social Science at the National Science Foundation, 1945–1991* (New Brunswick, NJ: Transaction, 1992), 62, 63.

18. Hines in "Director's Program Review: Social Sciences," Jan. 14, 1971, at 15, 19, NSF Library, NSF headquarters, Alexandria, VA.

19. On the Office, see NSF, *1959 Annual Report*, 44 ff; on the Division, see NSF, *1961 Annual Report*, 44.

20. Editorial, "U.S. Aids Research in Social Sciences," *New York Times*, Dec. 5, 1960.

21. Larsen, *Milestones and Millstones*, figures on 63.

22. Howard H. Hines to NSF Director, "Annual Report of Division of Social Sciences Fiscal Year 1968," Oct. 10, 1968, funding from other units on 5, Folder Social Science Annual Reports, Box 42, Series Historian Files, hereafter, NSF HF, NSF Collection Record Group 307, hereafter, RG 307, National Archives–College Park, MD.

23. Memo, Director Waterman to Members of the National Science Board, Aug. 12, 1959, at 2, in Board Books, 61st Meeting, Tab H, NSB Library, NSF headquarters, Alexandria, VA.

24. Henry W. Riecken to Leland J. Haworth, "Annual Review of the Division of Social Sciences, Fiscal Year 1963," July 15, 1963, at 21, Folder Social Science Annual Reports, Box 42, NSF HF.

25. See Gina Perry, *Behind the Shock Machine: The Untold Story of the Notorious Milgram Psychology Experiments* (New York: New Press, 2012), which notes (p. 27) that

Milgram received NSF funding in 1961, although a follow-up request for additional support was turned down (p. 213).

26. NSF, *1968 Annual Report*, 141.

27. NSF, *1963 Annual Report*, 32.

28. Figures on p. 35 in "Director's Program Review: Social Sciences."

29. U.S. Congress, Senate, Special Subcommittee on Science, Committee on Labor and Public Welfare, *National Science Foundation Act Amendments of 1968, Hearings*, hereafter, *1968 NSF Hearings*, 90th Cong., 1st Sess. (Washington, D.C.: GPO, 1968), figures on 121.

30. See Maurice Goldhaber and Gerald F. Tape, "Leland John Haworth, 1904–1979," in *National Academy of Sciences Biographical Memoir* (Washington, D.C.: NAS, 1985), 355–382, at 355.

31. NSF annual reports provide lists of NSB members.

32. Waterman quoted in Memo, Assistant Director for Social Sciences Henry Riecken to Director Waterman, "Annual Review of the Division of Social Sciences, FY 1961," July 15, 1961, pledged at 1, Folder Social Science Annual Reports, Box 42, NSF HF.

33. NSF, *1963 Annual Report*, 28–29.

34. Ralph L. Beals, *Politics of Social Research* (New Brunswick, NJ: Transaction, 1969), DSB membership on 101. Daniel S. Greenberg, *The Politics of Pure Science* (New York: New American Library, 1967), PSAC membership on 16.

35. John Walsh, "Behavioral Sciences: Report Bids for a Bigger Role," *Science* 161 (Sept. 13, 1968): 1112–1114, at 1113.

36. David Easton, *The Political System: An Inquiry into the State of Political Science* (New York: Knopf, 1953); Pendleton Herring, "Political Science in the Next Decade," *American Political Science Review* 39 (1945): 757–766; V. O. Key, "The State of the Discipline," *American Political Science Review* 52 (1958): 961–971.

37. Robert Adcock and Mark Bevir, "Political Science," in *The History of the Social Sciences since 1945*, ed. Roger E. Backhouse and Philippe Fontaine (New York: Cambridge University Press, 2010), 71–101, at 78. Also see Robert Adcock, "Interpreting Behavioralism," in *Modern Political Science: Anglo-American Exchanges since 1880*, ed. Robert Adock, Mark Bevir, and Shannon C. Stimson (Princeton, NJ: Princeton University Press, 2007), 180–208.

38. Ted V. McAllister, *Revolt Against Modernity: Leo Strauss, Eric Voeglin, and the Search for a Postliberal Order* (Lawrence: University of Kansas Press, 1996). Also see John G. Gunnell, *The Descent of Political Theory: The Genealogy of an American Vocation* (Chicago: University of Chicago Press, 1993); Nicholas Guilhot, ed., *The Invention of International Relations Theory: Realism, the Rockefeller Foundation, and the 1954 Conference on Theory* (New York: Columbia University Press, 2011).

39. Emily Hauptmann, "'Propagandists for the Behavioral Sciences': The Overlooked Partnership between the Carnegie Corporation and SSRC in the Mid-Twentieth Century," *Journal of the History of the Behavioral Sciences* 52 (2016): 167–187; Emily Hauptmann, "The Ford Foundation and the Rise of Behavioralism in Political Science," *Journal of the History of the Behavioral Sciences* 48 (2012): 154–173.

40. NSB Minutes, closed session, 61st Meeting, Aug. 27–28, 1959, Waterman on 12, NSB Library.

41. James D. Carroll, "Notes on the Support of Political Science Research Projects by the Division of Social Sciences of the National Science Foundation, Fiscal Years 1958–1965," in U.S. Congress, Senate, Committee on Government Operations, Research and Technical Programs Subcommittee, *The Use of Social Research in Federal Domestic Programs, Pt IV* (Washington, D.C.: GPO, April 1967), 81–106.

42. Duncan MacRae Jr., "The Development of Moral Judgment in Children," PhD diss., Harvard University, 1950.

43. Duncan MacRae Jr., with the collaboration of Fred H. Goldner, *Dimensions of Congressional Voting: A Statistical Study of the House of Representatives in the Eighty-first Congress* (Los Angeles: University of California Press, 1958), iv.

44. Memo, Assistant Director for Social Sciences [Henry Riecken] to Director Waterman, "Annual Review of the Division of Social Science, FY 1961," 8.

45. Anon, "NSF and Behavioral Science," *American Behavioral Scientist* 7 (Sept. 1963): 70–71, at 70.

46. Pooh-bah quote from Jacob Heilbrun, "Reagan's Athena," book review of Peter Collier, *Political Women: The Big Little Life of Jeane Kirkpatrick*, July 23, 2012, www.theamericanconservative.com, accessed June 7, 2016.

47. On Kirkpatrick, see David Binder, "Evron Kirkpatrick, 83, Director of Political Science Association," *New York Times*, May 9, 1995.

48. Gunnell, *Descent of Political Theory*, 258.

49. Mann quoted in Binder, "Evron Kirkpatrick."

50. Kirkpatrick, "The Impact of the Behavioral Approach on Traditional Political Science," 3, 14, 25, 29.

51. Evron M. Kirkpatrick, "Toward a More Responsible Two-Party System: Political Science, Policy Science, or Pseudo-Science?" *American Political Science Review* 65 (1971): 965–990, at 986–987.

52. Memorandum, Henry W. Riecken to Leland J. Haworth, Sept. 19, 1963, Folder Political Science 1963, Box 11, Series Haworth Director Files July 1963–Dec. 1964, RG 307.

53. Memorandum, Henry W. Riecken to Leland J. Haworth, 1.

54. Memorandum, Henry W. Riecken to Leland J. Haworth, 2, 3.

55. Evron M. Kirkpatrick to Leland J. Haworth, Nov. 6, 1963, Folder Political Science 1963.

56. Hubert H. Humphrey, *The Education of a Public Man: My Life and Politics* (Garden City, NY: Doubleday, 1976), 59, 62–63.

57. For an example, see Evron M. Kirkpatrick to Senator George S. McGovern, Nov. 26, 1963, Folder Political Science 1963.

58. See Carroll, "Notes on the Support of Political Science," Senate supporters on 87.

59. Plan to send letters noted in Kirkpatrick to Haworth, Jan. 2, 1964. Statement of Evron M. Kirkpatrick, Executive Director, The American Political Science Association, "Protesting the Discrimination against Political Science by the National Science Foundation," Dec. 12, 1963, attached to Memorandum, Evron M. Kirkpatrick to Leland J. Haworth, Jan. 2, 1964, Folder Political Science 1964, Box 11, Series Haworth Director Files July 1963–Dec. 1964, RG 307.

60. Kirkpatrick to Haworth, Jan. 2, 1964.

61. Riecken reported what he said to Kirkpatrick in Henry, W. Riecken, Diary Note, Jan. 7, 1964, Folder Political Science 1964.

62. Riecken, Diary Note, Jan. 7, 1964.

63. "Political Science and the NSF," *American Behavioral Scientist* 7 (Suppl., Mar. 1964): 2.

64. Leland J. Haworth, "Support of Political Science by the National Science Foundation," Aug. 26, 1964, Folder Political Science 1964.

65. Division Director [Howard H. Hines] to NSF Director [Leland Haworth], "Annual Report of Division of Social Sciences Fiscal Year 1968," Oct. 10, 1968, p. 21, Folder Social Science Annual Reports, Box 42, NSF HF.

66. Howard H. Hines, "Division of Social Sciences Annual Report, Fiscal Year 1970," p. 22, Folder Social Sciences Division, 1970, Box 8, Series Director Subject Files Q-Z, Acc. No. 307-75-053, National Archives, Washington National Record Center, Suitland, MD. When I did my research, this part of the NSF collection was not yet fully processed. After materials at Suitland are fully processed, they are usually moved to the National Archives in College Park, MD.

67. Fred R. Harris, "Political Science and the Proposal for a National Social Science Foundation," *American Political Science Review* 61 (1967): 1088–1095, figures on 1089.

68. "Annual Report, Division of Social Sciences, Fiscal Year 1968," figures on 7.

69. NSF, *1966 Annual Report*, 10.

70. NSF, *1967 Annual Report*, 106.

71. Robinson in *1967 NSSF Hearings*, 713, 714, 721. For full citation, see p. 341, n. 1.

72. Marvin Surkin and Alan Wolfe, "Introduction: An End to Political Science," in *An End to Political Science: The Caucus Papers*, ed. Marvin Surkin and Alan Wolfe (New York: Basic Books, 1970), 3–7, at 4.

73. Philip Green and Stanford Levinson, eds., *Power and Community: Dissenting Essays in Political Science* (New York: Random House, 1970), vii. Clyde W. Barrow, "The Political and Intellectual Origins of New Political Science," *New Political Science* 39 (2017): 437–472.

74. David Easton, "The New Revolution in Political Science," *American Political Science Review* 63 (1969): 1051–1061, at 1052.

75. Surkin and Wolfe, "Introduction," 6, 7. For further discussion, see David M. Ricci, *The Tragedy of Political Science: Politics, Scholarship, and Democracy* (New Haven, CT: Yale University Press, 1984), "The Decade of Disillusionment," 176–205.

76. On the NSF's science education efforts in the Cold War context, see John L. Rudolph, *Scientists in the Classroom: The Cold War Reconstruction of American Science Education* (New York: Palgrave, 2002).

77. Rudolph, *Scientists in the Classroom*, esp. chap. 5, "PSSC: Engineering Rationality."

78. Scott L. Montgomery, *Minds for the Making: The Role of Science in American Education, 1750–1990* (New York: Guilford, 1994), 212.

79. See Appendix IV to 77th NSB Meeting Minutes, May 17–18, 1962, at xv, NSB Library.

80. PSAC, "Strengthening the Behavioral Sciences," *Science* 136 (Apr. 20, 1962): 233–241, at 238.

81. NSF, *1962 Annual Report*, 26.

82. Assistant Director for Social Science [H. W. Riecken] to Director Alan T. Waterman, "Annual Review of the Division of Social Sciences, FY 1962," July 15, 1962, at 19, Folder Social Sciences Annual Reports, Box 42, NSF HF.

83. Riecken to Waterman, "Annual Review of the Division of Social Sciences, FY 1962," support from divisions noted on 19, quotes at 19 as well.

84. For a valuable insider's account, see Peter B. Dow, *Schoolhouse Politics: Lessons from the Sputnik Era* (Cambridge, MA: Harvard University Press, 1991).

85. On Bruner and Harvard's center, see Jamie Cohen-Cole, *The Open Mind: Cold War Politics and the Sciences of Human Nature* (Chicago: University of Chicago Press, 2014), 165–189, chap. 6, "Instituting Cognitive Science."

86. Jerome Bruner, *The Process of Education* (Cambridge, MA: Harvard University Press, 1960); Nance C. Lutkehaus, "Putting 'Culture' into Cultural Psychology: Anthropology's Role in the Development of Bruner's Cultural Psychology," *Ethos* 36 (2008): 46–59, copies sold on 51.

87. Jerome S. Bruner, *Toward a Theory of Instruction* (Cambridge, MA: Harvard University Press, 1966), 73–101, "Man: A Course of Study."

88. Bruner, *Toward a Theory of Instruction*, 74.

89. Bruner, *Toward a Theory of Instruction*, 76; on his college years, see Lutkehaus, "Putting 'Culture' into Cultural Psychology," 47.

90. Bruner, *Toward a Theory of Instruction*, 74.

91. Bruner, *Toward a Theory of Instruction*, 83, 84, 87.

92. Bruner, *Toward a Theory of Instruction*, 89.

93. Jamie Cohen-Cole, "The Politics of Psycholinguistics," *Journal of the History of the Behavioral Sciences* 51 (2015): 54–77, at 71.

94. Bruner, *Toward a Theory of Instruction*, 101.

95. Cohen-Cole, *The Open Mind*, 190–214, chap. 7, "Cognitive Theory and the Making of Liberal Americans."

CHAPTER 4: TWO CHALLENGES, TWO VISIONS

1. Harrington quoted in U.S. Congress, Senate, Committee on Government Operations, Subcommittee on Government Research, *National Foundation for Social Sciences, Hearings*, 90th Cong., 1st sess. (Washington, D.C.: GPO, 1967), hereafter, *1967 NSSF Hearings*, 642.

2. Harold Orlans, "Introduction," 1–19, at 2, in U.S. Congress, House, Committee on Government Operations, Research and Technical Programs Subcommittee, *The Use of Social Research in Federal Domestic Programs, Part 1: Federally-Financed Social Research—Expenditures, Status, and Objectives*, 90th Cong., 1st Sess. (Washington, D.C.: GPO, April 1967).

3. Vision and Kershaw quotes from "The Proper Study of Mankind," *Newsweek*, Aug. 15, 1966, 80–82.

4. Joy Rohde, *Armed with Expertise: The Militarization of American Social Research during the Cold War* (Ithaca, NY: Cornell University Press, 2013), 99.

5. "Appendix 3: Report of the Committee on Science and the Public Welfare," at 83, https://ia600207.us.archive.org/12/items/scienceendlessfr00unit/scienceendlessfr00unit.pdf, accessed Dec. 21, 2018.

6. On the problem of relevance before the mid-1960s, see Daniel Lee Kleinman and Mark Solovey, "Hot Science/Cold War: The National Science Foundation after World War II," *Radical History Review* 63 (Fall 1995): 110–139.

7. Haworth in NSF, *1966 Annual Report*, viii.

8. Daddario quoted in U.S. Congress, House, *Toward the Endless Frontier: History of the Committee on Science and Technology, 1959–79* (Washington, D.C.: GPO, 1980), 132.

9. Haworth quote of subcommittee recommendations in NSF, *1966 Annual Report*, xiii.

10. Brown in U.S., House, Committee on Science and Astronautics, Subcommittee on Science, Research, and Development, *Government and Science: Review of the National Science Foundation, vol. 1*, 89th Cong., 1st sess. (Washington, D.C., GPO, 1965), 123.

11. Herring in *Government and Science*, 443.

12. Haworth in U.S. Congress, House, Committee on Science and Astronautics, Subcommittee on Science, Research, and Development, *A Bill to Amend the National Science Foundation Act of 1950, Hearings*, 89th Cong., 2nd sess. (Washington, D.C.: GPO, 1966), 7.

13. Haworth quoted in NSF, *1966 Annual Report*, xix.

14. Director Haworth, letter of Jan. 14, 1966, in U.S. Congress, House, Committee on Government Operations, Research and Technical Programs Subcommittee, *The Use of Social Research in Federal Domestic Programs, Part 4: Current Issues in the Administration of Federal Social Research*, 90th Cong., 1st Sess. (Washington, D.C.: GPO, April 1967), 116.

15. "Commentary Subcommittee of the National Science Board Concerning H.R. 18696, A Bill to Amend National Science Foundation Act of 1950," in *A Bill to Amend the National Science Foundation Act of 1950, Hearings*, 86.

16. Guy Stever, *In War and Peace: My Life in Science and Learning* (Washington, D.C.: Joseph Henry Press, 2002), 227.

17. Hornig in U.S. Congress, Senate, Special Subcommittee on Science, Committee on Labor and Public Welfare, *National Science Foundation Act Amendments of 1968,* Hearings, 90th Cong., 1st sess. (Washington, D.C.: GPO, 1968), hereafter, *1968 NSF Hearings*, 64, 62. Douglas Martin, "Donald Hornig, Last to See First A-Bomb, Dies at 92," *New York Times*, Jan. 27, 2013.

18. On Daddario's establishment connections, see Daniel S. Greenburg, "Social Sciences: Progress Slow on House and Senate Bills," *Science*, Aug. 11, 1967, 660–662.

19. Daddario in *1968 NSF Hearings*, 124.

20. "House Discussion on Social Science Provision of National Science Foundation Bill, 1966," 106–109, in *The Use of Social Research in Federal Domestic Programs, Part 4*, Fulton at 106, Daddario at 107.

21. 1968 Public Law 90–407.

22. Toby Appel, *Shaping Biology: The National Science Foundation and American Biological Research, 1945–1975* (Baltimore: Johns Hopkins University Press, 2000), esp. 235–244.

23. Janet Abbate, "Applied Science as a Political Construct: Scientific Legitimacy and Social Utility at the US National Science Foundation," 8, unpublished manuscript.

24. Special Commission on the Social Sciences of the National Science Board, *Knowledge into Action: Improving the Nation's Use of the Social Sciences*, hereafter, *Brim Report* (Washington, D.C.: NSF, 1969), xi, xiii, xiv, xx.

25. NSF, *1970 Annual Report*, 5; "Notes on the Ninth Meeting of the Advisory Committee for Social Sciences May 7, 1969," 1, Folder Advisory Committee for Social Sciences—1968, Box 2, Series Haworth Director Files January–December 1969, NSF Collection Record Group 307, hereafter, RG 307, National Archives–College Park, MD.

26. On Snow, see John B. Phelps, "IRRPOS Looks for Relevance to Society's Problems," *Physics Today*, Nov. 1970, 61–62.

27. Daniel P. Moynihan, *The Negro Family: The Case for National Action* (Washington, D.C.: Office of Policy Planning and Research, U.S. Department of Labor, 1965). Daniel Geary, *Beyond Civil Rights: The Moynihan Report and Its Legacy* (Philadelphia: University of Pennsylvania Press, 2015).

28. Godfrey Hodgson, *The Gentleman from New York: Daniel Patrick Moynihan, A Biography* (New York: Houghton Mifflin, 2000), 12.

29. Daniel P. Moynihan, *Maximum Feasible Misunderstanding: Community Action in the War on Poverty* (New York: Free Press, 1969), 191, xxix–xxx. Nevertheless, Moynihan still saw a valuable role for social science (p. xxix) "not in the formulation of social policy, but in the measurement of its results." *Public Interest*, a journal founded in the mid-1960s to bring serious analysis to the study of social problems, became a main venue for criticisms of this sort, including from disenchanted liberal scholars and policy intellectuals who were moving rightward and became known as neoconservatives, such as Edward Banfield, Irving Kristol, Nathan Glazer, and Moynihan. In *The Unheavenly City: The Nature and Future of Our Urban Crisis* (Boston: Little, Brown, 1968), Banfield charged that Great Society programs and policies informed by social science perspectives had, in fact, made urban problems worse, not better.

30. M. Brewster Smith to Leland Haworth, "Report for the Year 1968," Nov. 15, 1968, at 2, in Board Books, 123rd meeting, Feb. 13–15, 1969, National Science Board Library, hereafter, NSB Library, NSF headquarters, Alexandria, VA. Craig Haney, "A Tribute to M. Brewster Smith," *SPSSI Forward 246* (Fall 2012): 28–29.

31. Bruce L. R. Smith, *American Science Policy since World War II* (Washington, D.C.: Brookings Institution, 1990), 81.

32. Haworth in NSF, *1968 Annual Report*, xv, xvi.

33. For recent discussions, see Kelly Moore, *Disrupting Science: Social Movements, American Scientists, and the Politics of the Military, 1945–1975* (Princeton, NJ: Princeton University Press, 2013). Jon Agar, "What Happened in the Sixties?" *British Journal for the History of Science* 41 (2008): 567–600.

34. For additional historical analysis, see Mark Solovey, "Senator Fred Harris's National Social Science Foundation Proposal: Reconsidering Federal Science Policy, Natural Science–Social Science Relations and American Liberalism during the 1960s," *Isis* 103 (2012): 54–82. On science boundary work in this episode, see Thomas Gieryn, "The U.S. Congress Demarcates Natural Science and Social Science (Twice)," 65–114, esp. 101–111, in Gieryn's book, *Cultural Boundaries of Science: Credibility on the Line* (Chicago: University of Chicago Press, 1998).

35. See Richard Lowitt, *Fred Harris, His Journey from Liberalism to Populism* (New York: Rowman & Littlefield, 2002).

36. Daniel S. Greenberg, "National Research Policy: Ambuscade for the 'Establishment,'" *Science* 153 (Aug. 5, 1966): 611–615, at 611. Dennis W. Brezina, "Rise and Demise of the Senate Subcommittee on Government Research," *Federation Proceedings* 29 (1970): 1821–1829.

37. Mark Solovey, "Project Camelot and the 1960s Epistemological Revolution: Rethinking the Politics–Patronage–Social Science Nexus," *Social Studies of Science* 31 (2001): 171–206; Rohde, *Armed with Expertise*, 63–89, chap. 3, "Deeper Shades of Gray: Ambition and Deception in Project Camelot." Fred Harris, "Project Simpatico," *Congressional Record—Senate* 112 (Feb. 7, 1966): 2281–2289.

38. Fred Harris, "National Foundation for the Social Sciences (S. 836)," *Congressional Record—Senate 112* (Oct. 11, 1966): 26028–26029, at 26028.

39. S. 836 reprinted in *1967 NSSF Hearings*, 1–5. In February 1967, Mondale introduced a separate bill, cosponsored by Harris, to create a Council of Social Advisers that would have been responsible, among other things, for developing an annual social report on the state of the nation and that would have provided the nation with a counterpart to the Council of Economic Advisers. Harris and Mondale considered their two bills to be companion pieces needed to realize the full potential of social science contributions to the national welfare. And Harris's subcommittee conducted hearings on Mondale's Full Opportunity and Social Accounting Act: U.S. Congress, Senate, Subcommittee on Government Research, *Full Opportunity and Social Accounting Act, pt. 1–3*, 90th Cong., 1st Sess. (Washington, D.C.: GPO, 1968).

40. *1967 NSSF Hearings*.

41. U.S. Congress, Senate, Committee on Government Operations, *Establishment of a National Foundation for the Social Sciences*, report to accompany S. 836, 90th Cong., 2nd sess. (Washington, D.C.: GPO, June 17, 1968), hereafter, *1968 Committee Print*.

42. Fred R. Harris, "The Case for a National Social Science Foundation," *Science*, Aug. 4, 1967, 507–509; "Political Science and the Proposal for a National Social Science Foundation," *American Political Science Review* 61 (1967): 1088–1095.

43. S. 836 text in *1967 NSSF Hearings*, 2.

44. *1967 NSSF Hearings*, 2, 4.

45. *1967 NSSF Hearings*, 5.

46. Rustow in *1967 NSSF Hearings*, 346.

47. Harrington in *1967 NSSF Hearings*, 646.

48. Dan Geary, *Radical Ambition: C. Wright Mills, the Left, and American Social Thought* (Los Angeles: University of California Press, 2009).

49. *1968 Committee Print*, 4. Problems in international relations associated with greater military support for the social sciences received extensive attention from Harris's subcommittee: U.S. Congress, Senate, Committee on Government Operations, Subcommittee on Government Research, *Federal Support of International Social Science and Behavioral Research, Hearings*, 89th Cong., 2nd sess. (Washington, D.C.: GOP, 1966).

50. Harris, "Political Science and the Proposal for a National Social Science Foundation," 1089, 1094, 1093.

51. *1968 Committee Print*, 7, 8.

52. Harris, "Political Science and the Proposal for a National Social Science Foundation," 1094.

53. Harris, "Political Science and the Proposal for a National Social Science Foundation," 1089.

54. *1968 Committee Print*, 7.

55. Eckstein in *1967 NSSF Hearings*, 776.

56. *1968 Committee Print*, 7, 8.

57. *1968 Committee Print*, 5, 7.

58. *1968 Committee Print*, 7, 12.

59. I. Bernard Cohen, *Interactions: Some Contacts between the Natural Sciences and the Social Sciences* (Cambridge, MA: MIT Press, 1994), "A Conversation with Harvey Brooks," 153–188, at 179. Lowitt, *Fred Harris*, 44. Rohde, *Armed with Expertise*, 104.

60. Harold Orlans, "Social Science Research Policies in the United States," *Minerva* 9 (1971): 7–31, at 8. Nathan Glazer, "Harold Orlans 1921–2007," *ASA Footnotes* 36 (April 2008), http://www.asanet.org/sites/default/files/savvy/footnotes/apr08/obit_0408.html, accessed June 7, 2016.

61. Daniel P. Moynihan to Senator Harris, Apr. 26, 1967, Folder 1, Box 84, Fred R. Harris Papers, University of Oklahoma Carl Albert Center Archives, Norman, Oklahoma. See Harris's mention of Mead and Schlesinger in *1967 NSSF Hearings*, 263, and statements in those hearings by Bohannan 686, Carter 298, Davis 265, Hays 379, Holton 620, Likert 733, Malone 446, McDougal 509, Miller 342, Millikan 606,

Murphy 647, Price 414. Kirkpatrick's strong support noted on p. 490 in Luther J. Carter, "Social Sciences: Where Do They Fit In?" *Science* 154 (Oct. 28, 1966): 488–491.

62. This division between the haves and have-nots, which is apparent from reading the relevant documents, is also evident in a telephone survey carried out by Harris's subcommittee that found "no consensus within the organized social science community" on the need for a new social science agency: Untitled document, Notes on telephone survey of professional association executive secretaries, Folder 13, Box 85, Fred R. Harris Papers.

63. Kaysen in *1967 NSSF Hearings*, 374, 376.

64. Wildavsky in *1967 NSSF Hearings*, 285, 288, 290.

65. Orlans, "Social Science Research Policies in the United States," 12.

66. Simon in *1967 NSSF Hearings*, 384, 392.

67. Haworth in *1967 NSSF Hearings*, 64, 66, 101, 67, 71, 94.

68. Haworth in *1967 NSSF Hearings*, 100, 72, 92, 96, 73, 101.

69. Minutes, Advisory Committee for Social Sciences Meeting, Feb. 2, 1967, at 1, Folder Division of Social Sciences 1968, Box 13, Series Haworth Director Files Jan.–Dec. 1969, RG 307.

70. Memorandum to Members of the National Science Board: "The Role of the National Science Foundation in Social Science Research," Jan. 12, 1967, at 6, in Documentary Supplement, Part 1, vol. 2, Box 1, Collection: Administrative History of the National Science Foundation, Lyndon B. Johnson Presidential Library Archives, Austin, TX.

71. Haworth in *1967 NSSF Hearings*, 101.

72. "S. 508—Introduction of Bill—A National Foundation for the Social Sciences," *Congressional Record—Senate*, Jan. 22, 1969, S699–S705.

73. NEH chairman Barnaby C. Keeney to Senator Pell, Dec. 14, 1967, included in *1968 NSF Hearings*, figures on 160.

74. Lowitt, *Fred Harris*, 54, 59, 62.

75. Ellen Herman, *The Romance of American Psychology: Political Culture in the Age of Experts* (Los Angeles: University of California Press, 1995), 166.

76. Otto N. Larsen, *Milestones and Millstones: Social Science at the National Science Foundation, 1945–1991* (New Brunswick, NJ: Transaction, 1992), says (p. 77) that "redundancy ... put a stop to the proposed secession of social science from NSF." Note that this statement is misleading also because it wrongly implies that success of the NSSF bill would have entailed separation of the social sciences from the NSF. More recently, Tiago Mata and Tom Scheiding, "National Science Foundation Patronage of Social Sciences, 1970s and 1980s: Congressional Scrutiny, Advocacy Network, and the Prestige of Economics," *Minerva* 50 (2012): 423–449, say (p. 425) that the NSSF proposal "floundered as redundant."

77. Michael D. Reagan, *Science and the Federal Patron* (New York: Oxford University Press, 1969), 151.

CHAPTER 5: LOSING GROUND

1. NSF, *1975 Annual Report*, 29.

2. John D. Plummer to Mr. President, undated but ca. 1975, Folder: AD/BBS: Social and Behavioral Sciences, General/Miscellaneous Background, Box 14, Series Historian Files, hereafter, NSF HF, NSF Collection Record Group 307, hereafter, RG 307, National Archives–College Park, MD.

3. David L. Sills, "Behavioral Field Widens Horizons," *New York Times*, Jan. 20, 1970.

4. Daniel J. Kevles, "Principles and Politics in Federal R&D Policy, 1945–1990: An Appreciation of the Bush Report," preface to the 40th anniversary ed. of Vannevar Bush, *Science the Endless Frontier* (Washington, D.C.: NSF, 1990), ix–xxv, xxii.

5. Godfrey Hodgson, *The World Turned Right Side Up: A History of the Conservative Ascendancy in America* (Boston: Houghton Mifflin, 1996); George H. Nash, *The Conservative Intellectual Movement in America since 1945* (Wilmington, DE: Intercollegiate Studies Institute, 2006 ed.); Justin Vaisse, *Neoconservatism: The Biography of a Movement* (Cambridge, MA: Harvard University Press, 2010).

6. For a contemporary discussion, see Henry J. Aaron, *Politics and the Professors: The Great Society in Perspective* (Washington, D.C.: Brookings Institution, 1978).

7. Kevles, "Principles and Politics in Federal R&D Policy, 1945–1990," xxiii.

8. See Toby Appel, *Shaping Biology: The National Science Foundation and American Biological Research, 1945–1975* (Baltimore: Johns Hopkins University Press, 2000), 235.

9. On OMB's power, see Anthony F. C. Wallace, chairman, Advisory Committee for the Social Sciences, "Annual Report, 1971," Board Books, 144th Meeting, Jan. 20–21, 1972, National Science Board Library, hereafter, NSB Library, NSF headquarters, Alexandria, VA, which says (p. 4) that "the OMB is in a position to determine effectively the balance between basic and applied research supported by federal money in various fields and in general the levels and kinds of training and research funds made available to various academic disciplines, to non-profit corporations and to industry."

10. The only substantial account of the NSF's involvement with the social sciences during this period comes from Otto N. Larsen's insider's history: *Milestones and Millstones: Social Science at the National Science Foundation, 1945–1991* (New Brunswick, NJ: Transaction, 1992), 91–127, chap. 5, "Relevance, RANN, Reorganization."

11. Roger L. Geiger, *Research and Relevant Knowledge: American Research Universities since World War II* (New York: Oxford University Press, 1993), 252.

12. NSF, *1970 Annual Report*, 53.

13. The second biologist was Rita Colwell, who began her appointment in 1998 and was the NSF's first female director.

14. See Milton Lomask, *A Minor Miracle: An Informal History of the National Science Foundation* (Washington, D.C.: NSF, 1976), 237–239. Also see Sylvia D. Fries, "The Ideology of Science during the Nixon Years: 1970–76," *Social Studies of Science* 14 (1984): 323–341; Cyril S. Smith and Otto N. Larsen, "The Criterion of 'Relevance' in the Support of Research in the Social Sciences: 1965–1985," *Minerva* 27 (1989): 461–482.

15. Glenn Bugos and Walter Vincenti, "Alfred John Eggers, Jr., 1922–2006," in National Academy of Engineering, *Memorial Tributes, vol. 15* (Washington, D.C.: National Academies Press, 2011), 60–64.

16. McElroy in NSF, *1971 Annual Report*, 2. U.S. Congress, House, Committee on Science and Technology, Subcommittee of Science, Research and Technology, *The Psychological and Social Sciences Research Support Programs of the National Science Foundation: A Background Report*, 95th Cong., 1st Sess. (Aug. 1977), hereafter, *1977 House Background Report*, RANN's objectives on 91–92. On RANN's rise and fall, see Dian Olson Belanger, *Enabling American Innovation: Engineering and the National Science Foundation* (West Lafayette, IN: Purdue University Press, 1988), esp. 76–123.

17. *1977 House Background Report*, 92.

18. Figures from Lomask, *A Minor Miracle*, 249.

19. Twenty-five percent noted in Larsen, *Milestones and Millstones*, 97.

20. Figures from Smith and Larsen, "The Criterion of 'Relevance,'" 465.

21. *1977 House Background Report*, 11.

22. *1977 House Background Report*, 97, 109.

23. NSF, *1976 Annual Report*, 67, 85. On RANN's social science work, see *1977 House Background Report*, esp. 91–142, "Problem-Oriented Applied Social Research in the Program of Research Applied to National Needs," three programs for productivity on 109.

24. Figures from Larsen, *Milestones and Millstones*, 97–98.

25. *1977 House Background Report*, RANN's applied social research accomplishments on 153–157, quotes at 154, 156.

26. NSB Minutes, 146th Meeting, Apr. 20–21, 1972, Appendix F: Highlights 1971 Annual Report, Advisory Committee for Research Application, at 32, NSB Library.

27. Wallace, "Annual Report 1971," 2.

28. Senator Edward M. Kennedy to Director Guyford Stever, Oct. 22, 1974, Folder Social Sciences, Division of, 1974, Box 8, Series Director Files, Subject Files, 1974 Alpha Files A-G, RG 307, Acc. No. 307-79-010, National Archives, Washington National Record Center, Suitland, MD. When I did my research, this part of the

NSF collection was not yet fully processed. After materials at Suitland are fully processed, they are usually moved to the National Archives in College Park, MD.

29. Criticisms noted in *1977 House Background Report*, 906–910.

30. Larsen, *Milestones and Millstones*, 95.

31. Committee on the Social Sciences in the National Science Foundation, *Social and Behavioral Science Program in the National Science Foundation, Final Report*, hereafter, *Simon Report* (Washington, D.C.: NAS, 1976), 64, 7, 8.

32. *Simon Report*, 8.

33. Robert Gillette, "1973 Budget: Administration Bets on Applied Science," *Science* 175 (Jan. 28, 1972): 389–392, at 392.

34. Figures from Larsen, *Milestones and Millstones*, 97. The total of $2,227 million is an approximation. To simplify, I have assumed that the $468 million is in 1975 dollars—although this amount was actually distributed over a few years—and then used the CPI inflation calculator, https://data.bls.gov/cgi-bin/cpicalc.pl, to derive the 2018 equivalent.

35. NSF, *1978 Annual Report*, 97.

36. Richard C. Atkinson, "The Golden Fleece, Science Education, and U.S. Science Policy," *Proceedings of the American Philosophical Society* 143 (1999): 407–417, esp. 415.

37. Richard J. Green and Will Lepkowski, "A Forgotten Model for Purposeful Science," *Issues in Science and Technology* 22 (2006): 69–73. For an earlier assessment, see Samuel J. Raff, "RANN Research at NSF: Some Results of an Evaluation," *Evaluation Review* 3 (1979): 497–512.

38. See Aaron, *Politics and the Professors*; Charles E. Lindblom and David K. Cohen, *Usable Knowledge: Social Science and Social Problem Solving* (New Haven, CT: Yale University Press, 1979); Robert F. Rich, *Social Science Information and Public Policy Making* (San Francisco: Jossey-Bass, 1981).

39. Carol H. Weiss and Michael J. Bucuvalas, *Social Science Research and Decision-Making* (New York: Columbia University Press, 1980), 25. "Carol (Hirschorn) Weiss," obituary, *Boston Globe*, Jan. 10, 2013.

40. Carol H. Weiss, "Introduction," in *Using Social Research in Public Policy Making*, ed. Carol H. Weiss (Lexington, MA: Lexington Books, 1977), 1–22, at 17, 18.

41. Weiss and Bucuvalas, *Social Science Research and Decision-Making*, 270.

42. Weiss, "Introduction," 9, 10.

43. NAS-NRC Assembly of Behavioral and Social Sciences, Study Project on Social Research and Development, *The Federal Investment in Knowledge of Social Problems*, hereafter, *FIKOSP* (Washington, D.C.: NAS, 1978), xii, 3–4. Related problems concerned the development of research agendas, which (p. 3) seemed to be "largely a reactive process, with few examples of systematic planning," and poor management

in the federal support system for social knowledge development and application, including (p. 4) a "rapid turnover of leadership."

44. Aaron, *Politics and the Professors*, 8.

45. *FIKOSP*, 8, 7.

46. Psychobiology grant noted in Appel, *Shaping Biology*, 107.

47. Dollar amounts mentioned in David Johnston, "Senator Proxmire Bars Race in 1988," *New York Times*, Aug. 28, 1987.

48. Senator William Proxmire, "Foreword: PPB, the Agencies and the Congress," v–xvii, at vi, in U.S. Congress, Joint Economic Committee, Subcommittee on Economy in Government, *The Analysis and Evaluation of Public Expenditures: The PPB System, a Compendium of Papers* (Washington, D.C.: GPO, 1969). Philip M. Boffey, "Love and Senator Proxmire," *Chronicle of Higher Education*, Mar. 24, 1975, 5–6, at 5.

49. Boffey, "Love and Senator Proxmire," 5.

50. Constance Holden, "Social Science at NSF Needs Pruning, Says Proxmire," *Science* 185 (Aug. 16, 1974): 597.

51. Proxmire quoted in "Note about Proxmire's Comments on NSF Projects from 7-1-74," Folder Social Sciences, Division of, 1974.

52. See discussion of awards in Johnston, "Senator Proxmire Bars Race in 1988."

53. Noted in Laurence J. Kotlikoff and Scott Burns, *The Coming Generational Storm: What You Need to Know about America's Economic Future* (Cambridge, MA: MIT Press, 2004), 118.

54. Proxmire quoted in Boffey, "Love and Senator Proxmire," 5.

55. Proxmire quoted in L. Stuart Ditzen, "Why Do Phila. Folk Talk That Way?" *Philadelphia Bulletin*, Mar. 2, 1977.

56. Proxmire press release, Sept. 27, 1979, http://content.wisconsinhistory.org/cdm/fullbrowser/collection/proxmire/id/241/rv/compoundobject/cpd/443, accessed Dec. 21, 2018. Also see the letters by Sherry B. Ortner (pp. 2, 8) and Roy A. Rappaport (pp. 8–9) with a reply by Senator Proxmire (p. 9), published under "Golden Fleece Award Sparks Protests" in *Anthropology Newsletter* 20:10 (Washington, D.C.: American Anthropological Association, 1979). There are slightly different versions of Ortner's letter and Rappaport's letter—Ortner archival letter, Rappaport archival letter—and a "Statement by Dr. Sherry B. Ortner re: 'Golden Fleece Award,' issued Sept. 27." These documents are in "Statements on the Golden Fleece Award for October, 1979," Folder 8219, Box 680, Sub-series 5, Series 3, RG 2, Social Science Research Council Archives, hereafter, SSRC Archives, Rockefeller Archive Center, Sleepy Hollow, NY.

57. Proxmire press release, 2.

58. Boffey, "Love and Senator Proxmire," 5.

59. Ortner archival letter, 2; Ortner letter, 8. On rethinking modernization theory, see Michael E. Latham, *The Right Kind of Revolution: Modernization, Development, and U.S. Foreign Policy from the Cold War to the Present* (Ithaca, NY: Cornell University Press, 2010), esp. 157–185, chap. 6, "Modernization under Fire: Alternative Paradigms, Sustainable Development, and the Neoliberal Turn."

60. "Statement by Dr. Sherry B. Ortner."

61. Rappaport archival letter.

62. Ortner letter, 2.

63. Rappaport letter, 8.

64. Daniel Schorr, "Cults and Charismatic Leaders Become a National Crisis Subject," *Arkansas Gazette* (Little Rock), Dec. 7, 1978.

65. Proxmire letter, 9.

66. Donald Lambro, *Fat City: How Washington Wastes Your Taxes* (South Bend, IN: Regnery/Gateway, 1980), xv, xviii.

67. Lambro, *Fat City*, 134–135.

68. Lambro, *Fat City*, chap. 17, 134–143, at 135, 136.

69. Lambro, *Fat City*, 140, 142.

70. Lambro, *Fat City*, 214.

71. Lambro, *Fat City*, 134.

72. Holden, "Social Science at NSF Needs Pruning, Says Proxmire," 597.

73. NSF, *1970 Annual Report*, 71.

74. NSF, *1970 Annual Report*, 71.

75. See Richard W. Burkhardt Jr., *Patterns of Behavior: Konrad Lorenz, Niko Tinbergen, and the Founding of Ethology* (Chicago: University of Chicago Press, 2005); Marga Vicedo, *The Nature and Nurture of Love: From Imprinting to Attachment in Cold War America* (Chicago: University of Chicago Press, 2013).

76. NSF, *1970 Annual Report*, 71.

77. NSF, *1970 Annual Report*, 71.

78. Peter Dow, *Schoolhouse Politics: Lessons from the Sputnik Era* (Cambridge, MA: Harvard University Press, 1991), 193. Dow worked at the Educational Development Center in Cambridge, MA, first as director of the Elementary Social Studies Program (1966–1967), then as director of the School and Society Programs (1967–1975).

79. NSF, *1970 Annual Report*, 71.

80. See Susan M. Marshner, *MAN: A Course of Study—Prototype for Federalized Textbooks?* (Washington, D.C.: Heritage Foundation, 1975), 2.

81. NSF, *1970 Annual Report*, 72.

82. NSF, *1970 Annual Report*, 72.

83. Dorothy Nelkin, *Science Textbook Controversies and the Politics of Equal Time* (Cambridge, MA: MIT Press, 1977), 31.

84. Nelkin, *Science Textbook Controversies*, 35, 34, 41. To derive the 2018 equivalent, I put $7.3 million in 1972 dollars into the CPI calculator, https://data.bls.gov/cgi-bin/cpicalc.pl, although that amount was actually distributed over many years.

85. See Dow, *Schoolhouse Politics*, 185 ff. For a recent account of the MACOS controversy by a historian of science, see Jamie Cohen-Cole, *The Open Mind: Cold War Politics and the Sciences of Human Nature* (Chicago: University of Chicago Press, 2013), 217–252, chap. 8, "A Fractured Politics of Human Nature."

86. Marshner, *MAN*, 10, 24, 26.

87. Marshner, *MAN*, 20, 43, 41, 20.

88. Marshner, *MAN*, 42.

89. Conlan quoted in Deborah Shapley, "Congress: House Votes Veto Power on All NSF Research Grants," *Science* 188 (Apr. 25, 1975): 338–341, at 341, 338.

90. Advertisement in *Washington Post*, Apr. 5, 1975.

91. Shapley, "Congress," 341.

92. Shapley, "Congress," 338–339, 341.

93. Harvey Brooks, "Knowledge and Action: The Dilemma of Science Policy in the '70's," *Daedalus* 102 (Spring 1973): 125–143, at 126.

94. Shapley, "Congress," 338–339.

95. Roth's proposal discussed in *1977 House Background Report*, Appendix III, 895.

96. Bills from Conlan and Helms noted in Nelkin, *Creationist Controversy*, 133–134.

97. Conlan quoted in John Walsh, "NSF Peer Review Hearings: House Panel Starts with Critics," *Science* 189 (Aug. 8, 1975): 435–437, at 435.

98. NSB Minutes, 174th Meeting, June 18–20, 1975, Appendix B: Policy Statement on Implementation of Science Curricula by the National Science Foundation Adopted by the National Science Board at Its 174th Meeting on June 20, 1975, and Appendix C: Policy Statement on Pluralism in Education Adopted by the National Science Board at Its 174th Meeting on June 20, 1975, NSB Library.

99. A 70 percent decline noted (p. 108) in Dorothy Nelkin, *Science Textbook Controversies and the Politics of Equal Time* (Cambridge, MA: MIT Press, 1978).

100. For a contemporary perspective, see Frances Fitzgerald, *America Revised: History Schoolbooks in the Twentieth Century* (Boston: Little, Brown, 1979), 185–205.

101. U.S. Congress, House Committee on Science and Technology, Subcommittee on Science, Research and Technology, *Secondary Math and Science Education,*

Hearings (Washington, D.C.: GPO, May 7, 1982), percentages from graph in testimony by Sarah E. Klein, p. 125.

102. Charles L. Heatherly, ed., *Mandate for Leadership: Policy Management in a Conservative Administration* (Washington, D.C.: Heritage Foundation, 1981), 239.

103. See Montgomery, *Minds for the Making*, 248–260.

104. Averch quoted in Dow, *Schoolhouse Politics*, 229.

CHAPTER 6: MOMENTUM LOST

1. Richard C. Atkinson, "Federal Support in the Social Sciences," *Science* 207 (Feb. 22, 1980), 829, at 829.

2. Percentages from Table 1: "Total Federal Obligations for Basic Research in Universities in the Social Sciences with Percent Distribution by Agency (FY 77–80)—$ in M," Folder BBS-Biological & Social Sciences, 1979, Box 3, NSF Collection Record Group 307, hereafter, RG 307, Acc. No. 307-87-223, National Archives, Washington National Record Center, Suitland, MD. When I did my research, this part of the NSF collection was not yet fully processed. After materials at Suitland are fully processed, they are usually moved to the National Archives in College Park, MD.

3. Percentages from Table 2: "Total Federal Obligations for Basic Research in the Social Sciences with Percent Distribution by Agency (FY 77–80)—$ in M," Folder BBS-Biological & Social Sciences, 1979.

4. H. Guyford Stever, *In War and Peace: My Life in Science and Technology* (Washington, D.C.: Joseph Henry Press, 2002), 238.

5. Patricia A. Pelfrey, *Entrepreneurial President: Richard Atkinson and the University of California, 1995–2003* (Los Angeles: University of California Press, 2012), 23–28 cover his NSF years.

6. "National Science Foundation Nomination of Richard C. Atkinson to be Director," April 21, 1977, http://www.presidency.ucsb.edu/ws/?pid+7374, accessed May 17, 2016. Also see Richard C. Atkinson Oral History Transcript, interviewed by Marc Rothenberg, Sept. 18, 2007, 1–3, NSF Oral Histories, NSF headquarters, Alexandria, VA.

7. Constance Holden, "NSF Gains Social Sciences Champion," *Science* 189 (Aug. 8, 1975): 436.

8. Pelfrey, *Entrepreneurial President*, 15–17. His curriculum vita, http://www.rca.ucsd.edu/docs/RCAVita.pdf, accessed Dec. 18, 2018.

9. Pelfrey, *Entrepreneurial President*, 17–19. Hunter Crowther-Heyck, "George A. Miller, Language, and Computer Metaphor of Mind," *History of Psychology* 2 (1999): 37–64.

10. Pelfrey, *Entrepreneurial President*, 19–20. On Stanford, see Rebecca S. Lowen, *Creating the Cold War University: The Transformation of Stanford* (Los Angeles: University of California Press, 1997).

11. R. C. Atkinson and R. M. Shiffrin, "Human Memory: A Proposed System and Its Control Processes," in *The Psychology of Learning and Motivation: Advances in Research and Theory*, Vol. 2, ed. K. W. Spence and J. T. Spence (New York: Academic Press, 1968), 89–195. Theory in "In Honor of … Richard Atkinson," https://fabbs.org/our_scientists/richard-atkinson-phd/, accessed May 24, 2016.

12. See Pelfrey, *Entrepreneurial President*, 21–23; Chizuko Izawa, ed., *On Human Memory: Evolution, Progress, and Reflections on the 30th Anniversary of the Atkinson-Shiffrin Model* (Hillside, NJ: Lawrence Erlbaum, 1999).

13. Ezra Bowen, "'Hello, Jimmie,' said the Machine, 'I've Been Waiting for You,'" *Life*, Jan. 27, 1967, 70, 72, 74, 75, 76, 78, 81; Richard C. Atkinson and Harlalee A. Wilson, "Computer-Assisted Instruction," *Science* 162 (Oct. 4, 1968): 73–77.

14. See Pelfrey, *Entrepreneurial President*, 21.

15. Richard C. Atkinson, *Human Memory and the Learning Process: Selected Papers of Richard C. Atkinson*, ed. by Y. Zabrodin and B. F. Lomov, trans. into Russian (Moscow: Progress Publishing House, 1980), quotes from Atkinson's preface.

16. Atkinson Oral History Transcript, 3. On the various critiques of scientific psychology, see Thomas Teo, *The Critique of Psychology: From Kant to Postcolonial Theory* (New York: Springer, 2005).

17. Stephen Cole, Leonard Rubin, Jonathan R. Cole, Committee on Science and Public Policy, National Academy of Sciences, *Peer Review in the National Science Foundation: Phase One of a Study* (Washington, D.C.: NAS, 1978).

18. See Pelfrey, *Entrepreneurial President*, 26–27.

19. Pelfrey, *Entrepreneurial President*, 27, 103.

20. Richard C. Atkinson, "The Golden Fleece, Science Education, and U.S. Science Policy," *Proceedings of the American Philosophical Society* 143 (Sept. 1999): 407–417, Rabi's visit on 415.

21. R. C. Atkinson, "Federal Support for Psychology and the Social Sciences: A Look Ahead," at 1, Folder: Remarks by Richard C. Atkinson, 1975–85, Box 46, Series: NSF Historian Files, hereafter, NSF HF, RG 307, National Archives–College Park, MD.

22. Atkinson, "Federal Support for Psychology and the Social Sciences," at 5.

23. Atkinson, "The Golden Fleece," 410.

24. Atkinson, "The Golden Fleece," 410.

25. Richard C. Atkinson, "Psychology and the Golden Fleece," transcript of invited address at Sixth National Institute on Teaching Psychology to Undergraduates, Jan. 4–7, 1984, at 5, https://files.eric.ed.gov/fulltext/ED242624.pdf, accessed Dec. 13, 2018.

26. Atkinson, "The Golden Fleece," 416. Pelfrey says much the same: *Entrepreneurial President*, 191, fn. 19.

27. List of awards: http://content.wisconsinhistory.org/cdm/ref/collection/tp/id/70852, accessed Dec. 20, 2018.

28. Atkinson, "The Golden Fleece," 412.

29. Atkinson, "The Golden Fleece," 412.

30. Atkinson, "The Golden Fleece," 413.

31. Atkinson, "The Golden Fleece," 413–414.

32. Peter Dow, *Schoolhouse Politics: Lessons from the Sputnik Era* (Cambridge, MA: Harvard University Press, 1991), 236.

33. Richard C. Atkinson, "Remarks," Temple University, Philadelphia, Oct. 30, 1975, at 6, Folder Remarks by Richard C. Atkinson, 1975–85, Box 46, NSF HF.

34. Atkinson, "Federal Support," 7.

35. Atkinson Oral History Transcript, 17.

36. On the 1975 reorganization, see U.S. Congress, House, Committee on Science and Technology, Subcommittee on Science, Research and Technology, *The National Science Board: Science Policy and Management for the National Science Foundation, 1968–1980*, hereafter, *NSB History 1968–1980*, 98th Cong., 1st sess. (Washington, D.C.: GPO, 1983), 646.

37. U.S. Congress, House, Committee on Science and Technology, Subcommittee of Science, Research and Technology, *The Psychological and Social Sciences Research Support Programs of the National Science Foundation, A Background Report*, hereafter, *1977 House Background Report*, 91st Cong., 1st Sess. (Washington, D.C.: GPO, Aug. 1977), 900.

38. Atkinson, "Federal Support," 9.

39. White House Press Release, July 2, 1976, www.fordlibrarymuseum.gov/library/document/0248/whpr19760702-011.pdf, accessed Dec. 13, 2018.

40. On Rabin, see Jeffrey Mervis, "Graham's Appointees Mirror His Credentials," *The Scientist*, June 29, 1987, www.the-scientist.com/news/grahams-appointees-mirror-his-credentials-63686, accessed Apr. 27, 2017.

41. Otto N. Larsen, *Milestones and Millstones: Social Science at the National Science Foundation, 1945–1991* (New Brunswick, NJ: Transaction, 1992), 126 fn. 34, 102, also see 112–113.

42.The historian Donald McGraw provides extensive analysis of the NSF's involvement with the biological sciences, including certain areas of the social and behavioral sciences, in *Millennial Biology: The National Science Foundation and American Biology, 1975–2005* (NY: Springer, forthcoming 2020).

43. Atkinson quoted in *Director's Program Review: Behavioral and Neural Sciences, Background Paper*, NSF Board Books, 249th Meeting, Nov. 17–18, 1983, Tab B, at 1,

National Science Board Library, hereafter, NSB Library, NSF headquarters, Alexandria, VA.

44. Figures from Larsen, *Milestones and Millstones*, 106.

45. Figures from Larsen, *Milestones and Millstones*, 106.

46. Figures from Larsen, *Milestones and Millstones*, 106.

47. Figures and percentages from Larsen, *Milestones and Millstones*, 106.

48. NSF, *1979 Annual Report*, 63.

49. Atkinson made the request noted in NSB Minutes, 184th meeting, Sept. 16–17, 1976, Appendix A, "Highlights of NSB Review of NSF Social Science Programs," p. 27, NSB Library.

50. Atkinson, "Federal Support," 9.

51. On NAS's involvement with the social and psychological sciences up through the early 1960s, see Rexmond C. Cochrane, *The National Academy of Sciences: The First Hundred Years, 1863–1963* (Washington, D.C.: NAS, 1978). For information about membership expansion and associated controversy during the 1960s and early 1970s, see Division of Behavioral Sciences, NRC-NAS, May 1972, Folder Behavioral Sciences: Proposed, Meetings: Annual, 1972, Central File Policy Files 1966–1972, National Academy of Sciences Archives, hereafter, NAS Archives, Washington, D.C. Also see I. Bernard Cohen, *Interactions: Some Contacts between the Natural Sciences and the Social Sciences* (Cambridge, MA: MIT Press, 1994), "A Conversation with Harvey Brooks," 153–188, esp. 167–177.

52. NSB, *Discussion Issues 1981, Social and Behavioral Sciences*, vol. 1, June 1981, percentage of NAS members on 5, doctorates on 4, NSB Library.

53. Members listed in Memorandum, Henry David, Executive Director, ABSS to Philip Handler, NAS President, June 12, 1973, Folder Assem Behavioral & Social Sciences, Membership, 1973, Central File Policy Files, 1966–1972, NAS Archives. The 1973 reorganization also created assemblies for the physical sciences, life sciences, and engineering sciences.

54. Constance Holden, "ABASS: Social Sciences Carving a Niche at the Academy," *Science* 199 (Mar. 17, 1978): 1183–1187, board supervision on 1185.

55. Herbert A. Simon, "Biographical," from *Nobel Lectures, Economics 1969–1980*, ed. Assar Lindbeck (Singapore: World Scientific Publishing, 1992), https://www.nobelprize.org/prizes/economic-sciences/1978/simon/auto-biography/, accessed May 24, 2016.

56. See Hunter Crowther-Heyck, *Herbert A. Simon: The Bounds of Reason in Modern America* (Baltimore: Johns Hopkins University Press, 2005).

57. Herbert A. Simon, *Models of My Life* (Cambridge, MA: MIT Press, 1996), 292.

58. Simon in *1967 NSSF Hearings*, 394. For a full citation, see p. 341, n. 1.

59. Memorandum, Herbert A. Simon, Chairman Division of Behavioral Sciences, to Committee on the Review of NRC, Jan. 21, 1970, at 3, 2, Folder Behavioral Sciences, General 1970, Central File Policy Files 1966–1972, NAS Archives.

60. *Simon Report*, 5. For a full citation, see p. 349, n. 31.

61. *Simon Report*, 28.

62. *Simon Report*, 5.

63. *Simon Report*, 17.

64. Simon quoted in U.S. Congress, House, Committee on Science and Technology, Subcommittee on Science, Research and Technology, *1977 National Science Foundation Authorization Hearings,* 94th Cong., 2nd Sess. (Washington, D.C.: GPO, 1976), 793.

65. See John D. Marks, *The Search for the "Manchurian Candidate": The CIA and Mind Control* (New York: Times Books, 1979).

66. Simon in *1977 NSF Authorization Hearings*, 796, 801.

67. Heyn's overview in NSB Minutes, 184th Meeting, Sept. 16–17, 1976, Appendix A: Highlights of NSB Review of NSF Social Science Programs, at 32, NSB Library. Atkinson quoted in Holden, "ABASS," 1183.

68. Atkinson, "The Golden Fleece," 415.

69. Dian O. Belanger, *Enabling American Innovation: Engineering and the National Science Foundation* (West Lafayette, IN: Purdue University Press, 1997), 118.

70. *1977 NSF Authorization Hearings*, Attachment E, "The Social Sciences as a Research Area in the National Interest," 511, 814.

71. NSF, *1979 Annual Report*, viii, 63.

72. P.M., "A Boost for Social Science in New Budget," *Science & Government Report*, March 1, 1978, 3–4, Carter administration on 3.

73. Larsen, *Milestones and Millstones*, 105–111.

74. Larsen, *Milestones and Millstones*, 106.

75. David Johnson, "Next to Nothingness and Being at the National Science Foundation: Part II," *Psychological Science* 3 (1992): 261–264, program changes on 263.

76. For Aborn's perspective, see "New Directions of Emphasis," 25–31, in *Director's Program Review: Social Sciences*, Jan. 14, 1971, NSF Library.

77. Nancy Burns, "The Michigan, Then National, Then American National Election Studies," 2006, 1–13, https://electionstudies.org/wp-content/uploads/2018/07/20060815Burns_ANES_history.pdf, accessed Dec. 13, 2018.

78. On the NES's development and NSF support, see Memorandum, Director Knapp to NSB Members, Oct. 19, 1983, in Board Books, 249th Meeting, Nov. 17–18, 1983, closed session, Tab 4, pp. 83-332-2 to 83-332-3, NSB Library. For additional

information, see Burns, "The Michigan, Then National, Then American National Election Studies," funding sources on 9, importance for political science on 10–11.

79. Memorandum, Director Knapp to NSB Members, Oct. 19, 1983, 83-332-3.

80. NSB Minutes, 190th Meeting, May 19–20, 1977, closed session, national resource at CS:190:21, NSB Library.

81. NSB, *Discussion Issues 1981, Social and Behavioral Sciences, vol. 2* (June 1981), at 2, NSB Library.

82. NSB 81–476, *Status of Science Reviews*, Nov. 1981, funding of all eleven facilities on 54, NSB Library.

83. Raymond A. Bauer, ed., *Social Indicators* (Cambridge, MA: MIT Press, 1966), "Foreword" by Earl P. Stevenson, vii.

84. U.S. Department of Health, Education, and Welfare, *Toward a Social Report* (Washington, D.C.: GPO, 1969).

85. Denis F. Johnston, "The Federal Effort in Developing Social Indicators and Social Reporting in the United States during the 1970s," in *Social Science Research and Government: Comparative Essays on Britain and the United States*, ed. Martin Bulmer (New York: Cambridge University Press, 1987), 285–302; Tim Booth, "Social Indicators and the Mondale Initiative," *Science Communication* 13 (1992): 371–398.

86. NSB, *Science and the Challenges Ahead: Report of the National Science Board*, 1974, at 8, https://www.nsf.gov/nsb/publications/1974/0574.pdfinadequate, accessed Dec. 20, 2018.

87. *1977 House Background Report*, 152.

88. See NSB, *Basic Research in the Mission Agencies: Agency Perspectives on the Conduct and Support of Basic Research* (Washington, D.C.: GPO, 1978), 207. Also Kenneth Prewitt, "Council Reorganizes Its Work in Social Indicators," *SSRC Items* 37 (1983): 74–78; Murray Aborn, "The Short and Happy Life of Social Indicators at the National Science Foundation," *SSRC Items* 38 (Sept. 1984): 32–41.

89. *1977 House Background Report*, 152.

90. Aborn, "The Short and Happy Life," 33.

91. Eleanor Bernert Sheldon and Robert Parke, "Social Indicators," *Science* 188 (May 16, 1975): 693–698, at 698.

92. Richard C. Rockwell to Murray Aborn, Oct. 14, 1982, at 6, Folder 6438, Box 536, RG 2, Series 1, Social Science Research Council Archives, hereafter, SSRC Archives, Rockefeller Archive Center, Sleepy Hollow, NY.

93. Larsen, *Milestones and Millstones*, 108.

94. Aborn, "The Short and Happy Life," 34.

95. NSB, *Status of Science Reviews, 1980* (NSB 79–370), Nov. 1979, at 144, NSF Library.

96. Atkinson quoted in P.M., "A Boost for Social Science in New Budget," 3.

97. Larsen, *Milestones and Millstones*, 104.

98. NSB, *Discussion Issues 1981, Social and Behavioral Sciences*, Vol. 2, 1.

99. NSF Office of Planning and Resource Management, *Status of Science Reviews 1981 Update*, Nov. 1980, at 55, 56, NSB Library.

100. For NSB membership up through Feb. 1981, see *NSB History 1968–1980*, 653–656.

101. David C. Leege, "Is Political Science Alive and Well and Living at NSF? Reflections of a Program Director at Midstream," *PS* 9 (1976): 8–17, at 12, 13.

102. Leege, "Is Political Science Alive and Well and Living at NSF?" 13, 14.

103. Herbert Costner to Otto Larsen, May 30, 1978, quoted in Larsen, *Milestones and Millstones*, 126–127, fn. 43.

104. Kelly G. Shaver to Richard C. Atkinson, Aug. 13, 1979, Folder BBS—Biological & Social Sciences, 1979. Shaver wrote a successful textbook, *Social Psychology* (1977), and became an expert on the psychology of management and entrepreneurship.

105. On Ashbrook's amendment, see U.S. Congress, House, *Toward the Endless Frontier: History of the Committee on Science and Technology, 1959–79* (Washington, D.C.: GPO, 1980), 541–547, at 541.

106. Ashbrook quoted in *Toward the Endless Frontier*, 546.

107. John M. Ashbrook, "A Critique of NSF," *Society* 17 (Sept./Oct., 1980): 12–14, at 13, 14.

108. Atkinson quoted in Daniel Schorr, "Cults and Charismatic Leaders Become a National Crisis Subject," *Arkansas Gazette* (Little Rock), Dec. 7, 1978, 3.

109. Atkinson, "Federal Support in the Social Sciences," 829.

110. Friedl comments noted in NSB Minutes, 219th Meeting, Sept. 18–19, 1980, at 14, NSB Library.

111. Discussion with Atkinson noted in Larsen, *Milestones and Millstones*, 120.

112. Donald N. Langenberg, Acting Director, Memorandum for Members of the National Science Board, Aug. 12, 1980 (NSB-80–325), at 14, NSB Library.

113. Langenberg quoted in NSB Minutes, 219th Meeting, at 14.

114. Slaughter's comments in NSB Minutes, 221st Meeting, Nov. 20–21, 1980, at 9, NSB Library.

115. Slaughter's comments in NSB Minutes, 221st Meeting, at 9. NSB Minutes, 220th Meeting, Oct. 16–17, 1980, at 26, 29, NSB Library. Also, see documents in Folder: Letters Pertaining to the Reorganization of the Biological, Behavioral, and Social Sciences Directorate, 1980–81, Box 16 Directorate of the Biological,

Behavioral and Social Sciences, General 1976–92 to Div. of Social and Economic Sciences 1974–81, Series Director Subject Files 1964–83, RG 307. For further discussion, see Larsen, *Milestones and Millstones*, 131–140.

116. Atkinson in "The Golden Fleece" claimed (p. 416) "sole credit" for establishing this award.

117. Vannevar Bush Award, May 21, 2003, Richard C. Atkinson, Citation, http://www.rca.ucsd.edu/speeches/Vannevar_Bush_Award_Citation_052103.pdf, accessed Oct. 27, 2018.

118. Atkinson, "The Golden Fleece," 415.

119. Atkinson Oral History Transcript, 26.

CHAPTER 7: DARK DAYS

1. Roger Witherspoon, "The Ax Falls Hard on Social Research," *Atlanta Constitution*, July 9, 1981.

2. Philip J. Hilts, "White House Uses Social Sciences, but Cuts Funding for Research," *Washington Post*, June 29, 1981.

3. Hilts, "White House Uses Social Sciences."

4. Roger E. Backhouse and Philippe Fontaine, "Toward a History of the Social Sciences," in *The History of the Social Sciences since 1945*, ed. Roger E. Backhouse and Philippe Fontaine (New York: Cambridge University Press, 2010), 184–223, at 197.

5. Charles A. Murray, *Losing Ground: American Social Policy, 1950–1980* (New York: Basic Books, 1984); Allan Bloom, *Closing of the American Mind: How Higher Education Has Failed Democracy and Impoverished the Souls of Today's Students* (New York: Simon and Schuster, 1987).

6. "Editor's Note," *Social Research* 35 (1968). To help rectify the imbalance, the same issue included a focus section on "Conservative Approaches in the Human Sciences."

7. George Gilder, *Wealth and Poverty* (New York: Basic Books, 1981).

8. David Dickson, *New Politics of Science* (New York: Pantheon, 1984), 3, 8. Also see John A. Remington, "Beyond Big Science in America: The Binding of Inquiry," *Social Studies of Science* 18 (1988): 45–72; Chris Mooney, *The Republican War on Science* (New York: Basic Books, 2005), esp. chap. 4; Genevieve J. Knezo, CRS Issue Brief, "Science and Technology Policy and Funding: Reagan Administration," updated Oct. 28, 1988, NSF Library, NSF headquarters, Alexandria, VA.

9. Dickson, *New Politics of Science*, 18, 13, 50. To see how these developments shaped studies on the causes and consequences of poverty and influenced antipoverty measures, see Alice O'Connor, "Dependency, the 'Underclass,' and the New

Welfare 'Consensus': Poverty Knowledge for a Post-Liberal, Postindustrial Era," in her book *Poverty Knowledge: Social Science, Social Policy, and the Poor in Twentieth-Century U.S. History* (Princeton, NJ: Princeton University Press, 2001), 242–283.

10. U.S. historians, historians of science, and historians of the social sciences have devoted little attention to NSF social science policies and programs in the 1980s. The best starting place remains the sociologist Otto N. Larsen's account: *Milestones and Millstones: Social Science at the National Science Foundation, 1945–1991* (New Brunswick, NJ: Transaction, 1992), chap. 6, "The 1980s: Threat, Unity, Promise," 129–226. Larsen offers many insights based on his experiences as an insider at the agency during this period. He also provides extensive excerpts from published and unpublished primary sources.

11. U.S. President, *America's New Beginning, A Program for Economic Recovery* (Washington, D.C.: The White House, Office of the Press Secretary, 1981), 4, 2.

12. Figures on proposed cuts from p. 573 of U.S. Congress, House, Committee on Science and Technology, Subcommittee on Science, Research and Technology, *1982 National Science Foundation Authorization Hearings*, hereafter, *1982 NSF Authorization Hearings*, 97th Cong., 1st Sess. (Washington, D.C.: GPO, 1981). I have rounded all figures to the nearest $100,000.

13. See science education figures in Robert Rienhold, "Social Scientists Fight for Government Funds," *New York Times*, July 29, 1981. Ronald Reagan, "Remarks at the Conservative Political Action Conference Dinner," Mar. 20, 1981, https://www.presidency.ucsb.edu/documents/remarks-the-conservative-political-action-conference-dinner-0, accessed Oct. 30, 2018.

14. Robert Reinhold, "Reagan's Plans on Research Cuts are Said to Aim at 'Soft' Sciences," *New York Times*, Feb. 9, 1981.

15. Constance Holden, "Dark Days for Social Research," *Science* 211 (Mar. 27, 1981): 1397–1398.

16. Figures from Table 7 "Federal Obligations for Research in Behavioral and Social Sciences by Discipline and Major Funding Agencies FY 1980 (est.)—$ Millions," in NSB, *Discussion Issues 1981: Social and Behavioral Sciences, Vol. 2*, 28, National Science Board Library, hereafter, NSB Library, NSF headquarters, Alexandria, VA.

17. John T. Wilson, *Academic Science, Higher Education, and the Federal Government, 1950–1983* (Chicago: University of Chicago Press, 1983), 93. Wilson had been the assistant director of the NSF's biological and medical sciences division from 1955 to 1961 and a deputy director from 1963 to 1968.

18. Kenneth Prewitt, "The Need for a National Policy for the Social Sciences," in *Social Science Research Council, Annual Report of the President, 1981–1982* (New York: SSRC, 1982), 1–15, at 3.

19. Bromley in *1982 NSF Authorization Hearings*, 751.

20. Ronald W. Reagan, "First Inaugural Address," Jan. 20, 1981.

21. OMB quote in Kenneth Prewitt and David L. Sills, "Federal Funding for the Social Sciences: Threats and Responses," *SSRC Items* 35 (1981): 33–47, at 33.

22. Mitchell C. Lynch, "Cuts Raise a New Social-Science Query: Does Anyone Appreciate Social Science?" *Wall Street Journal*, Mar. 27, 1981.

23. The two articles and the letter in Folder National Enquirer Articles, 1982, Box 49, Series Historian Files, NSF Collection Record Group 307, hereafter, RG 307, National Archives–College Park, MD.

24. Slaughter in *1982 NSF Authorization Hearings*, 574.

25. Larsen, *Milestones and Millstones*, 212, fn. 46.

26. Martin Anderson, *The Federal Bulldozer: A Critical Analysis of Urban Renewal, 1949–1962* (Cambridge, MA: MIT Press, 1964), 230.

27. See James Everett Katz, "Science, Social Science and Presidentialism: Policy during the Nixon Administration," in *The Use and Abuse of Social Science*, ed. Irving L. Horowitz (New Brunswick, NJ: Transaction, 1971), 265–283, esp. 272.

28. See William Greider, "The Education of David Stockman," *The Atlantic*, Dec. 1981.

29. William G. Wells Jr., "Politicians and Social Scientists," *American Behavioral Scientist* 26 (1982): 235–259, at 243. Tiago Mata and Tom Scheiding, "National Science Foundation Patronage of Social Sciences, 1970s and 1980s: Congressional Scrutiny, Advocacy Network, and the Prestige of Economics," *Minerva* 50 (2012): 423–449, also identifies Stockman (p. 435) as the "principal protagonist in the budget crisis."

30. John B. Slaughter Oral History Transcript, interviewed by Marc Rothenberg, Aug. 16, 2007, quote from Loweth—as recalled by Slaughter—at 9, NSF Oral Histories, NSF headquarters.

31. Stockman quoted in Hilts, "White House Uses Social Sciences."

32. Philip Handler, "Pruning the Federal Science Budget," *Science* 211 (March 20, 1981): 1261.

33. NAS Resolution, Folder 6470, Box 539, RG 2, Series 1, Social Science Research Council Archives, hereafter, SSRC Archives, Rockefeller Archive Center, Sleepy Hollow, NY.

34. Adams quoted in Holden, "Dark Days for Social Research," 1398.

35. Roberta B. Miller, "Social Science under Siege: The Political Response, 1981–1984," in *Social Science Research and Government: Comparative Essays on Britain and the United States*, ed. Martin Bulmer (New York: Cambridge University Press, 1987), 373–391, esp. 376.

36. Miller, "Social Science under Siege," 380.

37. Martin Bulmer, "The Social Sciences in an Age of Uncertainty," in Bulmer, ed., *Social Science Research and Government*, 345–352, at 345.

38. Anon, "Slicing through 'Soft' Since," *New York Times*, Apr. 4, 1981.

39. Shamansky in *1982 NSF Authorization Hearings*, 582.

40. Slaughter in *1982 NSF Authorization Hearings*, 544, 545. For a concise explanation of the curve by Laffer himself, see Arthur B. Laffer, "The Laffer Curve: Past, Present, and Future," June 1, 2004, https://www.heritage.org/taxes/report/the-laffer-curve-past-present-and-future, accessed Dec. 12, 2018.

41. Griliches in *1982 NSF Authorization Hearings*, 587. Number of his NSF grants noted (p. 204) in Daniel H. Newlon, "The Role of the NSF in the Spread of Economic Ideas," in *The Spread of Economic Ideas*, ed. David C. Colander and A. W. Coats (New York: Cambridge University Press, 1989), 195–228.

42. Griliches in *1982 NSF Authorization Hearings*, 589.

43. Griliches in *1982 NSF Authorization Hearings*, 589.

44. Klein in *1982 NSF Authorization Hearings*, 583.

45. Klein in *1982 NSF Authorization Hearings*, 585.

46. Robert E. Lucas Jr., "Supply Side Economics: An Analytic Review," *Oxford Economic Papers* 42 (1990): 293–316. On NSF funding for work on rational expectations, see Newlon, "The Role of the NSF in the Spread of Economic Ideas," 207.

47. Robert E. Lucas Jr., "Incentives for Ideas," *New York Times*, Apr. 13, 1981.

48. Kenneth Prewitt, "Summary of Symposium on Strategies for the Social Sciences, June 4, 1981," draft, June 18, 1981, at 2, Folder 6276, Box 515, Record Group 2, Series 1, SSRC Archives. Marshall Robinson, "Private Foundations and Social Science Research," *Society* 21 (May/June 1984): 76–80.

49. Prewitt, "The Need for a National Policy for the Social Sciences," 3.

50. Donna E. Shalala, "Politics and the Uses of Social Science Research," in *Political Science: The State of the Discipline*, ed. Ada W. Finifter (Washington, D.C.: American Political Science Association, 1983), 571–580, at 575.

51. On Miller's background, see David T. Kingsbury, AD/BBS Bulletin No. 84–8, July 31, 1984, Folder BBS-Directorate for Biological, Behavioral and Social Sciences—Divisional of Social and Economic Sciences 1982–1988 #3, Box 17, Series Director's Subject Files 1964–83, RG 307.

52. See Russell R. Dynes, "The Institutionalization of COSSA: An Innovative Response to Crisis by American Social Science," *Sociological Inquiry* 54 (1984): 211–229; Roberta B. Miller, "Crisis and Responses: The Politics of the Social Sciences in the United States (1980–1982)," *La Revue pour l'histoire du CNRS* (2002), https://journals.openedition.org/histoire-cnrs/548, accessed July 8, 2016. Also see COSSA's website, www.cossa.org. COSSA's first president was the anthropologist Dell Hymes. Early 1990s figures from Larsen, *Milestones and Millstones*, 217. The Federation included eight founding societies and was incorporated in March 1981, as noted (p. 326) in David Johnson, "Next to Nothingness and Being at the National Science Foundation: Part III," *Psychological Science* 3 (1992): 323–327.

53. Dynes, "The Institutionalization of COSSA," 211.

54. Roberta B. Miller, "The Social Science Lobby in the United States," in *The Human Sciences: Their Contribution to Society and Future Research Needs*, ed. Baha Abu-Laban and Brendan G. Rule (Edmonton: University of Alberta Press, 1988), 241–252, at 245. On COSSA's relations with congressional committees, also see Dynes, "The Institutionalization of COSSA," 223.

55. Winn in *Congressional Record—House*, July 21, 1981, H4619–H4631, at H4619.

56. Miller, "The Social Science Lobby in the United States," 247.

57. *Congressional Record—House*, July 21, 1981, Forsythe at H4620, Johnston at H4629.

58. Leach in *Congressional Record—House*, July 21, 1981, H4621.

59. Traxler in *Congressional Record—House*, July 21, 1981, H4623.

60. Final vote in *Congressional Record—House*, July 21, 1981, H4631.

61. Miller, "Social Science under Siege," 384.

62. LeBoutillier quoted in Robert Rienhold, "Social Scientists Fight for Government Funds," *New York Times*, July 29, 1981.

63. Heckler quoted (p. 246) in Miller, "The Social Science Lobby in the United States."

64. James L. McCartney, "Setting Priorities for Research: New Politics for the Social Sciences," *Sociological Quarterly* 25 (1984): 437–455, at 441.

65. Report from Michael S. Pallak, Alan Kraut, and Gary VandenBos to Research Support Network Members, "We Won This One … ," Aug. 4, 1981, Folder BBC—Biological, Behavioral and Social Sciences, Box 2, accession no. 307-87-220, RG 307, Washington National Record Center in Suitland, MD. When I did my research, this part of the NSF collection was not yet fully processed. After materials at Suitland are fully processed, they are usually moved to the National Archives in College Park, MD.

66. Doug Walgren, chair of Science, Research and Technology Subcommittee, to Roberta Miller, Oct. 30, 1981, in Folder: BBS-Directorate for Biological, Behavioral and Social Sciences—Division of Social and Economic Sciences 1982–1988 #3.

67. Figures from Larsen, *Milestones and Millstones*, 173.

68. See Dynes, "The Institutionalization of COSSA," 224–225.

69. Roberta B. Miller, "COSSA Progress Report," Aug. 13, 1982, in Folder BBS-Directorate for Biological, Behavioral and Social Sciences—Division of Social and Economic Sciences 1982–1988, Folder #3.

70. Miller, "The Social Science Lobby in the United States," 249.

71. Prewitt, "Summary of Symposium," 2; Prewitt, "The Need for a National Policy for the Social Science," 3.

72. Prewitt, "Summary of Symposium," 2.

73. "Summaries of Presentations Made by Drs. Blake, Converse, Rappaport and Simon at the Meeting of the National Science Board, April 16, 1981," in NSB, *Discussion Issues 1981: Social and Behavioral Sciences, vol. 2*, Section E, 2, 3.

74. "The Social and Behavioral Sciences" in NSB, *Discussion Issues 1981: Social and Behavioral Sciences, vol. 1*, 5.

75. Robert Pear, "Conservatives Demand a 'Defunding' of the Left," *New York Times*, Oct. 2, 1983.

76. "The Social and Behavioral Sciences," in NSB, *Discussion Issues 1981: Social and Behavioral Sciences, vol. 1*, 6, NSB Library.

77. "Statement on Social and Behavioral Sciences as Adopted by the National Science Board at Its 227th Meeting, June 17–19, 1981," included in Lewis M. Branscomb, Letter to the Editor "Social Science Research," *Science* 213 (Sept. 25, 1981): 1448.

78. Morrison quoted in National Research Council, *Behavioral and Social Science Research: A National Resource, Pt. 1*, hereafter, *BSSR, Pt. 1* (Washington, D.C.: National Academies Press, 1982), vi, vii.

79. *BSSR, Pt. 1*, 2, 3.

80. Slaughter in *1982 NSF Authorization Hearings*, 529, 547–549.

81. Slaughter in *1982 NSF Authorization Hearings*, 554.

82. Duncan MacRae Jr., *The Social Function of Social Science* (New Haven, CT: Yale University Press, 1976), xi, xiv. Also see his autobiography: *About an Academic Odyssey: Natural Science to Social Sciences and Policy Analysis* (Bloomington, IN: Xlibris, 2005).

83. David M. Ricci, *The Tragedy of Political Science: Politics, Scholarship, and Democracy* (New Haven, CT: Yale University Press, 1984), 209.

84. Raymond Seidelman, with the assistance of Edward J. Harpham, *Disenchanted Realists: Political Science and the American Crisis, 1884–1984* (Albany: State University of New York Press, 1985), 223.

85. Miller, "Crisis and Responses," 7.

86. Dynes, "The Institutionalization of COSSA," 219, 219–220.

CHAPTER 8: DEEP AND PERSISTENT DIFFICULTIES

1. Keyworth quoted in "Keyworth Rakes Social Sciences, USSR, Etc.," *Science and Government Report* 13 (Oct. 15, 1984): 5–6, at 5.

2. Otto Larsen quoted in "Footnotes," *Chronicle of Higher Education*, Sept. 12, 1984.

3. Bruce L. R. Smith, *American Science Policy since World War II* (Washington, D.C.: The Brookings Institution, 1990), 125.

4. Tiago Mata and Tom Scheiding, "National Science Foundation Patronage of Social Sciences, 1970s and 1980s: Congressional Scrutiny, Advocacy Networks, and the Prestige of Economics," *Minerva* 50 (2012): 423–449, at 426, 446–447, 447.

5. John A. Remington, "Beyond Big Science in America: The Binding of Inquiry," *Social Studies of Science* 18 (1988): 45–72, at 56.

6. Smith, *American Science Policy since World War II*, 126.

7. Figures in current dollars from Otto N. Larsen, *Milestones and Millstones: Social Science at the National Science Foundation* (New Brunswick, NJ: Transaction, 1992), Table 2.3, p. 24. Current dollars converted to 1982 dollars using cpi calculator from https://data.bls.gov/cgi-bin/cpicalc.pl.

8. Figures from Larsen, *Milestones and Millstones*, Table 2.3, p. 24.

9. See Constance Holden, "Reagan versus the Social Sciences," *Science* 226 (Nov. 30, 1984): 1052–1054, percentages on 1052.

10. F. Thomas Juster, Letter to the Editor, "Basic Research in the Social and Behavioral Sciences," *Science* 226 (Nov. 9, 1984): 610.

11. U.S. Congress, House, *Research Policies for the Social and Behavioral Sciences: Science Policy Study Background Report No. 6*, 99th Cong., 2nd sess. (Washington, D.C.: GPO, 1986), figures on 26.

12. Figures from Larsen, *Milestones and Millstones*, Table 6.1, p. 173. These figures include funding for the SES and BNS divisions. On funding for particular programs within these divisions, see pp. 174–180.

13. William Mishler, "Trends in Political Science Funding at the National Science Foundation, 1980–1984," *PS: Political Science and Politics* 17 (Fall 1984): 846–857, at 849.

14. Jean B. Intermaggio and Christine Ing, "Support for Research in Social Psychology at the National Science Foundation," *Personality and Social Psychology Bulletin* 15 (1989): 309–324, at 312.

15. Bloch in NSF, *1984 Annual Report*, vii.

16. Christina Rose, "Behavioral and Social Sciences Research: Reagan Administration" (Archives—04/08/87), updated Feb. 27, 1987, 2, NSF Library, NSF headquarters, Alexandria, VA.

17. Kim McDonald, "Social-Science Scholar Barred from Awards Aimed at Retaining Scientists at Colleges," *Chronicle of Higher Education*, Aug. 10, 1983.

18. Bloch in NSF, *1984 Annual Report*, vii.

19. See *Research Policies for the Social and Behavioral Sciences*, 31, 92. Also see Roberta B. Miller, "The Budget for Social Science Research," *Science* 224 (May 11, 1984): 561; Roberta B. Miller to Robert Rabin, May 14, 1984, Folder AD/BBS: Social

and Behavioral Sciences, General/Miscellaneous Background, Box 14, Series Historian Files, hereafter, NSF HF, NSF Collection Record Group 307, National Archives–College Park, MD.

20. Robert H. Haveman, *Poverty Policy and Poverty Research: The Great Society and the Social Sciences* (Madison: University of Wisconsin Press, 1987), 164.

21. Dean R. Gerstein, R. Duncan Luce, Neil J. Smelser, and Sonja Sperlich, eds., Committee on Basic Research in the Behavioral and Social Sciences, Commission on Behavioral and Social Sciences and Education, National Research Council, *The Behavioral and Social Sciences: Achievements and Opportunities* (Washington, D.C.: National Academy Press, 1988), figures on 2.

22. Figures from Larsen, *Milestones and Millstones*, 173.

23. On the 1986 law and NSF funding, see David Johnson, "Next to Nothingness and Being at the National Science Foundation: Part III," *Psychological Science* 3 (1992): 323–327, esp. 327.

24. Current dollars converted to 1982 dollars using cpi calculator from https://data.bls.gov/cgi-bin/cpicalc.pl.

25. Ronald J. Overmann, "Social & Behavioral Sciences at NSF, 1950–1990," 6, unpublished paper in author's possession.

26. Figures from Larsen, *Milestones and Millstones*, 171. Current dollars converted to 1982 dollars using cpi calculator from https://data.bls.gov/cgi-bin/cpicalc.pl.

27. Larsen, *Milestones and Millstones*, 24.

28. Larsen, *Milestones and Millstones*, 173.

29. Gerstein, et al., *The Behavioral and Social Sciences*, 2, 260, 205.

30. Gerstein, et al., *The Behavioral and Social Sciences*, 2.

31. Gerstein, et al., *The Behavioral and Social Sciences*, 246, 250.

32. "Implementing the National Academy of Sciences' Recommendations for Behavioral and Social Science Research at the National Science Foundation," undated but sometime in 1989, at 1, Folder AD/BBS: Social and Behavioral Sciences/NRC Study/Report, Box 13, NSF HF.

33. See Larsen, *Milestones & Millstones*, 119, 180–183, 180.

34. See Larsen, *Milestones & Millstones*, 119; Bruce Catton, "Otto Larsen's Grand View of Social Science," *Sociological Inquiry* 62 (1992): 1–10.

35. Otto N. Larsen, "Social Science out of the Closet," *Society* 22 (1985): 11–15, at 14, 12, 14.

36. Wil Lepkowski, "National Science Foundation Embarks on Conservative Course," *Chemical and Engineering News* 61 (May 16, 1983): 11–18, at 11.

37. Knapp in NSF, *1982 Annual Report*, viii.

38. Clarence Holden, "A Forceful New Hand on the Reins at NSF," *Science* 227 (Mar. 29, 1985): 1557–1558, at 1557. "Erich Bloch Honored with Vannevar Bush Award for Long-Running Contributions to S&T," NSF Press Release, Apr. 24, 2002, https://www.nsf.gov/news/news_summ.jsp?cntn_id=108854, accessed Nov. 8, 2018.

39. Keyworth quoted in Mervis, "Remembering Erich Bloch (1925–2016)," Dec. 1, 2016, https://www.sciencemag.org/news/2016/12/remembering-erich-bloch-1925-2016, accessed Oct. 25, 2019.

40. Knapp quoted in Lepkowski, "National Science Foundation Embarks on Conservative Course," 16.

41. Bloch in NSF, *1989 Annual Report*, 5.

42. For former NSB members and their terms, see https://www.nsf.gov/nsb/members/former.jsp, accessed Dec. 12, 2018.

43. Keith Schneider, "Working Profile; Scientist and Rule-Maker: Dr. David T. Kingsbury," *New York Times*, July 21, 1986. Kingsbury resigned under pressure due to a Justice Department investigation about his ties to a biotechnology company while holding public office. See Mark Crawford, "Kingsbury Resigns from NSF," *Science* 242 (Oct. 7, 1988): 28. About Clutter, see https://www.ascb.org/wp-content/uploads/2009/04/Mary_clutter.pdf, accessed Dec. 12, 2018.

44. Undated and untitled transcript of Kingsbury presentation on Jan. 28, 1985, Science Policy Forum Series, Washington, D.C., at 10, Folder: AD/BBS: Correspondence—Kingsbury 1984–1985, Box 12, NSF HF.

45. See Johnson, "Next to Nothingness and Being at the National Science Foundation, Part III," 327.

46. Slaughter Oral History Transcript, 17–18, NSF Oral Histories, NSF headquarters. Also see Larsen, *Milestones & Millstones*, 140–144.

47. NSF, *1984 Annual Report*, vii. Also Dian O. Belanger, *Enabling American Innovation: Engineering and the National Science Foundation* (West Lafayette, IN: Purdue University Press, 1988).

48. See Johnson, "Next to Nothingness and Being at the NSF, Part III," 327, which also notes that as the 1980s ended, SES division director Roberta Miller helped to establish a new NSF initiative called Human Dimensions in Global Change.

49. Gerstein, et al., *The Behavioral and Social Sciences*, 235.

50. Richard Rockwell to Kenneth Prewitt, June 9, 1981, Folder 6276, Box 515, RG 2, Series 1, Social Science Research Council Archives, Rockefeller Archive Center, Sleepy Hollow, NY.

51. See David Dickson, *New Politics of Science* (New York: Pantheon, 1984), 41–42; Philip J. Hilts, "Inside: The Science Agencies," *Washington Post*, May 13, 1983.

52. Roberta B. Miller, "Social Science under Siege: The Political Response, 1981–1984," in *Social Science Research and Government: Comparative Essays on Britain and the United States*, ed. Martin Bulmer (New York: Cambridge University Press, 1987), 377; Kenneth and David L. Sills, "Federal Funding for the Social Sciences: Threats and Responses," *SSRC Items* 35 (1981): 33–47, at 45.

53. Clarence Holden, "Science Board Cautiously Supports Social Research," *Science* 213 (July 31, 1981): 525. Larsen, "Social Science out of the Closet," 12.

54. David Price to Erich Bloch, May 1, 1989; Erich Bloch to David Price, June 6, 1989, both letters in Folder AD/BSS: Social and Behavioral Sciences, General/Miscellaneous Background.

55. "Remarks," Mr. Erich Bloch, NSF Director to COSSA, Annual Meeting Nov. 27, 1984, in Folder: "11-27-84 Consortium of Social Sci. Assoc. Annual Mts, Washington, DC," NSF HF.

56. Bloch to Price, June 6, 1989.

57. Erich Bloch, "Basic Research and Economic Health: The Coming Challenge," *Science* 232 (May 2, 1986): 595–599, at 597. Janet Abbate, "Applied Science as a Political Construct: Scientific Legitimacy and Social Utility at the US National Science Foundation," 16, unpublished manuscript.

58. Harvey Brooks to Doug Walgren, Mar. 9, 1981, 645–649, in U.S. Congress, House, Committee on Science and Technology, Subcommittee on Science, Research and Technology, *1982 National Science Foundation Authorization Hearings*, hereafter, *1982 NSF Authorization Hearings*, 97th Cong., 1st Sess. (Washington, D.C.: GPO, 1981), 646, 647.

59. Miller, "Crisis and Response: The Politics of the Social Sciences in the United States (1980–1982)," *La revue pour l'histoire du CNRS* 7 (Nov. 2002), 5, https://journals.openedition.org/histoire-cnrs/548, accessed Dec. 17, 2018.

60. Bulmer, "The Social Sciences in an Age of Uncertainty," in Bulmer, *Social Science Research and Government,* 345–352, at 347, 348.

61. On the development of economics as a profession and its importance in policy making and public discourse in the post–World War II era, see Roger E. Backhouse, "Economics," in *The History of the Social Sciences since 1945*, ed. Roger E. Backhouse and Philippe Fontaine (New York: Cambridge University Press, 2010), 38–70; Michael A. Bernstein, *A Perilous Progress: Economists and Public Purpose in Twentieth-Century America* (Princeton, NJ: Princeton University Press, 2001). On the contributions of economic thinking to American intellectual and social thought, especially in the post-1960s' period, see Dorothy Ross, "Changing Contours of the Social Science Disciplines," in *The Cambridge History of Science, vol. 7, The Modern Social Sciences*, ed. Theodore M. Porter and Dorothy Ross (New York: Cambridge University Press, 2003), 205–237.

62. Figures in current dollars from Larsen, *Milestones and Millstones*, Table 6.2, p. 175. Current dollars converted to 1982 dollars using cpi calculator from https://data.bls.gov/cgi-bin/cpicalc.pl.

63. Percentages from 1980 to 1989 in Larsen, *Milestones & Millstones*, 174. Figure 8.5 shows that during the 1980s Economics received slightly more than 40%, whereas my text says 39%, a minor discrepancy arising because I've relied on data from Larsen who used data in slightly different ways in different places.

64. Larsen, *Milestones and Millstones*, 65, 106.

65. Daniel H. Newlon, "The Role of the NSF in the Spread of Economic Ideas," in *The Spread of Economic Ideas*, ed. David C. Colander and A. W. Coats (New York: Cambridge University Press, 1989), 195–228, at 199.

66. Memo, Head, Office of Social Sciences [Henry Riecken] to Director [Waterman], "Annual Review of the Office of Social Sciences, FY 1960," July 15, 1960, support for economics discussed on pp. 9–11, first grant noted on 9–10, Folder Social Science Annual Reports, Box 42, NSF HF.

67. NSF, *1967 Annual Report*, 105. Also see NSF Economics program director James H. Blackman's essay, "The Outlook of Economics," *Southern Economics Journal* 37 (1971): 385–395, esp. 386–387.

68. Newlon, "The Role of the NSF in the Spread of Economic Ideas," 204.

69. Newlon, "The Role of the NSF in the Spread of Economic Ideas," 199.

70. On Newlon's relations with COSSA and AEA, see Mata and Scheiding, "National Science Foundation Patronage of Social Sciences, 1970s and 1980s," 444.

71. Walter W. Heller, *New Dimensions of Political Economy* (Cambridge, MA: Harvard University Press, 1966), 2, 9.

72. Angus Burgin, *The Great Persuasion: Reinventing Free Markets since the Great Depression* (Cambridge, MA: Harvard University Press, 2012), 5.

73. Roger E. Backhouse, "The Rise of Free Market Economics: Economists and the Role of the State since 1970," *History of Political Economy* 37 (Suppl., 2005): 355–392, at 355.

74. Backhouse, "The Rise of Free Market Economics."

75. Blackman in "Director's Program Review: Social Sciences," Jan. 14, 1971, at 4, NSF Library.

76. NSF, *1975 Annual Report*, 29–30, 30.

77. NSF, *1977 Annual Report*, 75.

78. Newlon, "The Role of the NSF in the Spread of Economic Ideas," 212.

79. NSB, *Discussion Issues 1981: Social and Behavioral Sciences, Vol. 2*, quote in "Significant Accomplishments" section p. 3, National Science Board Library, hereafter, NSB Library, NSF headquarters.

80. Feldstein quoted in Hilts, "White House Uses Social Sciences."

81. Griliches in *1982 NSF Authorization Hearings,* 590.

82. Newlon's efforts mentioned in Mata and Scheiding, "National Science Foundation Patronage of Social Sciences, 1970s and 1980s," esp. 444. Feldstein's 1981 grant noted in Christine Russell, "Cuts Hit Economic Research That Underpins Reagan's Programs," *Washington Star*, Mar. 22, 1981.

83. Mata and Scheiding, "National Science Foundation Patronage," 426.

84. On Anderson: http://web.stanford.edu/~andrsn/ https://www.hoover.org/profiles/annelise-anderson, accessed Nov. 10, 2018.

85. Ronald Reagan, "Nomination of John H. Moore to be Deputy Director of the National Science Foundation," Mar. 1, 1985, https://www.presidency.ucsb.edu/documents/nomination-john-h-moore-be-deputy-director-the-national-science-foundation, accessed Dec. 12, 2018.

86. Transcript of Kingsbury presentation, 5.

87. About this review, which was carried out by Otto Larsen, see his book *Milestones and Millstones*, 232–233.

88. *Emerging Issues in Science and Technology, 1981: A Compendium of Working Papers for the National Science Foundation* (Washington, D.C.: NSF, June 1982), "Contributions of Social Science to Innovation and Productivity," 37–48. Louis G. Tornatzky, et al., "Contributions of Social Science to Innovation and Productivity," *American Psychologist* 37 (1982): 737–746.

89. Tornatzky, et al., "Contributions of Social Science to Innovation and Productivity," 737.

90. Tornatzky, et al., "Contributions of Social Science to Innovation and Productivity," 743.

91. NSF official quoted in Constance Holden, "A Forceful New Hand on the Reins at NSF," *Science* 227 (Mar. 29, 1985): 1557–1558, at 1558; NSF Director Erich Bloch, "Remarks," 4, 7.

92. NSF, *1982 Annual Report*, 28.

93. Murray Aborn, "The Short and Happy Life of Social Indicators at the National Science Foundation," *SSRC Items* 38 (1984): 32–41, esp. 33–34.

94. Mishler, "Trends in Political Science Funding," 847. For further discussion of big social science, see Larsen, *Milestones and Millstones*, 105–111, 174–176.

95. On decentralization, see NSF, "Justification of Estimates of Appropriations, Fiscal Year 1985, to Congress," transfer on BBS-IV-11, NSF Library.

96. Memo, Director Knapp to NSB Members, Dec. 22, 1983, at 3, 5, 1, in Board Books, 250th Meeting, Jan. 19–20, 1984, closed session, Tab 2, NSB Library.

97. David C. Colander, "Money and the Spread of Ideas," in *The Spread of Economic Ideas*, ed. Colander and Coats, 225–230, at 232.

98. Board Books, 250th Meeting, Tab 2, funding history on 9, 16.

99. "Summary, Economics Advisory Panel, Nov. 18–19, 1983," in Board Books, 250th Meeting, Tab 2, at 20.

100. Board Books, 250th Meeting, Tab 2, 1979–1983 funding on 16.

101. Board Books, 250th Meeting, Tab 2, 16; Memo, Knapp to NSB Members, Dec. 22, 1983.

102. NSB Minutes, 268th Meeting, May 16, 1986, closed session, Appendix B, amounts on CS:5–86:4, NSB Library.

103. See Gerstein, et al., *Research Policies for the Social and Behavioral Sciences*, 47–48.

104. Ross, "Changing Contours," 236.

105. On the role of the social sciences in this shift, see Alice O'Connor, *Poverty Knowledge: Social Science, Social Policy, and the Poor in Twentieth-Century U.S. History* (Princeton, NJ: Princeton University Press, 2001).

106. Board Books, 250th Meeting, Tab 2, closed session, at 5, 6.

107. Haveman, *Poverty Policy and Poverty Research*, 128, further discussion of the PSID on 126–129.

108. Feldstein quoted in Hilts, "White House Uses Social Sciences, but Cuts Funding for Research."

109. Newlon, "The Role of the NSF in the Spread of Economic Ideas," 195.

110. Newlon, "The Role of the NSF in the Spread of Economic Ideas," 198.

111. Jason Stahl, *Right Moves: The Conservative Think Tank in American Culture since 1945* (Chapel Hill: University of North Carolina Press, 2016); John J. Miller, *A Gift of Freedom: How the John M. Olin Foundation Changed America* (San Francisco: Encounter, 2005).

112. Holden, "Reagan Versus the Social Sciences," 1054; call for proposals quoted in Robert Pear, "Scholars Charging Politics on Social Science Research," *New York Times*, Sept. 27, 1983.

113. Miller quoted in Pear, "Scholars Charging Politics on Social Science Research."

114. Assistant secretary quoted in Pear, "Scholars Charging Politics on Social Science Research."

CHAPTER 9: ALTERNATIVE VISIONS

1. Clifford Geertz, "School Building: A Retrospective Preface," in *Schools of Thought: Twenty-Five Years of Interpretive Social Science*, ed. Joan W. Scott and Debra Keates (Princeton, NJ: Princeton University Press, 2001), 1–11, at 4.

2. Henry W. Riecken, "The National Science Foundation and the Social Sciences," *SSRC Items* 37 (1983): 39–42, at 39.

3. Dean R. Gerstein, R. Duncan Luce, Neil J. Smelser, and Sonja Sperlich, eds., Committee on Basic Research in the Behavioral and Social Sciences, Commission on Behavioral and Social Sciences and Education, National Research Council, *The Behavioral and Social Sciences: Achievements and Opportunities* (Washington, D.C.: National Academies Press, 1988), 1.

4. Riecken, "The National Science Foundation and the Social Sciences," 39.

5. James A. Smith, *The Idea Brokers: Think Tanks and the Rise of the New Policy Elite* (New York: The Free Press, 1991), 159–165, at 165, 161. Andrew Rich, *Think Tanks, Public Policy, and the Politics of Expertise* (New York: Cambridge University Press, 2004) also notes (p. 47) that the Institute for Policy Studies made a striking break "with some of the institutional conventions associated with think tanks, particularly norms of neutrality and academic objectivity."

6. Marcus G. Raskin and Herbert J. Bernstein, eds., *New Ways of Knowing: The Sciences, Society, and Reconstructive Knowledge* (Totowa, NJ: Rowman and Littlefield, 1987).

7. John J. Miller, *A Gift of Freedom: How the John M. Olin Foundation Changed America* (San Francisco: Encounter Books, 2006). Lee Edwards, *The Power of Ideas: The Heritage Foundation at 25 Years* (Ottawa, IL: Jameson, 1997). On the AEI, see Jason Stahl, "From without to within the Movement: Consolidating the Conservative Think Tank in the 'Long Sixties,'" in, *The Right Side of the Sixties: Reexamining Conservatism's Decade of Transformation*, ed. Laura Jane Gifford and Daniel K. Williams (New York: Palgrave Macmillan, 2012), 101–118, at 102; and Stahl's recent book, *Right Moves: The Conservative Think Tank in American Culture since 1945* (Chapel Hill: University of North Carolina Press, 2016). Also see Thomas Medvetz, *Thank Tanks in America* (Chicago: University of Chicago Press, 2012).

8. Thomas S. Kuhn, *The Structure of Scientific Revolutions* (Chicago: University of Chicago Press, 1962).

9. Kuhn, *The Structure of Scientific Revolutions*, 160–161.

10. Steve Fuller, *Thomas Kuhn: A Philosophical History for Our Times* (Chicago: University of Chicago Press, 2000), chap. 5, "How Kuhn Saved Social Science," at 234. Also see Gary Gutting, "Introduction," in *Paradigms and Revolutions: Applications and Appraisals of Thomas Kuhn's Philosophy of Science*, ed. Gary Gutting (Notre Dame, IN: University of Notre Dame Press, 1980), esp. 12–13.

11. Gutting, "Introduction," 13, 14.

12. See, for example, Paul Rabinow and William M. Sullivan, eds., *Interpretive Social Science: A Second Look* (Los Angeles: University of California Press, 1987/1979).

13. Richard A. Shweder, "Clifford James Geertz, August 23, 1926–October 30, 2006," *NAS Biographical Memoir* (Washington, D.C.: NAS, 2010).

14. Geertz, "School Building," 4, 8, 9. Also see two important collections of Geertz's essays: *The Interpretation of Cultures* (New York: Basic Books, 1973) and

Local Knowledge: Further Essays in Interpretive Anthropology (New York: Basic Books, 1983).

15. U.S. Congress, House, Committee on Science and Technology, Subcommittee of Science, Research and Technology, *The Psychological and Social Sciences Research Support Programs of the National Science Foundation: A Background Report*, 95th Cong., 1st Sess. (Aug. 1977), Appendix III, 5, also see 22–24, 81–83.

16. Gina B. Kolata, "Social Anthropologists Learn to Be Scientific," *Science* 195 (Feb. 25, 1977): 770.

17. John E. Yellen and Mary W. Greene, "Archaeology and the National Science Foundation," *American Antiquity* 50 (1985): 332–341; Alice B. Kehoe, *The Land of Prehistory: A Critical History of American Archaeology* (New York: Routledge, 1998), "The New Archaeology," 115–132.

18. Richard N. Adams, "Letter to the Editor," *Science* 196 (Apr. 22, 1977): 372–374, at 372.

19. Adams, "Letter to the Editor," 374.

20. Thomas O. Beidelman et al., "Letter to the Editor," *Science* 196 (Apr. 22, 1977): 374.

21. Nancie L. Gonzalez, "Letter to the Editor," *Science* 196 (Apr. 22, 1977): 378.

22. Kenneth Prewitt and Haworth Editorial Submission, "Introduction to the Social Science Research Council's Contribution to the 5-Year Outlook for Science and Technology," *Behavioral and Social Sciences Librarian* 3 (1983): 47–61, at 55.

23. Prewitt, "Introduction," 55.

24. Robert McC. Adams et al., eds., *Behavioral and Social Science Research: A National Resource Part I* (Washington, D.C.: The National Academies Press, 1982), 36, 27, 28.

25. Harold Orlans, "Criteria of Choice in Social Science Research," *Minerva* 10 (1972): 571–602, at 601.

26. Richard P. Nathan, *Social Science in Government: Uses and Misuses* (New York: Basic Books, 1988), 10, 11. Also, Charles E. Lindblom, *Inquiry and Change: The Troubled Attempt to Understand and Shape Society* (New Haven, CT: Yale University Press, 1990), who proposed (p. 11) that in the social sciences, a form of research best understood as "inquiry" should "displace conventionally scientific investigation."

27. Neal Koblitz, "A Tale of Three Equations; or The Emperors Have No Clothes," *Mathematical Intelligencer* 10 (1988): 4–10, at 10.

28. Gene M. Lyons, "The Many Faces of Social Science," in *The Nationalization of the Social Sciences*, ed. Samuel Z. Klausner and Victor M. Lidz (Philadelphia: University of Pennsylvania Press, 1986), 197–208, at 201.

29. Lyons, "The Many Faces of Social Science," 204, 206.

30. Samuel Z. Klausner and Victor M. Lidz, "Nationalization and the Social Sciences," in Klausner and Lidz, *Nationalization of the Social Sciences*, 267–286, at 268, 269.

31. Klausner and Lidz, "Nationalization and the Social Sciences," 284.

32. Klausner and Lidz, "Nationalization and the Social Sciences," 285–286, 286.

CHAPTER 10: THE SOCIAL SCIENCES AT THE NSF

1. Robert M. Collins, *More: The Politics of Economic Growth in Postwar America* (New York: Oxford University Press, 2000), 233.

2. George Mazuzan, "The National Science Foundation: A Brief History," July 15, 1994, NSF88–16, at 20, https://www.nsf.gov/about/history/nsf50/nsf8816.jsp, accessed Nov. 18, 2018.

3. Otto N. Larsen, *Milestones and Millstones: Social Science at the National Science Foundation* (New Brunswick, NJ: Transaction, 1992), figures from Table 8.1 on 260.

4. Henry W. Riecken, "The National Science Foundation and the Social Sciences," *SSRC Items* 37 (1983): 39–42, at 40.

5. David L. Featherman, "Mission-Oriented Basic Research," *SSRC Items* 45 (1991): 75–77.

6. Roberta B. Miller, "Crisis and Responses: The Politics of the Social Sciences in the United States (1980– 1982)," *La revue pour l'histoire du CNRS,* 2002/online version 2006, https://journals.openedition.org/histoire-cnrs/548, accessed July 8, 2016.

7. Davis quoted in *1967 NSSF Hearings*, 270, 271. For full citation, see p. 341, n. 1.

8. There is little historical research about NSF social science during this period. However, for a discussion of developments from 1990 to 1991, see Larsen, *Milestones and Millstones*, "The 1990s: Continuities in Contested Legitimacy," 227–258; also Daniel S. Greenberg, *Science, Money, and Politics: Political Triumph and Ethical Erosion* (Chicago: University of Chicago Press, 2001), "Saving the Social Sciences," 451–456.

9. See Larsen, *Milestones and Millstones*, 239–249, Simon's testimony noted on 239.

10. Biological, Behavioral and Social Sciences Task Force, "Looking to the 21st Century," Executive Summary of the Task Force Report, June 1991 draft, in Folder BBS-Biological, Behavioral & Social Sciences Directorate, General Records 1976–92, although at 3, Box 16, Director Subject Files 1964–83, NSF Collection Record Group 307, hereafter, RG 307, National Archives, College Park, MD. Also see the final report: Courtland S. Lewis, *Adapting to the Future: Report of the BBS Task Force Looking to the 21st Century* (Washington, D.C.: NSF, 1991).

11. Figures from "Statements on Introduced Bills and Joint Resolutions—by Mr. Kerry: S1031," originally printed in *Congressional Record-Senate*, May 9, 1991, S5635, Folder AD/BBS: Social and Behavioral Sciences, General/Miscellaneous Background, Box 14, NSF Historian Files, RG 307.

12. Director Walter E. Massey, Memorandum [on appointment of Marrett], Jan. 30, 1992, Folder BBS: Biological, Behavioral, Social Sciences Directorate, General Records 1976–92; David Johnson, "Next to Nothingness and Being at the National Science Foundation: Part IV," *Psychological Science* 4 (1993): 1–6, Marrett's background on 4. Johnson also notes (pp. 5–6) that the agency sought to provide greater support for interdisciplinary efforts, implemented, for example, through the establishment of program clusters that cut across the main established disciplines. Among these clusters, there was one for social and political sciences and another for economic, decision, and management sciences.

13. See https://www.nsf.gov/nsb/, accessed Apr. 20, 2018.

14. David Johnson, "Next to Nothingness and Being at the National Science Foundation: Part 1," *Psychological Science* 3 (1992): 145–149, at 145.

15. https://www.nsf.gov/pubs/2004/nsf04219/nsf04219.pdf , https://www.nsf.gov/sbe/ses/soc/ISSQR_workshop_rpt.pdf.

16. Howard S. Becker, "How to Find Out How to Do Qualitative Research," *International Journal of Communication* 3 (2009): 545–553, at 546.

17. Personal communication, anonymous Ivy League sociologist to author, spring 2016.

18. Wendy A. Naus, "Advocating for Social Science Is a Team Sport," *Footnotes*, July/Aug. 2014, 3–4, at 3.

19. COSSA, *Social and Behavioral Science Research*, figures from unnumbered page with headline The National Science Foundation & Social & Behavioral Science, https://www.cossa.org/wp-content/uploads/2017/01/COSSA-Recommendations-for-Trump-Administration-Dec-2016.pdf, accessed Nov. 18, 2018.

20. Walker quoted on pp. 453–454 in Greenberg, *Science, Money, and Politics*. Also see Curt Suplee, "Fate of Science Is Unclear if Congress Abolishes Energy Department," *Washington Post*, May 12, 1995, A11.

21. See "Behavioral and Social Science Are under Attack in the Senate," *Inside Higher Ed.*, May 19, 2006.

22. Tatro and Coburn quoted in David Glenn, "Senator Proposes an End to Federal Support for Political Science," *Chronicle of Higher Education*, Oct. 7, 2009.

23. Coburn's suggestion noted in Patricia Cohen, "Field Study: Just How Relevant is Political Science?" *New York Times*, Oct. 20, 2009.

24. David Brooks, "The Unexamined Society," *New York Times*, July 7, 2011.

25. NSF, "FY 2008 Budget Request to Congress," Feb. 5, 2007, 61, including, predominant at SBE-2, NSF Library, NSF headquarters, Alexandria, VA.

26. COSSA, *Social and Behavioral Science Research*, percentages on unnumbered page with headline The National Science Foundation & Social & Behavioral Science.

27. Dylan Matthews, "Why Congressional Republicans Want to Cut Social Science Research Funding," *Vox*, Nov. 12, 2014, https://www.vox.com/2014/11/12/7201487/congress-social-science-nsf-funding, accessed Nov. 4, 2019.

28. See "Nobel Prizes—The NSF Connection: Economics," https://www.nsf.gov/news/special_reports/nobelprizes/eco.jsp, accessed November 18, 2018.

29. COSSA, *Social and Behavioral Science Research*, 2, emphasis in original.

30. Naus, "Advocating for Social Science Is a Team Sport," 3.

31. COSSA, *Social and Behavioral Science Research*, 2.

32. Sheila Jasanoff, "Back from the Brink: Truth and Trust in the Public Sphere," *Issues in Science and Technology* 33 (2017): 25–28, at 25.

33. Jasanoff, "Back from the Brink," 27, 28.

34. Theodore M. Porter, "Speaking Precision to Power: The Modern Political Role of Social Science," *Social Research* 73 (2006): 1273–1294, at 1273–1274.

35. For example, see David Price, *Weaponizing Anthropology: Social Science in Service of the Militarized State* (Petrolia, CA: Counterpunch, 2011). For further historical context, see Joy Rohde, *Armed with Expertise: The Militarization of American Social Research during the Cold War* (Ithaca, NY: Cornell University Press, 2013).

36. COSSA, *Social and Behavioral Science Research*, 4.

37. William A. Blanpied and Richard C. Atkinson, "Social Scientists' Contributions to Science Policy during the New Deal," http://www.rca.ucsd.edu/speeches/Social_Science_and_Science_Policy.pdf, accessed March 13, 2017, 4, 5.

38. Harry Alpert, "The Knowledge We Need Most," *Saturday Review*, Feb. 1, 1958, 36–38, at 38.

39. Fred R. Harris, "The Case for a National Social Science Foundation," *Science* 157 (Aug. 4, 1967): 507–509, at 508.

40. Michael D. Reagan, *Science and the Federal Patron* (New York: Oxford University Press, 1969), 189.

41. Anne Fausto-Sterling, *Myths of Gender: Biological Theories about Women and Men*, rev. ed. (New York: Basic Books, 1992/1985), 208.

INDEX

Page numbers in italics refer to figures.